Quality in Stored and Processed Vegetables and Fruit

Seventh Long Ashton Symposium 1979

Quality in Stored and Processed Vegetables and Fruit

Proceedings of a Symposium held at
Long Ashton Research Station
University of Bristol
8-12 April 1979

Edited by P.W. GOODENOUGH *and* R.K. ATKIN

1981

ACADEMIC PRESS
London New York Toronto Sydney San Francisco
A Subsidiary of Harcourt Brace Jovanovich, Publishers

ACADEMIC PRESS INC. (LONDON) LTD.
24/28 Oval Road
London NW1

United States Edition published by
ACADEMIC PRESS INC.
111 Fifth Avenue
New York, New York 10003

British Library Cataloguing in Publication Data

Quality in stored and processed vegetables and fruit.
1. Fruit, Canned - Congresses 2. Vegetables, Canned - Congresses 3. Fruit - Storage - Congresses 4. Vegetables - Storage - Congresses
I. Goodenough, P.W. II. Atkin, R.K.
664′.807′0212 TP368

ISBN 0-12-289740-4
LCCN 81-66382

Typeset by DMB Typesetting, Oxford
Printed in Great Britain by
St Edmundsbury Press, Bury St Edmunds, Suffolk

Preface

This book is a record of the proceedings of a Symposium organized jointly by Long Ashton Research Station and the Post Harvest Biology Group of the Association of Applied Biologists. In keeping with the aims of the Long Ashton Symposia, the event brought together workers in different disciplines, who do not normally meet regularly, to discuss problems associated with defining and maintaining "quality" in stored and processed vegetables and fruit. In this seventh symposium of the series, horticulturists, physiologists, biochemists, pathologists, flavour chemists, food technologists and others met to put each area into the context of high quality in foods.

Although advocated increasingly by producer and consumer alike, particularly in developed countries, "quality" *per se* is inevitably difficult to define in an objective, quantifiable manner. However, until quality attributes of foods can be defined more precisely and are more amenable to scientific measurement, progress towards the production of good quality produce, with consistency in standards and continuity of supply, will be impeded. Fundamental research has an important role to play in increasing our understanding of plant processes central to quality attributes, and indeed, the direct value of such basic work to producers is evident in several papers in this book.

As these proceedings also show, by reference to stored and processed horticultural crops, the final quality of produce bought by the consumer can be influenced by many factors, including the variety grown, and cultural and processing techniques used. Unravelling these many influences, whether in the field or factory, will clearly benefit from a multi-disciplinary discussion, which we hope this symposium has fostered. Such information will become increasingly necessary, particularly as the proportion of processed foods continues to increase, and with it demands to safeguard nutritional and quality standards. Another development necessitating more knowledge about "food quality" and ancillary attributes, such as nutritional value, could be the imposition of more stringent standards through legislation. We see evidence of this trend, and the consequent challenge to horticulturists and food technologists in Professor Kramer's stimulating Inaugural Address.

We hope that this book will encourage more research into the identification of characters, both genetic, biochemical and environmental, which must be evaluated if high quality produce is to be stored or processed. Moreover, we hope that a dialogue between those in university, research centres and industry will be maintained as each interest has an essential expertise to contribute in our endeavours to improve food production and storage, whilst also maintaining good "quality".

We thank all contributors for their co-operation in submitting their papers in time for precirculation before the symposium and their ready collaboration with the editors during the preparation of this book. We should also like to express our appreciation to the staff of Academic Press for their expert guidance given at all stages of publication. We acknowledge the permission granted by editors to use diagrams and tables from the journals noted in the relevant captions. Special thanks also go to Dr C.V. Cutting for his help in the organization of the Symposium, to Mr E.G.R. Chenoweth for redrawing many of the illustrations, to Mr I.M. Prosser for checking the references and the index, and to Mrs Beryl Courtice for her secretarial assistance.

March 1981

P.W.Goodenough
R.K.Atkin

Contents

SECTION II

The Foundations of Quality—Improving Quality in Breeding

The Foundations of Quality—Modifying Cultural Techniques to Develop Quality

SECTION III

Quality at Harvest—Physiological Aspects

Quality at Harvest—Harvesting and Handling

SECTION IV

Post-harvest Changes—Physiological Effects

Post-harvest Changes—Diseases

Post-harvest Changes—Control of Post-harvest Changes

SECTION V

Changes during Processing

INAUGURAL LECTURE

Qualities we Aim to Maintain

A. KRAMER

Dept. of Horticulture,
University of Maryland,
College Park, Maryland,
USA 20742

PALATABILITY OR NUTRITION

As late as one, certainly two decades ago, the "qualities we aim to maintain" would have been entirely self-evident. Everyone accepted that the aim of the food producer was to provide the consumer with what he wanted, that was to satisfy his sensory quality preferences. This approach was sound, since man, like other animals, developed an instinctive liking for those foods which coincidentally provided him with all his nutritional requirements. When he ate what he liked, he ate what he needed. Modern science and technology, however, has made it possible to separate the two associated but distinct factors of sensory and nutritional qualities (Ucko and Dimbleby, 1968). Thus in recent years the food and agricultural industry has been accused of doing what the public wanted it to do, namely, develop highly acceptable, safe, convenient foods to distribute on a mass basis at low prices. Nutrition was not a major factor as the system evolved. Now it is, but nutritional requirements may be too little and too late.

Although obesity is considered the major nutritional problem in America some authorities now claim that "good nutrition can prevent illness." Indirectly they are stating that the food industry should meet nutritional needs and also educate the public in its purchasing and eating habits so that a balanced diet is provided and the excesses that result in illness are avoided. By this method the industry would play a major role in reducing the national cost of medical care (Seligsohn, 1978).

FRESHNESS

Another concept assumed to control consumer behaviour is that "fresh is best." Yet this term is not readily understood. To some, "fresh" means just harvested or slaughtered, to others it means raw food that has not been preserved, except by refrigeration; while others use "fresh" as an adjective, as, for example, "fresh-baked" despite the fact that the flour may be several years old, or "fresh-frozen" when the produce may have been kept in the freezer for a year or longer. It should be recognized that such fresh-baked or fresh-frozen products are frequently of better quality than raw-fresh products. Nevertheless "open-dating" is a popular issue which at present is largely limited to short-life products, especially milk, but may soon affect all packaged foods, although the value to the consumer is dubious. The form of open-dating which may soon be required is the "pack-date" which will state the date the unit was processed and/or packaged. A more useful form is the "pull-date" which prints the date before which the unit may be sold, or a "use by" date. Perhaps the most useful form of open-dating is: "Best if used by ...."

There is already a bill in the US House of Representatives (H.R. 10358) intended to strengthen the Food, Drug and Cosmetics Act. While some of the language merely clarifies existing situations, this bill will put into law several practices in the food processing industry which are now only applied on a voluntary and partial basis. This bill will require coding of food for recall purposes, and in addition open-dating in the form of a "sell date". The listing of ingredients in order of decreasing quantity remains, except that the percentage of ingredients that may be significant because of their cost, nutritive value or acceptability, would be required.

CUSTOM AND CONVENIENCE

Contrary to the supposed wants and needs of consumers, custom and convenience are the primary criteria influencing consumer behaviour. Experiments with subjects of all ages showed that a child older than two or three years is already "sophisticated" enough to select foods to which he had become accustomed or that he had observed were being selected by his elders (Kramer, 1972).

Although man instinctively prefers foods that are sweet, sweet and sour, odourless or fragrant, and will not accept food that is even slightly bitter or salty or putrid, he has learned to accept large quantities of "bitters" (e.g. putrid fermented fish sauce) simply because originally such foods were a necessity. Later he became accustomed to these strange flavours, found that they did not harm him and grew to like these "unnatural" foods. We may surmise that people like and eat what they are accustomed to, not necessarily what "is good for them", to satisfy their nutritional requirements. They will even accept and consume foods that they do not like instinctively as a result of education, necessity and even the proselytizing efforts of advertisers and others.

This acceptance of custom explains many phenomena that we all experience in our attempts to "improve" quality. Two examples illustrating this pressure are (i) the rejection of vine-ripened tomatoes, because of their off-flavour, by a panel of college students who were accustomed to mature green tomatoes; and (ii) hand-shelled freshly-picked peas, because of lack of flavour.

The importance of convenience may be readily demonstrated by the success of so many "instant" dried foods and beverages, although the consumer may admit a preference for the food prepared from the freshly-harvested raw produce (Kramer, 1972).

WANTS OR NEEDS

In deciding on the qualities we aim to maintain, we are faced with the dilemma of having to decide whether we should cater for the consumer's wants, as expressed by his purchasing choice, which is not necessarily by his stated preference; or for his needs as determined by nutritionists, who themselves do not always agree. That this is a real dilemma is shown for many vegetables and some fruits where there is a negative correlation between sensory and nutritive quality (Kramer, 1953). For example, fancy quality *petit pois* are much lower in protein, carbohydrate and vitamins A, B_1, B_2 than more mature standard quality peas. Some plant breeders are attempting to block the pathway by which the red pigment lycopene is converted to the orange carotene, solely to produce redder tomatoes; coincidentally the pro-vitamin A (β-carotene) content is reduced giving a more attractive yet less nutritive product. There is little doubt that the consumer will prefer the redder tomato.

There could be a trend to discourage the consumer from purchasing food of low nutritive value. Some States and municipalities, in addition to taxing alcohol and tobacco, also tax soft drinks. Even this measure appears to have little effect on consumer purchasing habits, but is beneficial in cases where the tax revenue is appropriated for health research and education. The difficulty in providing such consumer protection is that the protection is imposed on the consumer, presumably "for his own good". In a democratic society this is very difficult to do, as demonstrated by the rise and fall of the Volstead Act, which legislated prohibition of alcoholic beverages in the United States.

Yet there are already several government guidelines that food producers may or must follow to maintain certain quality levels, as, for example, in the case of nutrients, though the main emphasis is admittedly in protecting the safety and wholesomeness of foods. We are all aware of regulations controlling the use and residual levels of pesticides, chemical additives and microbial numbers. Soon such regulations may conceivably affect new cultivars should they contain some undesirable constituents or inadequate amounts of essential nutrients. The consumers in America are now so thoroughly confused about all the conflicting claims of various food qualities that they invariably approve all proposed government regulations, but pay little attention to any label declarations because they assume the government will protect them. Thus consumers welcomed the addition of nutrient contents to labels, but are not rejecting products low in nutrient content. The mere presence of the nutrient statement satisfies them.

What then should be the position of the food researcher? We may anticipate that he will continue to produce foods similar to existing foods but higher in sensory quality, with a longer shelf-life, and more convenient, nutritious and safer than present foods. Such confidence may seem optimistic. The food industry argues that the opportunities to improve services and products for consumers are constrained by the increasing demands of mandatory regulations. Government agencies are proliferating, as are regulatory activities, and not surprisingly the consumer is confused and bewildered, invariably stating that he wants more protection whether he needs it or not (Kramer, 1979). However, we must not lose confidence in our abilities to improve the efficiency of food production and processing; the alternative is food shortage (Meadows *et al.*, 1972).

THE ENERGY PROBLEM

The energy crisis of the winter of 1973-74 elevated energy as the first limiting resource, and energy savings still dominate discussions on methods of food preservation. Recent findings, while indicating that energy usage, as well as total costs of fresh (chilled), canned and frozen foods are comparable, have shown that the most efficient method is freezing food. It has also been established that the major opportunities for energy savings are in the rapid utilization of the fresh product and reduction in energy required during home preparation. For canned products, the major savings are in the development of less expensive and energy demanding containers. For frozen foods, it has been assumed that low temperature storage is the major energy requirement. It has been demonstrated, however, that commercial storage even for a half year or longer requires a small part of the total energy, and it is the relatively short time of one or two weeks in the retail display cabinet and in the home appliance that provides the major opportunity for energy savings. Thus, regardless of the method of the initial food preservation, the most wasteful use of energy in connection with food production and consumption is in the home with the next opportunity for savings being in the retail outlet.

Energy saving is generally being accomplished by careful husbandry and recycling. Saving is also being made possible by new and improved packaging to permit the use of higher storage temperatures. Maintaining freezer temperatures at $-20°C$ ($-5°F$) or even $-18°C$ ($0°F$), instead of $-23°C$ ($-10°F$), and/or permitting a gradual rise in temperatures overnight, may permit a saving of approximately 20% in energy cost. Similar savings can also be accomplished by optimal management of heat resources. Yet despite the rising cost of energy and the economies already made, the total cost of refrigeration is increasing (Moleeratanond *et al.*, 1978).

The emphasis on energy savings has been boosted by a disenchantment with costly construction programmes for nuclear power plants. For example, the State of California now contends that its new minimum efficiency standards for refrigerators, which go into effect in 1979, could eliminate the need for two proposed nuclear reactors, thus saving over $2000 million.

While increased attention is being placed on energy savings in retail and home appliances, techniques such as "energy audits" and computerized control to limit energy used at peak load times are gaining momentum. Also, a new generation of power saving equipment, such as electric motors and conveyors, is being used when it is shown that substantial financial savings can be made as energy costs increase.

We may expect more attention to be given to energy saving and less to labour saving, although we cannot yet envisage a time when we will abandon mechanical harvesters, for example, to return to hand harvesting. In fact, developments in the uniform ripening of crops which may then be "once-over" machine harvested contribute both to labour and energy savings. The savings in energy must therefore come from:

(i) Increased yields in terms of food solids, i.e. nutrients.
(ii) Increased utilization of the plant materials grown.
(iii) Prompt stabilization of the harvested crops to extend shelf-life.
(iv) Reduced storage and transportation costs.

All this must, of course, be accomplished while maintaining if not improving sensory quality, safety and convenience of the foods when they reach the consumer.

INCREASED YIELDS

With the exception of a few crops such as potatoes or onions intended for dehydration, little attention has been given to horticultural crop yields in terms of solids. In fact, numerous studies may be cited where higher yields of new cultivars disappeared when expressed on a dry matter basis. Even tomatoes grown for processing, where concentrates are required to contain a minimum of tomato solids, were lower—not higher—in total solids. Being firm, red and ripening uniformly they satisfied certain sensory quality requirements and reduced harvesting costs, but provided less food and cost more to process (Watada, 1976).

Such an evaluation would not apply to tomatoes grown for fresh consumption because tomatoes, like many other fruits and some vegetables, are luxury items, satisfying certain aesthetic, sensory preferences of those who can afford them rather than supplying essential nutrients required by all consumers.

INCREASED UTILIZATION

Increased utilization is closely related to increased yield. What an important accomplishment it would be to develop a Brussels sprout plant 30% of which consists of edible sprouts. With total plant yield, fertilizer, and other inputs the same, yield of edible sprouts would be doubled. There are many opportunities for such "quantitative" quality improvements in vegetables and fruit, as only 20-30% of the plant is processed or packaged for human consumption; the rest, which is usually leaves and stems, is wasted. Efficient utilization of residual plant materials should be directed toward their nutritive value as food or feed. Such a procedure would not only solve a health and pollution problem but increase the value of the crop and the rate of food production per unit of cultivated farm land.

Ben-Sinai *et al.* (1965), for example, showed that pea plants harvested at a stage of maturity that would grade the peas as minimum fancy (Grade A) had a total plant yield of over 6 tons acre^{-1}. Yield of peas for human consumption was just a little over 1 ton acre^{-1} or 19% of the total pea plant yield. In terms of nutrients, 27% of the proteins and carbohydrates were found in the peas, 25% in the pods and 48% in the vines. It is evident, therefore, that a four-fold increase in food nutrients could be obtained (per acre) if the entire plant, instead of just the peas, were utilized for food or feed.

Until the early 1960s entire pea plants were harvested and transported to viner stations where the peas were separated from vines and pods. Some pea processors maintained livestock to which they fed the vines and pods. In recent years, however, mobile viners have come into use, so that pods and vines are left in the field and only the shelled peas are delivered to the plant for freezing or canning. Thus, where before about a quarter of the pea crop was used for food and the remainder for feed, three-quarters of the available pea plant nutrients are now wasted.

It would seem logical to utilize the nutrients in the entire plant and to use them directly as food rather than feed. This is technically feasible as demonstrated by Pirie (1966), who developed a pulper, sometimes called the "mechanical cow". This pulper extracts up to 75% of the protein and also other nutrients from leafy crop material fed into the machine at the rate of about 1 ton per hour.

Utilizable wastes are also produced at the processing/packaging plant, mainly due to sorting, peeling, trimming and blanching operations. Some of these wastes are uneconomical to recover, usually because of the high cost of drying. Utilization of dehydrated vegetable waste is largely confined to tomato, asparagus and spinach wastes (Mercer and Rose, 1968); pea and sweet corn wastes are seldom wasted, but fed directly or ensiled for livestock feed.

Problems in utilization of vegetable and fruit processing or production wastes as foods or feeds may arise because of the concentrations of chemical residues found in them. Whenever the need for more food and/or the problems of waste disposal become sufficiently serious, either solutions which are less expensive than dehydration and involve omission of pesticides will be found, or less expensive recovery techniques will be developed.

For example, a process for fully utilizing tomato cannery waste was developed. In this process the waste was compressed and the presscake, containing two-thirds of the solids, but only 10% of the protein, was used as cattle feed, while the expressed juice containing one-third of the solids and 60% of the protein, was precipitated as a tomato protein concentrate. This concentrate can be extracted with acetone to form an 85% protein isolate, plus a tomato flavour and pigment residue (Fig. 1). This tomato protein concentrate has good nutritional properties, and a Protein Efficiency Ratio similar to soy protein. If produced on a large scale, and the waste provided at no cost, a pound of tomato protein could be produced for about 10¢ (Kramer and Kwee, 1977).

Utilizing solar energy for the production of nutrients is a new concept that may well be the focus of agricultural research for the next 20 years and which incidentally may result in increased requirements for refrigerated storage and other methods of extending shelf-life. The idea is to crop land for maximum nutrient production rather than gross crop yields, only a small fraction of which consists of digestible nutrients. For example, it may be found that two crops of barley harvested in the milk stage contain more nutrients than one crop allowed to mature and dry in the field. But this would be only the beginning of the required research and development.

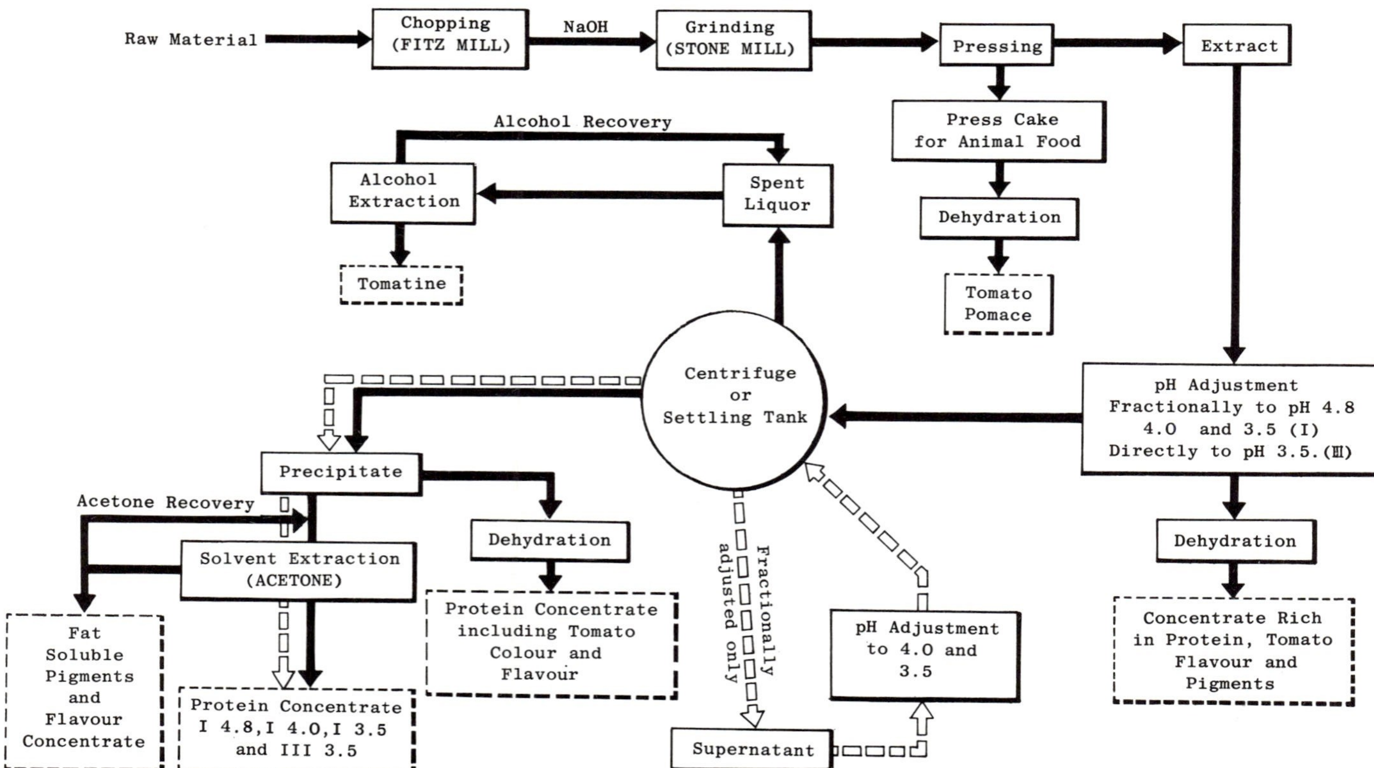

Fig. 1. Schematic flow chart of a process for the total utilization of tomato waste.

What is one to do with milk stage barley seeds? How will the nutrients in the other parts be utilized?

STABILIZATION AFTER HARVEST

After many years of neglect, the international community is beginning to accept that vast food resources are wasted simply because of inadequate handling facilities after harvest (National Academy of Sciences, 1978). The approximate estimates now available suggest that 25-50% of all crops are lost—which, if avoided, would eliminate current food shortages and provide adequate supplies of superior quality for years to come. Although the major crops of concern are the cereals, nearly a quarter of the losses are in vegetable and fruit crops (if roots and tubers are included) upon which millions subsist (Pentzer and Kramer, 1976). Jul (1971) estimated that only 1-2% of the world food supply requiring refrigeration is refrigerated.

Assuming that maintenance of freshly harvested quality of "natural" products is the goal, there should be general agreement among post-harvest physiologists, that this could be accomplished by the "instant" reduction of the harvested product temperature to 0°K or −273°C. It is also recognized that this is impractical, if not unattainable. However, prompt temperature reduction without frost damage, with or without a preliminary high temperature "curing" or "blanching" treatment is the usual treatment recommended and is frequently carried out successfully by various pre-cooling techniques. Unless the vegetable or fruit products are then promptly consumed or further processed (and so stabilized), they are then stored at appropriate temperatures and humidities, to extend high quality shelf-life long enough for the distribution of the product to the consumer.

For that fraction of the product which cannot be so handled, a compromise is made, the material is further processed, usually by more vigorous heating and/or cooling combined with special packaging, but at a cost in modifying the product from its "fresh-like" condition. Thus canned, frozen or dehydrated products are manufactured which can gain at times a dominant share of the market. All these post-harvest techniques for food preservation add to the cost of the product and reduce its "fresh-like" quality. Our efforts were, and will continue to be, to develop or modify these techniques of extending "fresh-like" quality at a lower cost.

CANNING

As stated already, a major opportunity in reducing the cost of canning is to develop an inexpensive container; in fact, a major reason for predicting a slowing down in the rapid growth of frozen foods is the appearance of the retort pouch/tray. Robe and Andres (1978) prepared a comprehensive report on the development of the retortable pouch, which is a flexible container, and of the retort tray. They emphasize the savings in energy achieved during processing but fail to point out that little overall saving is made because of the need for an expensive container and much more expensive and slower processing equipment.

There are some claims that foods processed in a pouch are superior to canned or frozen, and equal to raw foods prepared from raw ingredients. While we are awaiting more definitive studies, I have assumed that products such as most raw fruits, vegetables and other products, which are overcooked during conventional canning processes, are superior when not overcooked by the pouch processing method to conventionally canned products. However, they lose this advantage with storage unless they are held in cold storage. Furthermore, such items, even if short-cooked in a retortable pouch, are not likely to be superior to the equivalent frozen product. Thus the superior quality of the retort pouch, and particularly the retort tray, if it exists at all, is probably limited to prepared foods of the stew or soup-type where overcooking is not a problem and may add to the "blending" of flavours and improve the flavour quality. It is interesting that many of the products that are now distributed in Europe, and are about to appear on the American market in retortable pouches or trays are the "stew-type". Thus, before the "shelf-stable retortable pouch" has a real impact on frozen/chilled/ambient vegetable and fruit storage mix, it will be necessary to (i) develop a cheaper, rigid, rapidly sealable container which can be sterilized more efficiently (perhaps by microwave), and (ii) determine limits, if any, for time-temperature of storage.

DEHYDRATION

Recent advances in improving quality of dehydrated foods seem to have culminated in the lyophilization (or freeze-drying) process. Cost again is a problem. The US Army Research and Development Command has achieved some cost reduction by a new process. Raw food materials are first freeze-dried to 25-35% moisture content, at which point they are compressed to less than half volume before further drying by conventional methods. The fully dehydrated and compressed product retains all the qualities of a completely freeze-dried material. However, this new process is not expected to affect refrigerated storage requirements as the demand for high quality dehydrated foods will probably remain limited. Also, if energy is to be saved, all dehydrating techniques, except direct sun-drying, are not likely to be encouraged.

MODIFIED ATMOSPHERES

The original studies of Kidd *et al.* (1927) on stored apples, which were assigned the name of controlled atmosphere (CA) storage, were based on slowing respiratory rates by storing in a gas-tight chamber under reduced oxygen (O_2) and increased carbon dioxide (CO_2) concentrations. New findings are that CO_2 has the additional benefit of reducing the growth of micro-organisms, as well as in interfering with oxidation. Other reports indicate than an application of 20% or more of CO_2 even for one or two weeks benefits a variety of plant and animal products. Thus the original concept of increasing CO_2 and reducing O_2 concentrations has been extended to other fruits and vegetables, and also to many meat products, particularly fresh-cut red meats, poultry and fish.

Another extension of the CA concept is the use of gases other than O_2 and CO_2. We have demonstrated the possibility of extending high quality shelf-life of grapefruit segments, sliced peaches and mushrooms from 1-3 days to about a month by evacuating the container and breaking the vacuum with carbon monoxide (CO) (Besser and Kramer, 1972). Other studies have shown similar beneficial results with the use of CO in combination with CO_2, on red meat. While CO and CO_2 are effective in arresting aerobic enzymatic and microbial activity, they appear to have little effect on the anaerobes. We therefore embarked on the GASPAK process which consists of treating fresh products with gases, which should leave the product entirely free of aerobic or anaerobic biological activity. To date we have succeeded in maintaining "fresh-like" quality of raw potato strips, apple slices and grapefruit segments by a sequential SO_2-CO-CO_2 treatment. After two months at ambient storage in sealed Saran-polyethylene pouches, the samples retained original colour, texture and flavour, less than 10^2 g^{-1} micro-organisms, less than 60 ppm SO_2 and no detectable toxins. Equally satisfactory results were obtained by the use of ethylene oxide (C_2H_2O-CO-CO_2), but it was not possible to remove entirely the ethylene chlorohydrins and glycols, as would be required by the United States Food and Drug Authority.

Research into long-term harmful effects, using different species of animals, will be needed before any substantial oxygen reduction procedures will be considered as entirely safe by regulatory agencies, unless the process renders the product entirely sterile or incapable of supporting *Clostridium botulinum*. This could affect the speed and cost of applying a process such as the GASPAK and may delay or limit hypobaric storage, or simple vacuum packing of low-acid vegetables and fruit.

MODIFIED PRESSURES

I do not intend to review the hypobaric storage technique, but would like to mention hyperbaric, or high pressure processing and storage.

It has been demonstrated by Charm (personal communication) that at pressures of over 250 atmospheres even microbial spores are damaged. Exposing raw foods to these pressures may be a means of sterilization without the application of heat or chemical preservatives. Building pressurized vessels, introducing products into them and removing, and aseptically packaging the product may be very expensive. Shalkowski (personal communication) accomplished the same bactericidal effect by passing milk through a "jet-gun", i.e. ultra-high pressure nozzle. This process may also have its problems, particularly in reducing particle size.

Even if the high pressures could be applied economically, with little or no change in temperature and render the product practically sterile, it is not known what effect the pressures have on specific/particular enzyme systems. Thus, unless accompanied by a heating or cooling treatment, off-flavours and other undesirable effects may be found to delay use of high pressures—as was found with irradiation-preservation.

CONCLUDING REMARKS

The quality the consumer believes he wants is freshness, and prompt handling after harvest is the only way to extend fresh quality from the field or orchard to the consumer's table.

Government imposition of open dating is of questionable value, and the conventional methods of presentation by canning, freezing or drying change fresh-like quality. What remains is expanding CA storage by introducing additional gases, or modifying pressures as a treatment, with the treated raw product promptly transferred to an appropriate package which will also serve as the CA storage.

REFERENCES

Ben-Sinai, I.M., Ben-Sinai, M., Ahmed, E.M. and Kramer, A. (1965). *Fd Technol., Champaign* **19**, 856-859.

Besser, T. and Kramer, A. (1972). *J. Fd Sci.* **37**, 830-823.

Jul, M. (1971). *ASHRAE. J.* **12**, (8), 32-37.

Kidd, F., West, C. and Kidd, M.N. (1927). Gas storage of fruit. *Spec. Rep. Fd Invest.*, DSIR, No. 30.

Kramer, A. (1953). *The Food Packer* **34** (September), 11-16.

Kramer, A. (1972). "Food and the Consumer". Avi Publishing Co., Westport, Connecticut, USA.

Kramer, A. (1979). *Quick Frozen Foods* **41**, (7), 74.

Kramer, A. and Kwee, W.H. (1977). *J. Fd Sci.* **42**, 212-215.
Meadows, D.H., Meadows, D.L., Randers, J. and Behrens, W.W. (1972). "The Limits to Growth". Potomac Books, Washington, D.C.
Mercer, W.A. and Rose, W.W. (1968). Integrated Treatment of Liquid Wastes from Food Canning Operations. Publ. No. D3015. Nat. Canners Assoc. Washington, D.C.
Moleeratanond, W., Kramer, A., Ashby, B.W., Baily, W.A. and Bennet, A.M. (1978). *In* "How Ready Are Ready-to Serve Foods" (K. Paulus, ed.), pp. 294-307. S. Karger, Basel and New York.
National Academy of Sciences (1978). *In* "Postharvest Food Losses in Developing Countries", pp. 110-139. National Academy of Sciences, Washington, D.C.
Pentzer, W.T. and Kramer, A. (1976). First report, U.S. Growers Fellowship. Refrig. Res. Found. 7315 Wisconsin Ave., Washington, D.C. p. 34. (Obtainable from The Refrigeration Research Foundation).
Pirie, N.W. (1966). *Science, N.Y.* **152**, 1701-1705.
Robe, K. and Andres, C. (1978). *Fd Processing* **39**, 70-74.
Seligsohn, M.R. (1978). *Fd Eng.* **50**, (1), 24-26.
Ucko, P.J. and Dimbleby, G.W. (eds) (1968). "The Domestication and Exploitation of Plants and Animals". Aldine-Atherton, Chicago.
U.S. Congress (1978). H.R. 10358. Govt Printing Office, Washington, D.C.
Watada, A.E. (1976). *In* "Proceedings Second Tomato Quality Workshop", 166-170. Univ. Calif., Davis, Veg. Crops Series No. 178.

Section I
What is Quality?

I.1 Relating Sensory Aspects to Quality

A.A. WILLIAMS

Long Ashton Research Station, University of Bristol, Bristol BS18 9AF, England

INTRODUCTION

According to the Concise Oxford Dictionary, quality is defined as "the degree of excellence"; as quality in foods normally implies some measure of acceptability it is this definition which is normally taken when dealing with such products. Many factors can contribute to quality, ranging from those concerned with various functional properties to the various sensory attributes contributing to the final enjoyment or otherwise of the product. Although one can usually compile a list of such factors, the difficulty is in evaluating their relative importance, particularly when this can change depending on where one is in the food chain.

Irrespective of whether a horticultural product is eaten fresh or after processing, success or failure ultimately depends on the consumer's opinion. It is the criteria used by the consumer, therefore, which should be paramount when deciding whether to introduce a new cultivar for the grower or to modify some piece of equipment in the case of the processor. Unlike the grower and processor, however, the consumer is influenced largely by sensory properties although other factors such as advertising, habit, quality, variation in use, convenience, varietal name, how grown, nutritive value and, of course, price all play their part.

To understand fully what is meant by sensory quality in fruits and vegetables requires a combination of approaches (Williams, 1978a,b,c, 1979). It is first necessary to define what attributes contribute to appearance, aroma, flavour and texture of the product. By monitoring these aspects and relating them to the product's acceptability it is possible to discover how they relate to quality. Although such information helps to define what is meant by quality, to completely understand it, we must be able to relate the sensory information to factors independent of the human assessor, that is, the chemical and physical basis of these attributes in the fruit or vegetable being assessed. An understanding of the origin of particular sensory characteristics is also important if we are to improve the sensory properties of new cultivars or of a processed product scientifically rather than empirically.

This three-directional approach of obtaining descriptive information on a product and relating it to preference information on the one hand and chemical and physical information on the other has been widely used at Long Ashton Research Station (Williams and Carter, 1977; Williams *et al.*, 1977; Williams *et al.*, 1979). The problems associated with the relationship and interrelationships of the two sensory aspects are discussed in this paper although, as will become apparent, even this cannot be viewed independently of the chemical and physical data.

OBTAINING SENSORY INFORMATION

DESCRIPTIVE SENSORY ASSESSMENT OF FRUITS AND VEGETABLES

The key to evaluating sensory quality in horticultural products is an accurate description of their sensory properties. The profile procedure first advocated by the A.D. Little Corporation (Caul, 1957) offered the best approach and this has since been developed and improved by others (von Sydow and Karlsson-Ekstrom, 1971; Persson *et al.*, 1973; Clapperton, 1973, 1974; Williams, 1975; Williams and Carter, 1977). To keep terminology as rational as possible, particularly with respect to aroma characteristics, vocabularies have been based on the glossary of 44 terms advocated by Harper *et al.* (1968a,b), although to get adequate coverage specific odour terms have had to be expanded.

The questionnaire developed at Long Ashton for the assessment of apples (Williams and Carter, 1977) is typical of the vocabulary necessary to describe the sensory properties of horticultural products. The questionnaire is divided into ten main sections (Table I). Each section contains descriptions covering the various aspects of that particular attribute; a typical example of the items scored is given in Table II. Terms are defined either by reference to a standard, a photograph in the case of visual attributes or by association with commonly known foods in the case of texture. In developing any vocabulary it is essential that all terms are precise, otherwise valuable information will be lost. For example, in the assessment of an apple, scoring for the general term "fruitiness" might in one situation mean "apple-like" and in another "banana-like". Both these attributes are "fruity"; if this is the only term available for scoring, therefore, the two assessments would appear similar, whereas they are quite different.

By using standards for training, it is hoped that panelists will use the

TABLE I

Divisions of laboratory profile questionnaire and aspects of fruit on which the consumer was asked for preference information

Laboratory profile questionnaire		Consumer survey questionnaire
Main attribute	No. of descriptions available for scoring	
Appearance: External	5	Appearance
Internal	4	”
Colour: External	4	
Internal	1	
Aroma: External	21	Aroma
Internal	21	”
Taste (mouth flavour)	25	Flavour, sharpness, sweetness, juiciness
Texture	4	Texture, crispness
Feel to hand	2	
Overall quality		Overall quality

TABLE II

Attributes assessed in external aroma profile for Cox's Orange Pippin apples (Score range 0-5)

Attribute	Score	Attribute	Score
1. Overall intensity		9. Estery wax-like	
2. Complexity/depth of aroma		10. Fruity (other than apple) - specify	
3. Dried leaves/powdery		11. Banana-like	
4. Green		12. Sugary	
5. Sharp		13. Spicy	
6. Alcoholic winey		14. Fatty greasy	
7. Scented		15. Rancid	
8. Estery synthetic		16. Cox-like	

same word to describe the same sensory property. But even this may be in doubt. It is known that people can react differently to the same stimuli, particularly when giving preference information (Harper *et al.*, 1968a). People also vary in their threshold for perception of compounds (Meilgaard and Reid, 1979) and it is likely that the power function relating the concentration of a compound to its perceived intensity will also vary from individual to individual. This has obvious implications when trying to devise standards, although effects due to these factors can be minimized by using simple standards and concentrations well above their thresholds. When it comes to assessing foods and beverages, however, where one has a multistimulus system, the problem is much greater and although panelists might be trained to describe a standard with the same term, one, again, cannot guarantee they are perceiving the same sensation in the food.

Ignoring any potential interactive effects (Williams, 1978a), Fig. 1 illustrates three responses one might get from an equal concentration mixture of two stimuli, one with a rancid odour (X) and the other fruity (Y). The figures show the relationships between concentration and perceived intensity for the two stimuli for the three respondents. The solid line superimposed on these graphs shows how the equal concentration stimulus mixture intercepts these lines. For person (*i*) the concentration of the stimulus with the rancid odour is greater than its threshold value (X) and that of the stimulus with the fruity odour less than its threshold value (Y). The odour perceived is rancid, no contributions being made from the fruity component. For person (*ii*) both components are present in concentrations greater than their thresholds and both stimuli have an approximately equal power relationship between concentration and intensity of response. The rancid note is therefore stronger than the fruity note but the latter will have some effect giving a fruity background. For person (*iii*) both stimuli are present above threshold but the factor relating stimulus concentration to response is greater for the fruity stimulus than for the rancid one. Although the fruity component has a higher threshold concentration, at the concentration of the mixture it therefore dominates. Other variations can, of course, be postulated, the whole system becoming more complex when interactive effects are included and a multicomponent system, as present in a real product, is being considered.

Despite these reservations the use of detailed descriptive assessments has proved a valuable scientific and quality control tool enabling one to find out just what are the sensory differences between cultivars (Williams *et al.*, 1977) or between different manufacturing processes (Brown *et al.*, 1974; Williams, unpublished results). By relating variation in the scores given to descriptions to the analytical data (Karlsson-Ekstrom and von

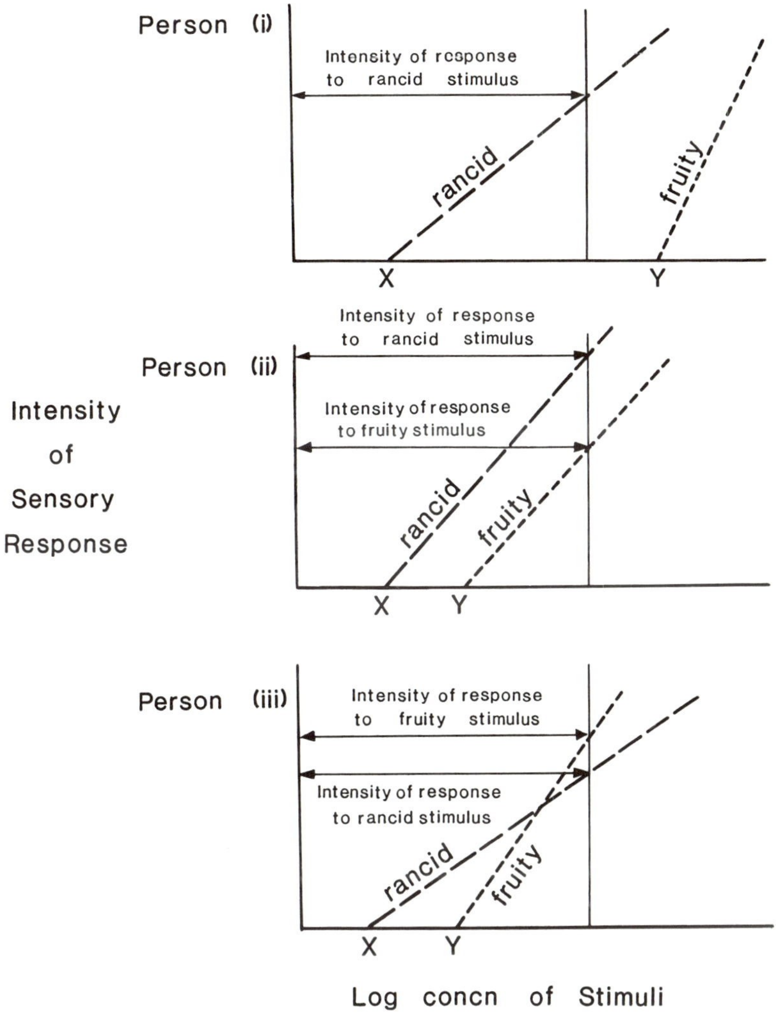

Fig. 1. Effect of variation in threshold concentration and the value of the power function relating concentration of stimuli to responses on the perceived sensation in a binary mixture of equal concentration.

— — — — — — = relationship between concentration of stimulus to intensity of sensory response for compound with "rancid" aroma; X = threshold value.

------------------ = relationship between concentration of stimulus to intensity of sensory response for compound with "fruity" aroma; Y = threshold value.

____________ = concentration of each stimulus.

Sydow, 1973), such languages have also given insight into the origin of flavour.

PREFERENCE ASSESSMENT OF FRUITS AND VEGETABLES

To relate the description of a product to quality one must have preference information. It is possible to obtain this by using the same panel as used for profile assessment, but because people vary much more in what they like and dislike than in their ability to describe a product, although useful, such information cannot be considered as representative of the consumer at large. Table III lists some approaches for obtaining such information from the consumer. The method chosen depends very much on the funds available. In our research at Long Ashton much use has been made of audiences at national agricultural shows.

TABLE III

Types of consumer tests

Information required from consumer	Place where information obtained	Approach
1. Answer to questionnaire	1. At a Centre	1. Issue and collect questionnaire and/or sample
2. Assessment of sample and answer to questionnaire: (a) monadic (b) paired comparison (c) ranking	2. In the home	2. In-depth interview 3. Group discussion

TABLE IV

General questions asked of consumers

General character		
Is the appearance of Apple A	unacceptable	☐
	just acceptable	☐
	fair	☐
	good	☐
	very good	☐

Similar questions asked of aroma, flavour and texture

It would be ideal if the consumer and profile panel could be asked similar questions. The consumer is obviously not as experienced in the assessment of products and the use of descriptive vocabulary as the trained panelist; therefore one has to be careful regarding the choice of questions asked to ensure that meaningful replies are obtained. Surveys on apples have been used to discover which terms are meaningful to consumers and untrained panelists (Holland, 1977). In Long Ashton questionnaires, in order to keep terms simple and meaningful and to be able to relate them to the laboratory assessments, questions were restricted to those given in Table I and the type of questions asked shown in Table IV.

RELATIONSHIPS WITHIN THE SENSORY DATA

The sensory data both from the trained panel and the consumer reveals various interrelationships which can be investigated, and which all have a bearing on overall quality. They can be divided into three types:

(i) The interrelationships of preference data.
(ii) The relationships between preference and objective descriptive data.
(iii) The interrelationships of objective descriptive data.

Each of these groups will now be discussed, together with the types of relationships which potentially exist, the problems in interpreting the information and their importance to understanding quality. In most assessments, the complete interdependence of many terms being scored adds to the difficulties in resolving these relationships.

INTERRELATIONSHIPS OF PREFERENCE DATA

Such relationships cover those that exist between the rating of a particular attribute, such as appearance, taste, texture, etc., and overall quality. They inherently give the relative importance of the individual attributes to overall quality and must always be positive, although the slope and shape of the curve may vary (Fig. 2). In (*i*) the influence of attribute quality on overall quality is greater with lower quality than with higher quality products, whereas with (*iii*) the reverse applies. In (*ii*) the influence of attribute quality is independent of overall quality. An increase in the quality of an attribute, however, may not have an effect on overall quality, but by its nature can never have an adverse effect. Determining these relationships in fruits or vegetables is not easy because it is virtually impossible to vary the attributes independently. Overall quality may appear to be very sensitive to changes in a relatively unimportant attribute merely through its association with one of more significance. Typical relationships of this sort are given by the regression equations of overall quality against internal colour and texture in apples (Fig. 3), the former having a much steeper slope than the latter. If the relationships could be obtained independently of each other, it is probable that internal colour would have less effect on overall quality than these graphs tend to indicate. In such an interrelated system one must try to find the effects of each variable independently of the others, the use of coefficients of determination of the linear regression equation (Table V), for example, giving a better indication of the importance of the various attributes.

Apart from the interrelationship of sensory properties in the product itself, there is evidence in the literature that variations in one attribute can influence a person's perception of others. Maga (1974), for example, reports that thresholds for basic taste sensations are affected by colour.

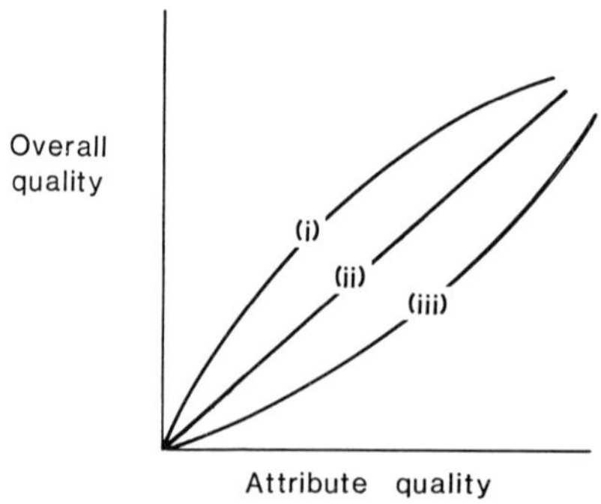

Fig. 2. Relationships between preference data.
(i) Overall quality = $k[Qa]^n + C \quad n < 1$
(ii) Overall quality = $k[Qa] + C$
(iii) Overall quality = $k[Qa]^n + C \quad n > 1$

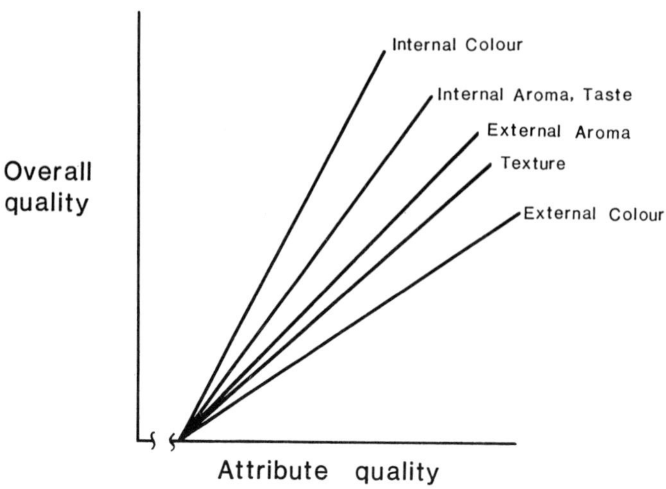

Fig. 3. Relationships between attribute quality and overall quality in apples as indicated by regression equations.

TABLE V

Coefficients of determination (%) from regression of attribute quality against overall quality for fresh and stored apples

Attribute	%
External colour	16
Internal colour	61
External aroma	28
Internal aroma	57
Taste (with peel)	74
Texture (with peel)	74

Similarly, when scoring preference, the rating given to one attribute may well influence criteria for assessing another. Given products with identical flavour characteristics for example, a good appearance could lead one to expect a good flavour; if, when tasted, it does not come up to expectations, scoring will probably be lower than if the appearance was comparable with, or inferior to, the flavour. Evidence for such interrelationships has been found in wine where the score given to flavours and aroma quality depends on whether the sample is assessed in a clear or coloured glass. In this case, when the wine had a good colour, higher scores were given for aroma and flavour when the colour of the wine was visible, the converse occurring when the wine had a relatively poor colour (Timberlake, 1976).

Such psychological effects are, however, probably small in comparison with the effect of one quality rating on the relationship of another to overall quality (Williams, 1979). This is most apparent at the low quality end of the scale where if one attribute is particularly bad, no matter how good other aspects might become the product never becomes acceptable. At higher levels of quality a similar, although less dramatic, interdependence occurs. For example, with apples it is believed that texture is the most influential attribute over the quality range as a whole, the effect of the other general attributes, such as taste and aroma, depending on the level of textural quality. With fruit whose textural quality is just acceptable (Fig. 4) (i), improvements in taste and aroma have only minimal influence on quality; improvements in texture (the major attribute) being the only factor which significantly influences overall quality. As textural quality improves (ii), its influence on overall quality becomes less while the influence of the other attributes increases. Once textural quality has reached its optimum, significant improvements in overall quality can only be achieved by increasing taste and aroma quality (iii).

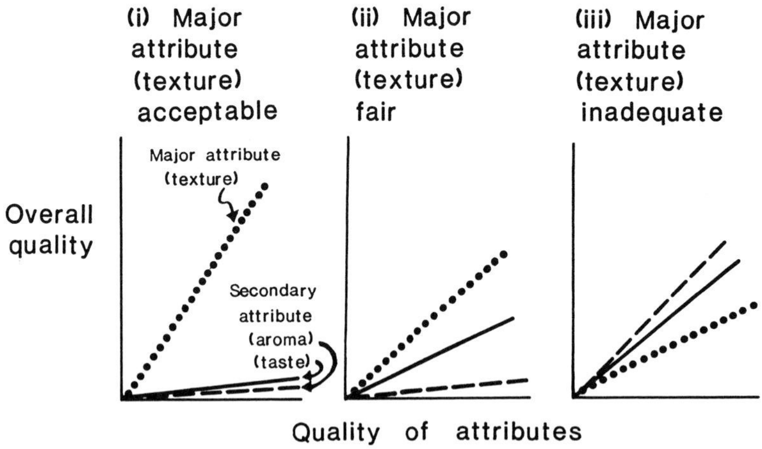

Fig. 4. Effect of one attribute on the influence of others on overall quality. Overall quality = $f_1(Qa) + f_2(Qb)(Qa) + f_3(Qc)(Qa)(Qb)$

Work at Long Ashton on Cox's Orange Pippin apples bears out these types of interdependence, particularly with respect to taste and texture (Williams, 1979). With apples given a low textural rating, there was little relationship between overall quality and taste quality. As texture improved, the influence of taste on overall quality increased.

In the case of apples, it seems that once above a critical minimum quality in all attributes, texture is the dominant sensory character. In other fruits and processed products other characteristics, such as appearance, aroma, taste or flavour character, may dominate. Alternatively, or as must exist below the point at which the major attribute begins to dominate, a more stepwise interrelationship can exist, when no one attribute can improve quality beyond a certain level until the qualities of all the other sensory attributes have been improved. In this case the relationship becomes even more complex:

$$\begin{aligned}\text{Overall quality} &= K_1 [f_1 (\text{quality } a) (\text{quality } b) (\text{quality } c) \ldots] \\ &+ K_2 [f_2 (\text{quality } b) (\text{quality } a) (\text{quality } b) \ldots] \\ &+ K_3 [f_3 (\text{quality } c) (\text{quality } a) (\text{quality } b) \ldots] \\ &+ \ldots\ldots\ldots\ldots\ldots\ldots\end{aligned}$$

RELATIONSHIPS BETWEEN PREFERENCE AND OBJECTIVE DESCRIPTIVE DATA

Preference-preference relationships are complex enough, but when relating objective descriptions of a particular attribute to quality, e.g. the intensity of sweetness, sharpness and the various flavour character descriptions to taste and mouth flavour quality, the situation is even more difficult.

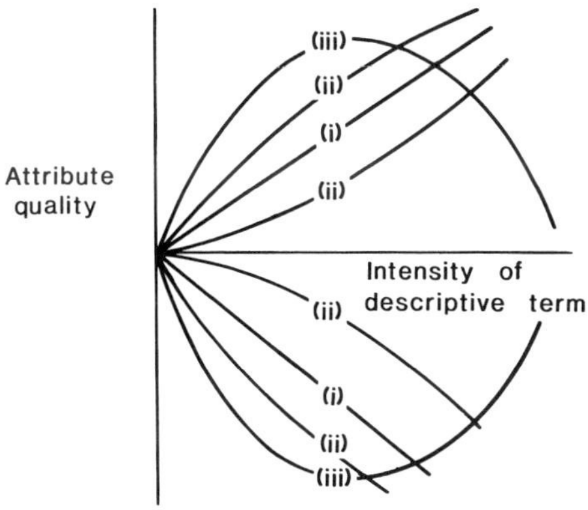

Fig. 5. Relationships between descriptive and preference data.
(i) Quality $= \mathrm{ka} + \mathrm{C}$
(ii) Quality $= \mathrm{ka}^{\mathrm{n}} + \mathrm{C}$
(iii) Quality $= \mathrm{k}[\mathrm{a} - \mathrm{B}]^{2\mathrm{x}} + \mathrm{C}$
$\mathrm{x} =$ integer

Relationships can, of course, be similar to preference-preference relationships but in this case they can be both negative and positive, for example in apples an increase in the scoring of "flouriness" will decrease textural quality, as will an increase in the sensory character rancidity decrease flavour quality. Examples of such relationships are given in Fig. 5. It is also possible to have attributes which initially relate positively to overall quality eventually reaching an optimum level above which their effect either levels out or starts to decrease quality (e.g. Fig. 5 (*iii*)).

To represent this mathematically, square terms are introduced into the equations. If a correlation improves when an objective description is used in this type of equation, there is good indication that the relationship is of this form. In cases where a linear effect dominates, the use of simple linear and systematic multiple regression techniques can help to unravel some of the relationships which exist. For example, in apples textural quality can be explained largely in terms of "crispness" and "flouriness" (Table VI), the variation in other attributes of quality being partially accounted for by variations in various other descriptive terms. Like preference-preference relationships, however, terms may be interrelated. One set of attributes may be important in defining quality at one level and others at another level. Mathematically such relationships can be expressed by a dependent equation of the type,

$$\text{Quality} = K_1[f_1\,(\text{attributes } a, b, c)] + K_2[f_2\,(\text{attributes } d, e, f)\,(\text{attributes } a, b, c)].$$

One implication of such relationships is that relative importance of attributes will change depending on the comparison being made, or, in a buying situation, the choice available. The aroma of Cox apples can be defined

TABLE VI

Information obtained from multiple regression of attribute quality against scoring of objective descriptions—aroma, texture and taste

General attribute	Most important terms in relating to general attribute quality
External aroma	Cox-like, dried leaves
Internal aroma	Cox-like, lack of potato
Texture (without peel)	Crispness, flouriness, toughness
Texture (with peel)	Crispness, flouriness, hardness
Taste (without peel)	Acidity, sweetness, depth of flavour
Taste (with peel)	Juiciness, depth of flavour

by three sensory characteristics described as "dried-leaf", "spicy" and "scented". As Cox apples are considered by many people superior to other cultivars, these attributes are obviously important when comparing Cox quality with such cultivars. When we compare Cox apples with one another, however, comparisons are being made at the top end of the quality scale. Here it seems that an over-riding "esteriness' superimposed on the basic three-attributes of Cox is the major criterion of quality.

When examining various quality relationships, preference-preference or preference-objective, it is essential to produce as wide a range as possible of all the variables, otherwise anomalous information will result. If apples vary only in colour or aroma inevitably these will be the only aspects appearing to relate to quality. Texture and taste would in this instance appear insignificant. Multivariate techniques are commonly used to examine the factors which account for the bulk of the variations in a particular set of data and often permit a reduction in the number of attributes being assessed. Care must be taken however, that attributes are not inadvertently dropped in any scoring system merely because they are not varying in the particular sample batch being examined. It is possible that in another set of samples they could be important.

THE INTERRELATIONSHIPS OF "OBJECTIVE" DESCRIPTIVE DATA

When dealing with any objective descriptive data, whether for relating to preference information or to analytical data, interrelationships of the terms used must be known. Many terms, although clear to the assessors, can overlap. The nature of such overlaps must be known. Principal component plots on the data obtained when using a vocabulary or from scoring similarities of terms give useful information in this respect. For example, Fig. 6 shows how various wine terms classified as "woody" are related.

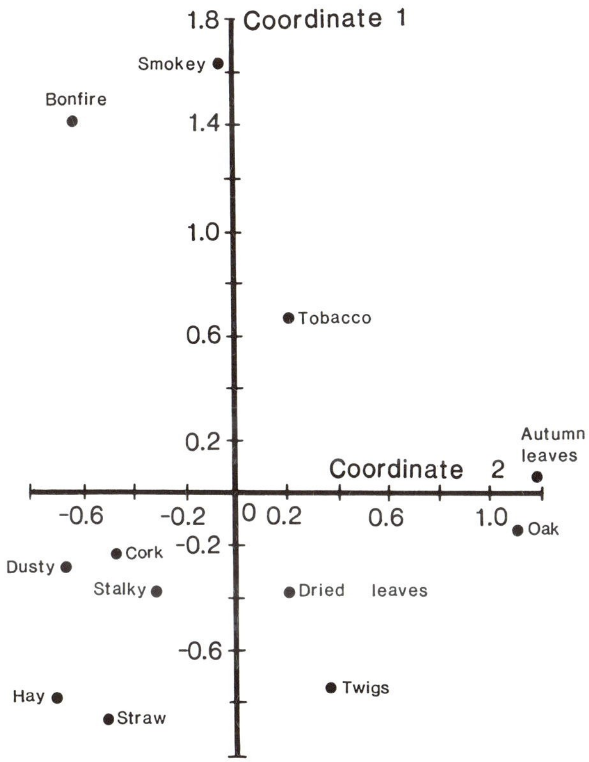

Fig. 6. First two principal co-ordinates of similarities of wine terms showing the relationship between wine terms considered to have "woody" type aromas.

In developing all vocabularies it is often necessary to include terms which the panel members feel characterize the product but which are difficult to define precisely. Such terms are varietal names for fruits e.g. "Cox-like", "Victoria plum-like" and "black currant-like" and, type names for wines, e.g. "claret-like" and "Beaujolais-like". Discussions alone do not break down such terms further, but by examining their relationship to the more precisely defined descriptions attempts at more accurate interpretations can be made. A typical example is the relationship between the scoring of "Cox-like" within the internal aroma of the apple to the other internal aroma attributes (Fig. 7). To reduce interactive effects within descriptive terms, regressions were carried out against principal component vectors; those contrasting "overripe", "rancid", "potato-like" and "cucumber-like" with "scented", "estery", "synthetic" and possibly "spicy" proved to be related significantly. Even so, these terms account for only a relatively small portion of the variation in the term "Cox-like". A possible

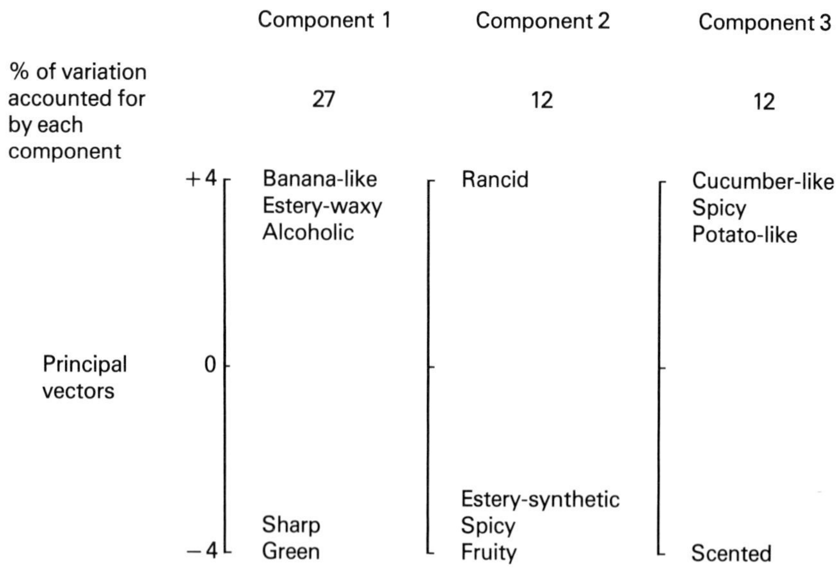

Regression of Cox-like on principal components

Cox-like = 3·6 + 0·02 (Component 1)
− 0·23 (Component 2) − 0·30 (Component 3)
+ 0·12 (Component 4) + 0·01 (Component 5)

Fig. 7. Relationship between principal components of internal aroma attributes and "Cox-like", as obtained from assessments of stored Cox apples.

explanation is that "Cox-like" is indeed a unique term, but a further explanation may be found in the nature of the task being requested of the panelist.

In all descriptive assessments of products, the panel is assessing the intensity of a descriptive term against a background of other attributes. The panelist is not just merely scoring the attributes but also his or her ability to recognize individual attributes amongst the other stimuli. Sensory attribute scores therefore will depend both on their inherent intensities and on a recognition factor which is related to the background against which the attribute is being assessed. When the product is relatively poor and therefore out of sensory balance, attributes are relatively easy to pick out, but as the product becomes more "harmonious" they become more difficult to perceive, although their actual aroma intensity may have increased. Figure 8 illustrates this effect diagramatically using three attributes, *a*, *b* and *c*, potentially important to an ill-defined character, e.g.

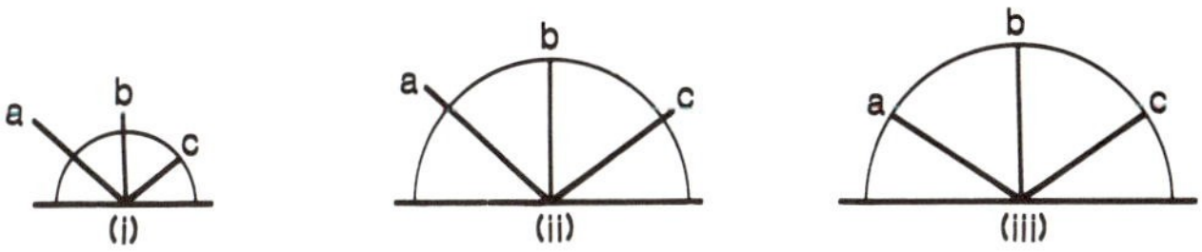

Fig. 8. Relationships of specific attributes to general attributes.

Intensity of general attribute $= f_1(\min f_2(a,b,c))$

"Cox-like"; this character comprises three terms, indicated by the hemicircular region of balance (it is assumed for this exercise that all three attributes, *a*, *b* and *c*, have equal weight). In Fig. 8 (i) the product is out of balance. The amounts of the attributes, a, b, c, extending beyond this hemicircular area are easily recognizable. As the quality improves (i to iii) the area of overall harmony increases, but the actual true levels of the attributes *a*, *b*, *c*, may also alter, some increasing and others decreasing. Due to the increased intensity of the harmonious region, however, these attributes, even though they might have increased in actual intensity, are now becoming more difficult to perceive and may well be scored more weakly than before. In Fig. 8 (iii) the product is completely harmonious and, although the attributes contributing to the ill-defined character are all stronger than in (i), they are virtually undiscernable over the background.

Postulated models of this nature clearly indicate the difficulties in unravelling the true interrelationships of descriptive sensory data and, in particular, of discerning the pattern of attributes giving rise to a general ill-defined term. The problem is that although the overall intensity of a product can be evaluated sensorily, the inherent intensity of individual attributes cannot. To understand just what happens when attributes merge to form a pattern, it is necessary to have some fixed reference point to which to relate perceived intensity and from which one can estimate the potential attribute intensity. This fixed point can only be the chemical components or physical attribute responsible for the sensation in the fruit or vegetable itself. Once the basis of the various sensations is understood, one can then explore relationships amongst them and from this information deduce the true interrelationships which are occurring amongst the sensory attributes themselves.

VARIATION WITHIN POPULATIONS

When examining sensory information, whether preference or objective, it is usual to average scores for the whole panel and deal with panel means. This implies that everyone is inherently the same and that variations are all due to random error in the assessment. Unfortunately, as already seen, people vary in their likes and dislikes and, to some extent, in what they actually perceive. Rather than dealing with mean scores (or the average consumer), scientists should be investigating response patterns within the community, and evaluating the magnitude of the various subsets and the relationships of attributes within the most important of these. An examination at Long Ashton of relationships between attribute rating and overall ratings for individual members of an apple assessment panel has shown that people can be segregated on the importance they appear to attach to particular attributes (Williams, 1979). Within this panel, the two major groups are most influenced by taste and texture, but differed in the importance they attached to the other attributes. The third group related all attributes equally to quality. These groupings were based on preference relationships; when it comes to objective-preference relationships and explaining the ill-defined descriptions, do the same attributes have equal importance to everybody or do groupings again exist? While further work into various response patterns will help to resolve this question, to truly understand what is determining such groupings it will again be necessary to resort to physical and chemical information.

CONCLUSIONS

Sensory methods, both using trained panels or the consumer, can produce useful information. Trying to identify interrelationships within this data, their meaning and what ultimately gives rise to quality, is not, however, a straightforward task. Even where good mathematical relationships exist, care must be taken in dissociating those which are purely mathematical from those which have meaning in terms of quality rating. However, whether investigating the reasons for the variation in a particular population or the interrelationships within descriptive terms, the chemical constituents and physical properties of the fruit or vegetable giving rise to the various sensations are the final key as to what is happening. To understand sensory phenomena, sensory responses must be related to such data.

Despite the problems outlined, if quality in horticulture is to be improved to the benefit of the consumer, processor and grower, it is essential that we understand exactly what is meant by quality in sensory, chemical and physical terms.

REFERENCES

Brown, D.G.O., Clapperton, J.F. and Dalgliesh, C.E. (1974). *Proc. Amer. Soc. Brew. Chem.* **32**, 1-4.
Caul, J.F. (1957). *Adv. in Fd Res.* **7**, 1-40.
Clapperton, J.F. (1973). *J. Inst. Brew.* **79**, 495-508.
Clapperton, J.F. (1974). *J. Inst. Brew.* **80**, 164-173.

Harper, R., Bate Smith, E.C., Land, D.G. and Griffiths, N.M. (1968a). *Perfumes-sent. Oil. Rec.* **55**, 22-36.
Harper, R., Land, D.G., Griffiths, N.M. and Bate Smith, E.C. (1968b). *Br. J. Psychol.* **59**, 231-252.
Holland, D.A. (1977). *J. Sci. Fd Agric.* **28**, 436-442.
Karlsson-Ekstrom, G. and von Sydow, E. (1973). *Lebensm.-Wiss.-Technol.* **6**, 86-89.
Maga, J.A. (1974). *Chem. Senses and Flavor,* **1**, 115-119.
Meilgaard, M.C. and Reid, D.S. (1979). *In* "Progress in Flavour Research" (D.G. Land and H.E. Nursten, eds), pp. 287-305. Applied Science, London.
Persson, T., von Sydow, E. and Åkesson, C.A.J. (1973). *J. Fd Sci.* **38**, 682-689.
Timberlake, C.F. (1976). Proceedings of the 1st Wine Subject Day, Long Ashton Research Station, 17 Sept. 1976, 25-37.
von Sydow, E., and Karlesson-Ekstrom, G. (1971). *Lebensm. Wiss.-Technol.* **4**, 152-157.
Williams, A.A. (1975). *J. Sci. Fd Agric.* **26**, 567-582.
Williams, A.A. and Carter, C.S. (1977). *J. Sci. Fd Agric.* **28**, 1090-1104.
Williams, A.A. and Knee, M. (1977). *Ann. appl. Biol.* **87**, 127-131.
Williams, A.A. (1978a). *Int. Flavour Fd Addit.* **9**, 80-85.
Williams, A.A. (1978b). *Int. Flavour Fd Addit.* **9**, 131-133.
Williams, A.A. (1978c). *Int. Flavour Fd Addit.* **9**, 171-175.
Williams, A.A. (1979). *In* "Progress in Flavour Research" (D.G. Land and H.E Nursten, eds), pp. 287-305. Applied Science, London.
Williams, A.A., Lea, A.G.H. and Timberlake, C.F. (1977). *In* "Flavour Quality, Objective Measurement" (R.A. Scanlan, ed.) Am. Chem. Soc. Symp. Ser., No. 51, 71-88.
Williams, A.A., Carter, C.S. and Langron, S. (1979). Proceedings of the 4th Wine Subject Day, Long Ashton Research Station, 6-7 September, 1978, 57-67.

I.2 Aroma in Fruit and Vegetables

T.R. GORMLEY

The Agricultural Institute
Kinsealy Research Centre, Malahide Road, Dublin, Eire

INTRODUCTION

Quality of food can be divided into sensory and hidden quality. In the sensory area there is appearance, texture, and flavour. Of these characteristics, appearance is the most important at point of purchase. This is followed by texture and then by flavour, i.e. aroma and taste when the food is consumed (Kramer, 1973). While it has been established that taste is "four dimensional", aroma is much more complex and has many dimensions.

The importance of aroma in fruit and vegetables in relation to consumer acceptability is a matter of debate. For example, there is no doubt that with onions aroma is important, as it is with certain apple cultivars and in wine and fruit juices. It can be countered that aroma of many foods is only of major importance to the flavourist who wishes to blend chemical compounds to produce a flavour and is of secondary importance to the consumer who may buy on price, appearance and texture.

Kazeniac (1977) has pointed out that an important criterion for the success of a food flavour appears to be recognition and compatability of the flavour of the food with flavours familiar to the consumer. The increased interest in home-grown foods offers evidence of changing habits towards natural food flavours. In the UK there are signs that the domestic purchaser is selecting fruit on sensory properties (Williams and Knee, 1977) especially Cox's Orange Pippin apples. However, the position with Golden Delicious is different and both Belgian, British and Irish consumers seem to buy on the basis of colour (Monin, 1974; Centre Français du Commerce Exterieur, 1975; Gormley and Egan, 1977).

This paper reviews some of the work on the identification of aroma compounds in fruit and vegetables, gives an account of aroma related studies at Kinsealy Research Centre and touches on other aspects of aroma including off-flavours and aroma in processed foods.

AROMA COMPOUNDS IN FRUIT AND VEGETABLES

The separation and identification of aroma giving compounds in fruit and vegetables is complex, involving gas liquid chromatography (g.l.c.) and mass spectrometry (m.s.) studies and the contribution of the compounds thus identified to human sensory perception is even more difficult to quantify. There is extensive literature on this topic and only some review articles are referred to here.

Von Sydow (1971) suggested a system of odour classification, with 40 odour qualities, and also model correlation techniques using single or multiple correlations. Flavour chemistry in the 1970s has been outlined by Rohan (1972), while Nursten (1975) has discussed the chemistry of flavour, past, present and future.

VEGETABLES

Macleod (1970) has reviewed the chemistry of vegetable flavours pointing out that less attention had been given to vegetables than to fruits, juices and alcoholic beverages. A survey has also been carried out on vegetable volatiles by Johnson *et al.* (1971,a,b). The genus *Allium* contains the most highly flavoured vegetables, i.e. onion, garlic, and chives, and their flavour chemistry has been outlined by Abraham *et al.* (1976) and Whitaker (1976). The most important flavour compounds in the *Allium* species include thiols, sulphides, disulphides, trisulphides, and thiosulphinates. In brassicas, sulphur compounds are again important. Brassicas contain sinigrin which under the action of enzymes can produce allyl isothiocyanate as a potential flavour compound. In blanched products, e.g. frozen Brussels sprouts, the enzymes are destroyed and some of the flavour compounds are not formed. The complexity of the situation in relation to the number of volatile compounds present can be seen as follows; in French beans 24 volatile substances were identified by g.l.c. of which only two seemed to contribute to flavour (Kramer and Fox, 1966); in peas about 80 compounds have been identified (Johnson *et al.*, 1971a), while in tomatoes about 110 have been found (Johnson *et al.*, 1971b).

FRUIT

Information on volatiles in fruit is extensive and Nursten (1970) has reviewed fruit volatiles under the headings of nature of the volatiles and their biosynthesis. He lists 160 different volatile compounds for apples, 56 for bananas, 155 for oranges, and 143 for strawberries. Most of these compounds are acids, esters, carbonyls, ethers, or acetals. He suggested that methanol may come from pectin, branched alcohols from the appropriate amino acids as may aldehydes. Acetals are probably formed from aldehydes and alcohols, organic acids are usually formed from other organic acids or sugars while the formation of esters may be more complex than a straight interaction between alcohols and acids. Salunkhe and Do (1976) have reviewed the biogenesis of aroma constituents of fruits and vegetables.

IMPORTANT VOLATILE FLAVOUR COMPONENTS

With regard to the number of compounds producing aromas in fruit and vegetables, the question arises as to the practical significance of these. This can be approached in two ways; their significance to the flavourist and their significance to the consumer or buyer.

Fruit can be placed in four categories on the basis of their flavour volatiles: (i) fruits whose aroma is largely given by one compound, i.e. a character impact compound; (ii) fruits in which a small number of compounds are involved; (iii) fruits where a large number of compounds are required to reproduce the aroma; and (iv) fruits whose aroma cannot be reproduced even by a large number of compounds (Nursten, 1977). In addition to hydroxy compounds, aldehydes, ketones, acids, esters, and sulphur-containing compounds, oxygen heterocyclics have also been found to be important (Flament, 1975). Pyrazines are important components in vegetables especially those with a methoxyl group (Murray and Whitfield, 1975).

Some foods whose aroma is largely given by one compound or group of similar compounds are as follows: pear (*trans*-2, *cis*-4-decadienoates), cucumber (*trans*-2, *cis*-6-nonadienal), green pepper (2-isobutyl-3-methoxypyrazine), raw potato (2-isopropyl-3-methoxypyrazine), cooked mushrooms (1-octen-3-one), and beetroot (geosmin). In Delicious apples, raspberries, tomatoes and celery, aroma is given by a character impact compound plus a number of other contributory flavour compounds. Williams *et al.* (1977) have shown that 4-methoxyallylbenzene is an important compound in Cox apple aroma.

FLAVOURISTS, FOOD PROCESSORS, AND CONSUMERS

The basic work described above is of considerable significance for flavourists, who are continually seeking new and better synthetic and extracted flavours. It also has application for food processors as it may enable them to modify a process in order to conserve a flavour. Plant production specialists and plant breeders also benefit as they may be able to grow or select for increased amounts of character impact or aroma compounds while the "fresh" food storage expert may be able to improve storage conditions.

When the aroma of a food is due to a character impact compound, correlation with sensory evaluation can be tested by splitting the carrier gas into several streams on exit from the g.l.c. where panelists can smell the aromatics. When a number of compounds rather than a chemical impact compound is involved the correlation with sensory evaluation is more difficult and complex techniques as described by von Sydow *et al.* (1970) must be used. A scheme for establishing the organoleptic importance of the individual components of a flavour in a systematic fashion has been suggested by Parliment and Scarpellino (1977) using modified flavour profile techniques.

Broderick (1974a,b, 1975a,b) in a series of articles on the aroma components in fruits outlines how basic g.l.c. work may assist the flavourist. He lists aroma components identified in the fruits prior to 1952 some of which were used by flavourists. This is followed by those flavour components used commercially but not identified in the fruits. He then describes the additional results obtained since instrumental methods became standard procedures and points out that with apples, research results on aroma to date—for the flavourist—have not warranted the cost. In contrast, the research worker and flavourist have combined forces to produce good banana flavours. With pineapple, researchers have some distance to go before the results are likely to influence commercial pineapple flavours. In raspberries it has been established that α- and β-ionone are present, but flavourists had used ionones prior to their identification. As for strawberries, the extracted volatiles give a poor commercial flavour and the flavour must still be altered to suit the end use. Broderick concludes that the combination of the flavourist and aroma researcher has on balance been rewarding.

AROMA-RELATED RESEARCH AT KINSEALY RESEARCH CENTRE

Aroma-related research is carried out at Kinsealy Research Centre on apples, onions, and mushrooms. Most of the work on apples relates to Golden Delicious as it is the most important apple cultivar in Ireland. However, before giving details of this work it is useful to describe some concepts which are relevant to all food research at Kinsealy.

THE FOOD TRIANGLE, THE FOOD SYSTEM, THE QUALITY SPECTRUM

The food triangle employs the concept of a triangle with food in the centre, and the sides of the triangle representing the consumer, food quality, and human nutrition. While these three aspects are interrelated, a separate emphasis is given to each in all our food research. While aroma is a major aspect of sensory quality and falls on the quality side of the triangle, it must also be looked at from the consumer point of view—in the final analysis what does it mean to the consumer?

The food system is the second concept. Food is produced in a system which is continuous from the soil to the point of consumption and its quality including aroma, can be influenced by many factors along the way. One of the most important influences is intensification—with the motto "Yield Rules All"—being applied and this has resulted in a lowering of quality in some crops.

The third concept is the quality spectrum which is a list of quality factors for a given food in order of importance and also the point in the food chain where each quality factor is important. In this context, aroma of fruit and vegetables is important at the point of consumption and its degree of importance varies according to the cultivar in question.

PRACTICAL AROMA MEASUREMENTS

Because of the high cost of equipment for complex aroma analysis of foods many laboratories must use more practical aroma measurements. Two possible approaches are considered below.

Linking to another index

Taste panels at Kinsealy Research Centre have noted that there is a relationship in Golden Delicious apples between skin colour and aroma. Golden Delicious apples that have developed a degree of blush have a rich aroma while green fruit have little or no aroma. It is possible, therefore, to gain considerable information on aroma through linking it to colour data which is more readily measured. This has only been tested for Golden Delicious and may only apply in Ireland where there is considerable variation in skin colour of Golden Delicious from season to season and within cultural treatments in a given season.

These observations by taste panels were confirmed by selecting four groups of Golden Delicious apples which varied in colour from blush/yellow to yellow to green/yellow to green. The correct placing of the apples in four categories was confirmed by taking Hunter "a" value measurements on each apple on the blush and shaded sides and obtaining a mean for the two values. A different set of four apples was used for each of 12 panelists who were asked to place the apples in descending order from the most to least aromatic, on the basis of smell. The tests were carried out in a dark room where panelists were not influenced by the colour of the fruit. The results (Table I) show the mean Hunter "a" values for the four groups of apples and also the aroma ranking given by the panelists.

TABLE I

Aroma panel rank scores for 48 Golden Delicious apples placed in four categories on the basis of skin colour

Apple category	Hunter "a" values (mean)[a] for skin	Rank totals for 12 panelists[b]
Blush/yellow	+5·1	20 (most aromatic)
Yellow	−3·3	24
Green/yellow	−6·6	29
Green	−8·9	47 (least aromatic)

[a]Skin colour, means for blush and shaded side (Hunter "a" values, red (+), green (−)).
[b]Range for significance, 12 panelists × 4 samples, 21-39 ($P<0{\cdot}05$)

The blush/yellow samples were rated most aromatic and the green least aromatic with the yellow and green/yellow apples also being placed in the correct order. The results confirm the previous observations and suggest that skin colour can be used as a convenient index of aroma in Golden Delicious apples.

It is possible that linking to another index could also apply to other fruits and vegetables, e.g. a high soluble solids (SS) content probably indicates good husbandry and possibly a warm growing season which could also be reflected in a large amount of aroma giving compounds.

Saucepan-cone test system

This system employs a set of four to six stainless steel saucepans each fitted with a lid in the form of a cone. The saucepans, 19 cm in diameter, 10 cm tall, have a cone 10 cm from apex to base. There is a 4-cm aperture at the top of the cone which can be closed or opened by a stopper. This system enables six samples of food to be tested by panelists. A weighed quantity of sample is used and the panel members cannot see the sample. Aroma can be tested at different temperatures by placing the saucepans in a water bath or on a cooker in the case of foods during cooking. This system may have application for the aroma testing of soft fruit, jams, fruit pulps, soups, apples, and cooked foods, either during or after cooking. Samples are normally ranked from the most to the least aromatic or from the most to the least preferred aroma. The system can be used to evaluate the effect of production, storage, packaging, and processing techniques on the aroma of a wide range of fruit and vegetables. Loss of aroma in a given sample over a period of weeks or months could also be tested provided some form of calibration can be devised, possible using a known concentration of a synthetic or natural aroma which is characteristic of the food being tested.

APPLE AROMA

Flavour versus aroma of some apple cultivars/selections

The aroma and flavour of 26 cultivars/selections was compared using aroma and taste panels. Five samples were tested in each of five separate panels with Golden Delicious being included as the sixth sample in each panel (Table II). The samples were tested in December, 1978 (panels 1-5) and again in February, 1979 (panels 1A-5A) after post-harvest storage at 2°C. There were ten panelists in each panel. The apples were smelt unpeeled and were then evaluated for flavour as peeled slices with different codes the following day by the same panelists. Each panelist was asked to rank the samples from most (1) to least (6) aromatic; the rank totals were added and were referred to rank total tables (Kramer and Twigg, 1966; Gormley and Sherington, 1978) to determine statistical significance. In the flavour test each panelist was asked to rank samples from best (1) to worst (6) for flavour and the data analysed as above.

The rank totals in each panel for flavour and aroma were in turn ranked

TABLE II

Apple cultivars/selections tested for aroma, flavour and % soluble solids

Panels 1 and 1A	Panels 2 and 2A	Panels 3 and 3A
Sinta	Empire[a]	Nero 26[a]
Golden Delicious	Melrose	Jonathan
Suntan	Starkspur	Gala
Karmijn	Moss' Seedling[a]	Charden
Ivette	Golden Delicious	Golden Delicious
Kent	Crispin	Goldjon

Panels 4 and 4A	Panels 5 and 5A
Aori No. 1[a]	GD 85
Aori No. 2[a]	GD Smoothee
GD 88	GD C446
Gloster 69	GD 4E-23-10
Idared 68A	Golden Delicious
Golden Delicious	GD Stark

Panels 1-5 carried out December, 1978.
Panels 1A-5A carried out February, 1979.
[a]Tested in December, only. GD = Golden Delicious.

in order of magnitude with the lowest total being given 1 and the highest 6. Results show that there was a good relationship between aroma and flavour ranks for five cultivars/selections and between flavour rank and SS rank for nine cultivars/selections. Close agreement between the ranks for all three factors, i.e. aroma, flavour and SS was found for fruit of the cultivars Empire, Idared and the Japanese selections from Aomori Research Station, Aori No. 1 and Aori No. 2. In the samples tested in February, 1979 a good relationship was obtained between the ranks for aroma and flavour for six cultivars/selections and between flavour and SS ranks for ten cultivars/selections. Close agreement between ranks for all factors was obtained for the cultivars/selections Ivette, Starkspur, Goldjon, Idared, and GD Smoothee. These data suggest that aroma may be contributing significantly to the flavour of certain cultivars. However, this did not apply to all cultivars. These observations are made bearing in mind the limitations of the experiment in that panelists were asked to rank samples from most to least aromatic—and not for aroma preference—whereas they were asked to rank for flavour preference. It must also be recognized that it is not possible to have all cultivars/selections at their optimum for eating at the same time as some have longer high quality storage life than others.

French versus Irish Golden Delicious

A study (Gormley and Egan, 1977) in the seasons 1974-75 and 1975-76 compared the quality of Irish- and French-grown Golden Delicious apples. Samples were purchased at retail outlets in Dublin and the French Goldens were on average 48 and 28% more expensive per kg than the Irish in 1974-75 and 1975-76 respectively. The main quality difference between fruit from the two sources was in colour, the French fruit being green, and the Irish green/yellow to golden.

Thirty one samples were taken in 1974-75 and 27 in 1975-76. Each sample constituted 12 Irish and 12 French Goldens. The data (Table III) show that the Irish apples were yellow/golden in colour while the French apples were green which suggests on the basis of the data in Table I that the Irish apples had more aroma. This was borne out at the time of testing when taste panelists often remarked that the Irish Golden Delicious apples had a pleasant aroma while the green French apples had practically no detectable aroma. Yet, over a total of 58 paired comparison taste panels using peeled apples during the course of the experiment, the Irish apples were preferred significantly on six occasions, the French on nine occasions while on the remaining 43 occasions there was no significant preference. The grand mean SS values for Irish and French apples in 1974-75 were 12·4 and 12·2% respectively and in 1975-76 the values were 13·0 and 12·0%. These data suggest that aroma seemed to have little effect on flavour acceptance and

TABLE III

Mean skin colour[a] data (Hunter "a" values[b]) for Irish and French Golden Delicious apples purchased in November-March, 1974-75 and 1975-76

Month	French	Irish	Month	French	Irish
Nov. 74	−6	+7	Nov. 75	−10	+6
Dec. 74	−9	+7	Dec. 75	−7	+1
Jan. 75	−9	+4	Jan. 76	−8	−1
Feb. 75	−10	+5	Feb. 76	−9	−3
Mar. 75	−5	+7	Mar. 76	−9	0

[a]Measured on blush or most ripe side.
[b]a(−) = green, a = 0, yellow, a(+) = red.

the consumer seemed to be prepared to buy more expensive green French Golden Delicious apples. This may be because consumers equate the green colour with Granny Smith apples imported prior to the Golden Delicious season. The attitude of British consumers to French Goldens is similar (Centre Français du Commerce Extérieur, 1975); they also tend to think of green apples as crunchy and juicy and of yellow ones as over-mature and soft. Because French Goldens are picked at an immature stage they are greener than the Irish fruit. However, under Irish climatic conditions the natural colour for Golden Delicious apples is green/yellow to golden; if allowed to reach this stage of ripeness they will have good eating quality but if consumed when green they are of poor eating quality with little aroma (Gormley, 1975). The results of this study also agree with those from Belgium where a pink flush in Golden Delicious apples was consistent with high eating quality (Monin, 1974). The Belgian study also showed that home-grown Goldens were selling for a much lower price than those imported from France or the Cape.

Soil management and apple aroma

The effect of method of soil management on the quality of Golden Delicious apples was reported by Gormley *et al.* (1973) and it was shown that soil treatments giving high yields, e.g. overall herbicides, gave fruit of lower quality than plots which had been herbicide-treated or clean cultivated for three years followed by grassing down. Testing of fruit from this experiment was continued up to and including 1975 with similar results. The treatments were as follows:

(1) Clean cultivation.
(2) Clean cultivation for three years followed by grassing down.
(3) Overall herbicides.
(4) Overall herbicides for three years followed by grassing down.

(5) Overall herbicides plus straw mulch.
(6) Grassing down from planting time.

The data (Table IV) show mean values for skin colour and yield of Golden Delicious fruit from this experiment over the period 1973-75. The "grassing down" treatments (2 and 4) gave yellower fruit than the overall herbicide treatments and the rank correlation coefficient between skin colour and yield was −0·71. This suggests that fruit from high yielding treatments have less aroma than those from low yielding treatments. This may be related to the leaf/fruit ratio as discussed by Gormley *et al.* (1973). In the seasons 1969-71 and 1973-75 the rank correlation coefficients between skin colour and yield were negative and values ranged from −0·27 to −0·70. These data have application in relation to the effect of "intensification" on quality as discussed above. Gormley *et al.* (1973) have also indicated that yellowness in Golden Delicious apples may be associated with high SS values, so a favourable taste panel response to yellow Irish Golden Delicious apples could be due to a high SS content or a more intense aroma. However, it is more likely to be associated with a combination of both.

TABLE IV

Mean skin colour data and yield for Golden Delicious apples grown under different soil managements (1973-75)

Soil management treatment	Skin colour[a] (Hunter "a", blush side)	Yield (kg/tree)
1	−1·1	48·8
2	+3·2	42·8
3	−3·5	58·2
4	+0·6	41·7
5	−3·4	48·9
6	−1·6	42·1

[a] a(−) = green, a = 0, yellow, a(+) = red

Soil/climate and aroma

A study was begun in 1972 (O'Kennedy and McDonnell, 1974) to evaluate the performance of Golden Delicious apples in seven orchards representing a range of soil and climatic conditions. Data for skin colour (non-blush side), yield and SS over the seasons 1973-75 are presented in Table V; once again there was a high inverse relationship between skin colour and yield indicating that fruit from high yielding trees would have less aroma. The corresponding SS figures also correlated negatively with yield.

These experiments show that Golden Delicious, when grown in Ireland,

TABLE V

Mean skin colour data, % soluble solids and yield for Golden Delicious apples grown at seven different orchard locations over three seasons (1973-75)

Orchard location	Skin colour (C) (Hunter "a", non-blush side)	% Soluble solids	Yield (kg/tree) (Y)
Leixlip	−6·9	14·2	14
Tullow	−8·2	13·1	22
Cuffesgrange	−8·0	13·6	24
Bennettsbridge	−5·0	14·6	12
Piltown	−8·7	13·8	23
Dungarvan	−7·7	13·7	14
Clonroche	−6·9	14·1	10

Rank correlation coefficients: $C \times Y = -0{\cdot}78$
$\% SS \times Y = -0{\cdot}83$

appears sensitive to factors such as soil management, climate, and others. Whereas green Irish GD apples are of very poor flavour with low levels of SS and practically no aroma, if allowed to develop a degree of blush they have an excellent aroma and a high level of SS.

ONIONS

Onions have long been popular and remain widely used. They are usually graded on the basis of pungency and lachrymatory factor. There is extensive literature on the nature, estimation and origin of the volatiles in onions including recent papers by Tewari and Bandyopadhyay (1977), and Freeman and Whenham (1974, 1976). The chemistry of onion aroma will not be discussed here except to note that the volatile flavour compounds are not present in the intact vegetable but are formed by enzyme (alliinase) action on flavour precursors—derivatives of L-cysteine—to give pyruvic acid, ammonia, and sulphenic acid. Sulphenic acid undergoes a further change to give the lachrymatory compound, thiopropanol-S-oxide.

In Ireland onions valued at about £600,000 were imported in 1977 in addition to £35,000 worth of pickled onions. Research at Kinsealy Research Centre (Kenny, personal communication) aims to produce home-grown onions with good pungency and dry matter (DM) characteristics. A DM value equal to or greater than 12% is required for dehydration. Some aspects of this research are outlined below.

Pungency measurement

The Chemical Oxygen Demand (C.O.D.) method of Saguy *et al.* (1970), which has a correlation factor of $r = -0{\cdot}98$ with odour threshold, was used. The method requires that the onion/water macerate be left for 15 min before distillation. However, with many samples it would be convenient to macerate them and distil them in batches. This would result in samples being left standing for different lengths of time between maceration and distillation. Pungency C.O.D. tests on fresh and powdered onions left standing for up to 18·5 h between maceration and distillation show that the C.O.D. figures (i.e. the amount of water soluble reducing substances) increase with time of standing of the onion/water mixture prior to distillation (Table VI).

Standing overnight gave a two-fold increase in C.O.D. value for both fresh and dehydrated onion, and it is therefore necessary to adhere to a definite blending/distillation time interval whether it be 15 min or a conveniently longer period.

TABLE VI

Changes in pungency values (C.O.D. method) with varying time interval between blending and distilling the onion macerate

Interval between macerating and distillation (h)	C.O.D. values (ppm sucrose)	
	Fresh onion	Powdered onion
0·7	1844	1249
2·0	2781	1816
17·0	4235	2990
18·5	4371	2964

Onion pungency—effect of peat and mineral soils

Pungency and dry matter content in onion cultivars grown in peat and mineral soil was tested six times between October 1972 and May 1973 (Table VII). The results show that dry matter (DM) values were suitably high for dehydration and all cultivars had similar pungency values. The effect of soil type was small. However, when the same cultivars and some additional ones were grown in peat in 1973-74 (Table VIII), DM and pungency values were very much lower. The differences between the two seasons were linked with prevailing climatic conditions. These data indicate that a payment scheme relating DM × pungency × ha would be desirable. Extensive tests on the pungency of dehydrated onion powder over a four-year-period showed that cv. Dura gave powder with a mean C.O.D. value

TABLE VII

Mean pungency (C.O.D. method) and dry matter (DM) values for onions grown in peat and mineral soil[a]

Cultivar	Peat soil		Mineral soil	
	C.O.D. (ppm sucrose)	% DM	C.O.D. (ppm sucrose)	% DM
Dura	2370	14·0	2712	15·1
Hygrow	2374	13·1	2236	12·6
Wijbo	2098	12·4	2384	13·3
Produdure	2466	12·7	2085	13·9
Mean	2327	13·1	2354	13·7

[a]Samples tested six times between October 1972 and May 1973.

TABLE VIII

Mean pungency (C.O.D. method) and dry matter (DM) values for onion cultivars grown in peat[a]

Cultivar	%DM	C.O.D. (ppm sucrose)	Cultivar	%DM	C.O.D. (ppm sucrose)
Buan	12·0	1657	Laco	9·9	2167
Hyduro	10·6	1627	Hygro	9·7	1816
Dura	10·8	1451	Wijbo	9·7	1824
Hyprodo	9·8	1860	Robusta	9·5	1859
Solidor	10·0	2027	Augusta	9·5	2004

[a]Samples tested four times between December 1973 and May 1974.

(ppm sucrose) of 1757 compared with 2166 for powder from cv. Robusta. The DM content of the cultivars was in the reverse order, Dura 13·1% and Robusta 10·5%.

MUSHROOMS

Mushrooms, like onions, have long been used to garnish foods. Aroma in the common field mushroom resides mainly in the pileus and also the central portion of the stipe exclusive of the cuticle (Bernhard and Simone, 1959).

Observations at Kinsealy Research Centre have suggested that freshly-picked mushrooms (*Agaricus bisporus*) have a poor flavour which improves during subsequent storage. This was confirmed by a taste panel in which 17 tasters ranked a series of five samples of cooked mushrooms for flavour of age zero, one, two, and three days post-harvest. The mushrooms had been stored at 15°C. The fifth sample was freshly-harvested flat mushrooms. All samples were sautéed in butter. The freshly-harvested sample was significantly downgraded ($P<0.01$) for flavour, there being no difference between the other samples. In subsequent aroma assessments the panel preferred the aroma of the one- and two-day post-harvest sautéed mushrooms and downgraded that of the zero and three-day post-harvest samples. A combination of the flavour and aroma data above suggests that mushrooms held at about 15°C have an optimum flavour about two to three days after harvest.

OTHER AROMA STUDIES

Finally, other examples where aroma of fruit and vegetables is important are also considered briefly.

FRESH FRUIT AND VEGETABLES

Chill storage will generally reduce rate of loss of aroma. However, while chill storage maintains strawberries in good condition for two or three days as far as appearance and texture was concerned, loss of aroma was very noticeable. This was less evident in raspberries (Kenny, 1975). Certain forms of packaging, e.g. pre-packaging with synthetic films, may conserve aroma.

PROCESSED FRUIT AND VEGETABLES

Blanching results in loss of aroma in fruit and vegetables and may also destroy enzymes which are responsible for flavour development. In canned and dried vegetables, aroma is lost during heating and other odours may develop during the heating process. This does not imply that canned and dried products are inferior, but only that aroma and taste may be very different from those of the fresh fruit and vegetables (Guadagni, 1969). In canned fruits and vegetables the aroma is diluted by the syrup or brine respectively; in food dehydration volatiles are lost during drying, while in frozen foods a gradual loss of volatiles during storage occurs.

FRUIT AND VEGETABLE PRODUCTS

These include vegetable powders, jams, fruit juices, fruit pulps, wine and other items.

The aroma of vegetable powders and pieces (kibble), especially onion, is of particular importance as they are used as flavour ingredients based largely on volatile components (Kenny, 1973). Celeriac (turnip root celery) powder is used as a substitute for celery powder in some countries.

The importance of aroma in jam flavour may be greater than is generally realized. The soluble solids and natural acidity of soft fruit is modified in jam-making by the addition of sucrose and citric acid. The flavour as sensed by the consumer may be largely due to the aroma producing compounds. Bearing in mind that volatiles are lost during the boiling process, more attention should be given to the aroma of new strawberry cultivars/ selections intended for jam-making using a simple test system such as the saucepan/cone system described in p. 40. Tests on confectionary jams using artificial flavouring materials (flavour boosters) obtained from two sources showed that in general, amounts recommended for addition were too high. Apricot, raspberry, and blackcurrant flavours were best but strawberry flavours from both sources gave jam with a very synthetic flavour (Gormley, 1973).

Aroma is important in fruit juice flavour and aspects of apple juice flavour have been reviewed by O'Beirne (1976). Kenny (personal communication) showed that apple juices made from Bramley's Seedling, Golden Delicious, Cox's Orange Pippin, and Laxton's Superb apples had low flavour acceptability when consumed individually. Combinations of Bramley's Seedling with other juices gave an improved flavour to the product while the addition of Cox juice gave a blend with a high level of aroma. Aroma is also important in the flavour of wine.

OFF-FLAVOUR

Many of the off-flavours in foods are caused by aroma compounds produced by changes in the food itself or by the interaction of outside agents, e.g. micro-organisms, with the food. A frequent occurrence in fruit is the production of alcohol to give a beer-like taste. Catty taints occur as sporadic outbreaks of objectionable odours in a range of foods including carrots, processed peas, raspberries, lemon juice and others. This off-flavour is due to mesityl oxide which is a contaminant found in solvents and elsewhere and may enter the food-chain in can lacquers, through the use of solvents for solubilizing colours and flavours, and by migration from packaging materials (Spencer, 1969).

REFERENCES

Abraham, K.O., Shankaranarayana, M.L., Raghavan, B. and Natarajan, C.P. (1976). *Lebensm.-Wiss. Technol.* **9**, 193-200.

Bernhard, R.A. and Simone, M.J. (1959). *Fd Res.* **24**, 165-166.

Broderick, J.J. (1974a). *Flavour Ind.* **5**, 184-185.

Broderick, J.J. (1974b). *Flavour Ind.* **5**, 284-285.

Broderick, J.J. (1975). *Int. Flavours Fd Addit.* **6**, 351.

Broderick, J.J. (1976a). *Int. Flavour Fd Addit.* **7**, 27,30.

Broderick, J.J. (1976b). *Int. Flavours Fd Addit.* **7**, 51-52.

Centre Français du Commerce Extérieur. (1975). Gouts et attitudes des consommateurs britanniques à l'égard des pommes et en particulier de la Golden Delicious. Paris. pp. 37.
Flament, I. (1975). *In* "Aroma Research", Proceedings of the International Symposium on Aroma Research (H. Maarse and P.J. Groenen, eds), pp. 221-237. Centre for Agricultural Publishing and Documentation, Wageningen.
Freeman, G.G. and Whenham, R.J. (1974). *J. Sci. Fd Agric.* **25**, 499-515.
Freeman, G.G. and Whenham, R.J. (1976). *Int. Flavours Fd Addit.* **7**, 222-227, 229.
Gormley, T.R. (1973). *Res. Rep. An Foras Talunt. (Hort.).* 75-76.
Gormley, T.R. (1975). *Lebensm.-Wiss. Technol.* **8**, 168-171.
Gormley, T.R. and Egan, J.P. (1977). *Ir. J. Fd Sci. Technol.* **1**, 23-32.
Gormley, T.R. and Sherington, J. (1978). *Ir J. Fd Sci. Technol.* **2**, 59-66.
Gormley, T.R., Robinson, D.W. and O'Kennedy, N.D. (1973). *J. Sci. Fd Agric.* **24**, 227-239.
Guadagni, D.G. (1969). *In* "Quality and Stability in Frozen Foods" (W.B. von Arsdel, M.J. Copley and R.L. Olsen, eds), pp. 84-116. Wiley-Interscience.
Johnson, A.E., Nursten, H.E. and Williams, A.A. (1971a). *Chemy Ind.* **21**, 556-565.
Johnson, A.E., Nursten, H.E. and Williams, A.A. (1971b). *Chemy Ind.* **43**, 1212-1224.
Kazeniac, S.J. (1977). *Fd. Technol., Champaign* **31**, 26-33.
Kenny, A. (1973). *Fm Fd Res.* **4**, 85-87.
Kenny, A. (1975). *Fm Fd Res.* **6**, 64-66.
Kramer, A. (1973). "Food and the Consumer". Avi Publishing Co., Westport, Connecticut, USA.
Kramer, A. and Fox, M. (1966). *Fd Technol., Champaign* **20**, 1594-1595.
Kramer, A. and Twigg, B.A. (1966). "Fundamentals of Quality Control for the Food Industry". Avi Publishing Co., Westport, Connecticut, USA.
Macleod, A.J. (1970). *Flavour Ind.* **1**, 665-672.
Monin, A. (1974). *Acta Hort.* **40**, 467-469.
Murray, K.E. and Whitfield, F.B. (1975). *J. Sci. Fd Agric.* **26**, 973-986.
Nursten, H.E. (1970). *In* "The Biochemistry of Fruits and their Products" (A.C. Hulme, ed.), Vol. 1, pp. 239-268. Academic Press, New York and London.
Nursten, H.E. (1975). *Int. Flavours Fd Addit.* **6**, 75-82.
Nursten, H.E. (1977). *In* "Sensory Properties of Food" (G.G. Birch, J.G. Brennan and J.K. Parker, eds), pp. 151-166. Applied Science, London.
O'Beirne, D. (1976). "Flavour and Processing Aspects of Irish Apple Juice." M. Agr. Sc. Thesis, University College, Dublin.
O'Kennedy, N.D. and McDonnell, P. (1974). *Res. Rep. An Foras Talunt. (Hort.) 1973*, 47-48.
Parliment, T.H. and Scarpellino, R. (1977). *J. agric. Fd Chem.* **25**, 97-99.
Rohan, T.A. (1972). *Fd Process. Ind.* **41**, 33-36.
Saguy, M., Mannheim, C.H. and Peleg, Y. (1970). *J. Fd Technol.* **5**, 165-171.
Salunkhe, D.K. and Do, J.Y. (1976). *Crit. Rev. Fd Sci. Nutr.* **8**, 161-190.
Spencer, R. (1969). British Food Manufacturing Industries Research Association Laymans guide. Catty Taints in Food, August 1-6.
Tewari, G.M. and Bandyopadhyay, C. (1977). *Lebensm.-Wiss. Technol.* **10**, 94-96.
von Sydow, E. (1971). *Fd Technol., Champaign* **25**, 40-44.
von Sydow, E., Andersson, J. Anjou, K., Karlsson, G., Land, D. and Griffiths, N.M. (1970). *Lebensm.-Wiss. Technol.* **3**, 11-17.

Whitaker, J.R. (1976). *Adv. Fd Res.* **22**, 73-133.
Williams, A.A. and Knee, M. (1977). *Ann. appl. Biol.* **87**, 127-131.
Williams, A.A., Tucknott, O.G. and Lewis, M.J. (1977). *J. Sci. Fd Agric.* **28**, 185-190.

I.3 Cell Wall Structure in Relation to Texture

M.A. HALL

Department of Botany and Microbiology,
University College of Wales, Aberystwyth,
Aberystwyth, Dyfed SY23 3DA, Wales

INTRODUCTION

Any attempt to relate the biochemical, biophysical and chemical properties of higher plant cell walls to textural characteristics must, of necessity, resort to hypothesis and speculation. The reasons for this involve an admission of the existence of enormous gaps in our knowledge of the nature of cell walls. This situation has improved considerably recently but much remains to be done. This paper will, in the first place, set out what has been done and what it is now feasible to undertake, and will then attempt to provide a basis for further work on cell walls in relation to texture.

Clearly, no progress can be made in this direction unless and until we have rigorously defined models for the cell walls in question and it is at this point that the gaps in our knowledge begin to become apparent. In fact, the cell walls of only four plant species have been examined in any detail in this respect, namely those of sycamore callus (Keegstra *et al.*, 1973), pea epicotyl (Gilkes and Hall, 1977), maize coleoptile (Darvill *et al.*, 1977, 1978) and red kidney bean (Nusbaum and Albersheim, personal communication). This is not to say that we lack knowledge about the precise structure of cell wall components from many plant species and different plant parts (see e.g. McNeil *et al.*, 1979), but, as discussed below, the way in which the components of the cell wall interact and are interconnected is at least as important as a knowledge of the structure of the components themselves.

GENERAL STRUCTURE OF PLANT CELL WALLS

Cell walls, from whatever source, can be said to comprise a three phase system, namely a crystalline or semi-crystalline fibrillar component, cellulose; a typically amorphous component, the matrix, consisting largely of non-cellulosic polysaccharides but also including significant quantities of protein; and lastly, a packing component. In the primary cell wall the packing component is water and this may make up more than half the total fresh weight of the wall. In secondary cell walls the water is replaced by other materials, lignin in xylem tissue, cellulose in other cases. In some instances various substances such as suberin, cutin and tannins are deposited in the cell wall.

When considering the mechanical properties of cell wall structure, Northcote (1972) has pointed out that analogies may be drawn between the properties of plant cell wall structures and those of various man-made materials such as steel-reinforced concrete and glass fibre-reinforced plastic. Fortunately, we know a good deal about the factors upon which the properties of these man-made materials depend. Thus, the tensile strength is dependent on (a) the length and cross-section of the fibrillar component; (b) the relative amounts of fibre and matrix; (c) the elastic moduli of the components; (d) the arrangement of the fibres in the matrix; (e) the interaction between the matrix component and the microfibrillar component and (f) the interactions between components of the matrix. Consideration of factors (e) and (f) makes it clear why we need to know not only the structure of cell wall components but also how they interact. It is also germane to observe that whereas processing of plant material is unlikely to affect either (a) or (b), the converse is true of (e) and (f).

ANALYSIS OF CELL WALL COMPOSITION

Modern methods for the analysis of cell wall structure owe much to the pioneering work of Albersheim and his co-workers (see Talmadge *et al.*, 1973). The fundamental techniques rely on modifications of the well tried method of methylation analysis whereby the position of the glycosidic linkages may be elucidated. This type of analysis is illustrated in Fig. 1. While the overall linkage distribution in a cell wall may be determined in this way, such "whole wall" determinations do not provide information on the distribution of such linkages within individual polysaccharides or indeed how the latter are interconnected. To achieve this end it is desirable to remove specific components of the wall. In the past this was done by extraction with such treatments as dilute acid or alkali. While work on the polysaccharides so extracted has given valuable information, the methods of extraction are relatively unspecific in their action and furthermore may generate artefacts such as alteration in degree of polymerization. A more desirable approach is to use enzymes capable of hydrolysing *specific* glycosidic linkages. By the judicious selection of such enzymes it is possible to obtain specific polysaccharides intact and also much information on their connections to other cell wall components.

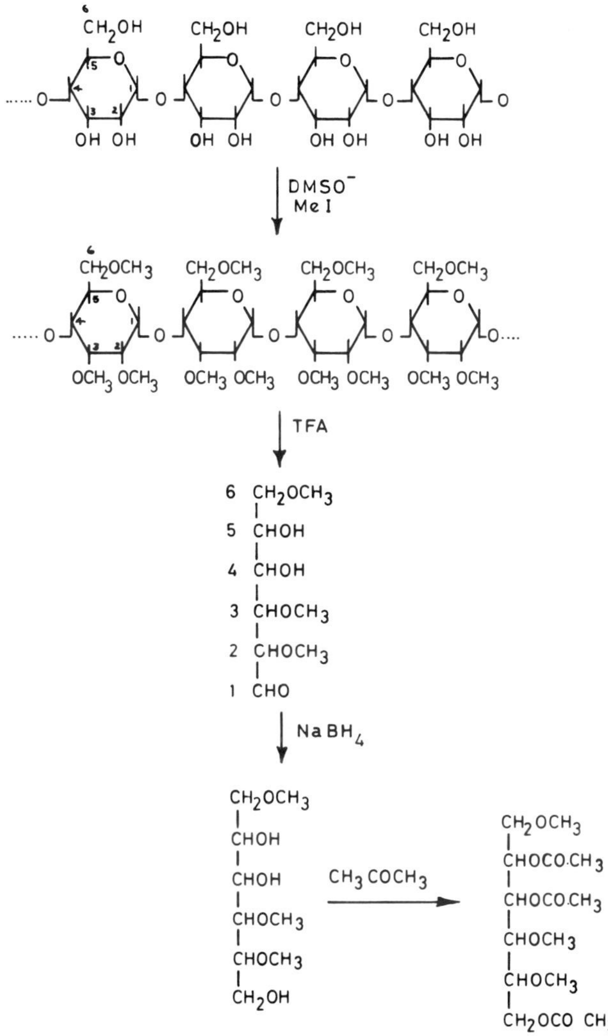

Fig. 1. Procedure for determining glycosidic linkage composition of a polysaccharide. The final products—methylated acetylated alditols—are separated and identified by gas liquid chromatography—mass spectrometry. DMSO$^-$—methyl sulphinyl anion; MeI—methyl iodide; TFA—trifluoroacetic acid; $NaBH_4$—sodium borohydride; CH_3COCH_3—acetic anhydride.

A number of such enzymes have now been used in cell wall studies, including endopolygalacturonase (Talmadge *et al.*, 1973; Darvill *et al.*, 1977), glucanase (Bauer *et al.*, 1973), arabanase (Weinstein and Albersheim, personal communication) and galactanase (Labavitch *et al.*, 1976). It is crucial to purify such enzymes before use, as they are usually derived from micro-organisms and crude preparations are contaminated with other polysaccharide hydrolases and glycosidases. At present, most work on cell walls is concerned with the problems referred to above but researchers are increasingly turning their attention to the secondary and tertiary structures of cell wall polymers (e.g. Rees and Welsh, 1977) and some of these aspects will also be dealt with below.

THE NATURE OF CELL WALL COMPONENTS

POLYSACCHARIDES

The structure and properties of the microfibrillar component of the cell wall, namely cellulose, are so well known that a detailed description is unnecessary. Suffice to say that cellulose is a $\beta 1 \rightarrow 4$-linked glucan, the aggregation of which into microfibrils is a consequence of inter- and intramolecular hydrogen bonding. Such is the degree of order of the aggregated glucans within a fibre that they are in fact crystalline. It is this property of cellulose which confers on it the property of high tensile strength.

Although further work is needed, it is now possible to categorize the main types of polysaccharide present in the cell wall matrices of mono- and of dicotyledons and further to put forward reasonable hypotheses concerning their arrangement *in muro*.

The original grouping of matrix polysaccharides into a pectic fraction and a hemicellulosic fraction was formulated on the basis of extractability in various solvents. This has led to some confusion and to apparently contradictory findings since the extractants are far from specific for particular polysaccharide groupings. The modern classification (Bauer *et al.*, 1973) defines the pectic polysaccharides as those found in covalent association with galacturonosyl-containing polysaccharides, namely rhamnogalacturonan I and II, homogalacturonan, arabans, galactans and arabinogalactans; whereas hemicelluloses are defined as those polysaccharides non-covalently associated with cellulose, namely the xyloglucans and glucuronoarabinoxylans. Examples of these various types of polysaccharide are given in Fig. 2.

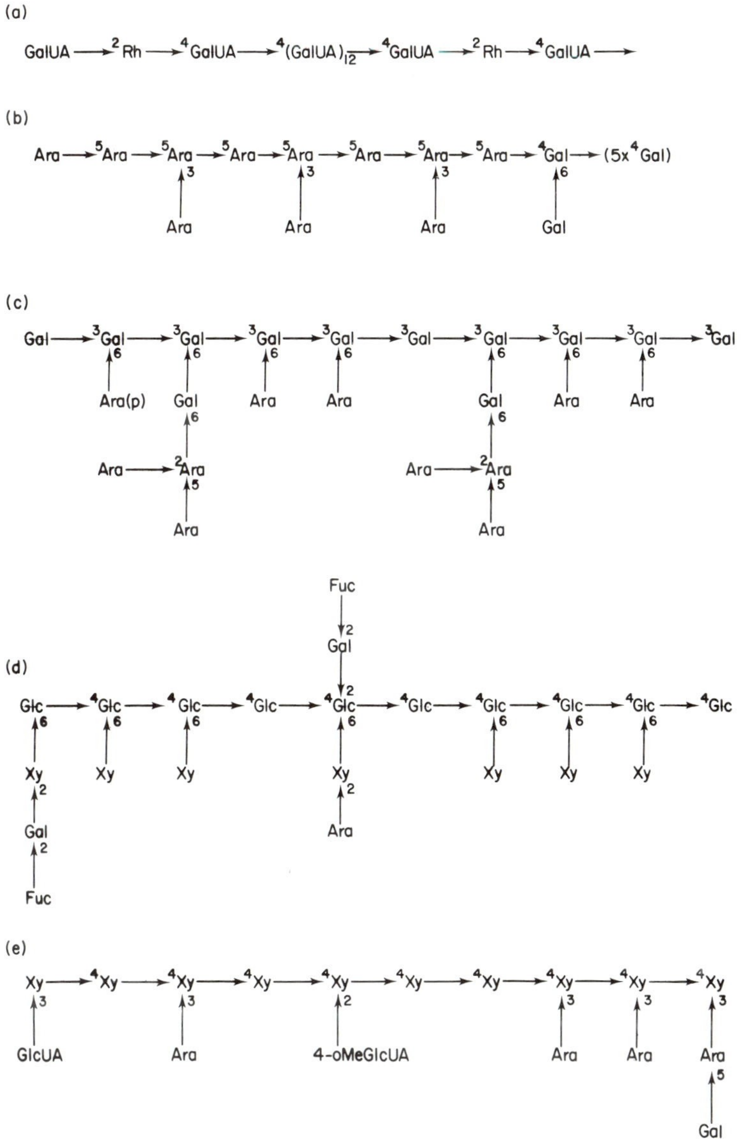

Fig. 2. Matrix polysaccharides of higher plants: a. Rhamnogalacturonan I; b. Arabinan/Galactan; c. Arabinogalactan; d. Xyloglucan; and e. Glucuronoarabinoxylan. Glc—glucose; Ara—arabinose; Xy—xylose; Rh—rhamnose; Gal—galactose; GalUA—galacturonic acid; Fuc—fucose; GlcUA—glucuronic acid.

(Fig. 2. a—d, after Talmadge *et al.*, 1973; Bauer *et al.*, 1973; and Keegstra *et al.*, 1973. Fig. 2. e, after Darvill *et al.*, 1977; Darvill, 1976; Smith, 1977.)

CELL WALL PROTEIN

Primary cell walls contain 5-10% protein (e.g. Lamport, 1970) which is particularly rich in hydroxyproline (up to 20%), alanine, serine and threonine. Fragments of such proteins always contain arabinosyl and galactosyl residues. It is now established that the hydroxyproline residues are substituted with tetra-arabinosides (Lamport and Miller, 1971) and that the galactosyl residues are attached to serine residues (Lamport *et al.*, 1973) (Fig. 3).

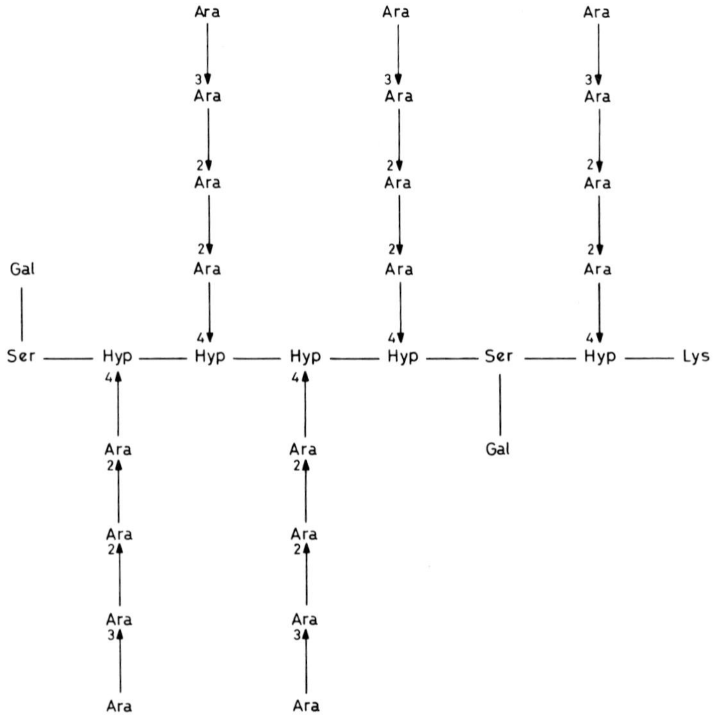

Fig. 3. Proposed structure of hydroxyproline-rich glycoprotein from cell walls. Ara—arabinose; Gal—galactose; Hyp—hydroxyproline; Ser—serine; Lys—lysine (Akiyama and Kato, 1977; Lamport *et al.*, 1973).

INTERCONNECTIONS BETWEEN COMPONENTS OF THE CELL WALL

COVALENT INTERCONNECTIONS

It is now well established that the neutral pectic polysaccharides are covalently attached to the acidic pectic polysaccharide rhamnogalacturonan I and it is generally accepted that homogalacturonan is attached to rhamnogalacturonan. It seems likely that rhamnogalacturonan II is covalently attached to other pectic polysaccharides but definitive proof is lacking.

Some evidence has been adduced to suggest that covalent attachments exist between xyloglucan and pectic polysaccharides in the wall (Bauer *et al.*, 1973; Keegstra *et al.*, 1973) but the same authors also demonstrated that this was only true of a portion of the xyloglucan. It is not known whether glucuronoarabinoxylan is covalently attached to pectic polysaccharide.

It was thought initially that the hydroxyproline-rich protein was connected to polysaccharide components of the matrix through the galactosylserine linkage but this is now open to doubt (Monro *et al.*, 1976).

NON-COVALENT INTERACTIONS

It has now been demonstrated that both xyloglucan and glucuronoarabinoxylan bond strongly to cellulose via multiple hydrogen bonds (Bauer *et al.*, 1973; Darvill *et al.*, 1978; McNeil *et al.*, 1979). The $\beta 1 \rightarrow 4$-linked glucan backbone of xyloglucan hydrogen bonds to cellulose in the same way as pairs of cellulose molecules, namely via the hydroxyl groups at the 6-position and glycosidic oxygens in adjacent chains. In xyloglucan the xylose substituents at the 6-position of many parts of the glucan backbone prevent further association. In sycamore and pea cell walls there is just sufficient xyloglucan to completely cover the existing cellulose microfibrils. In maize and other monocots on the other hand, glucuronarabinoxylans or arabinoxylans appear to constitute the major polysaccharide component of the matrix and are therefore present in excess relative to cellulose. There is evidence, however, that such components may aggregate to form mixtures of random coils and linear chains (Dea *et al.*, 1973, 1977) which in turn may lead to gel formation and may be a means of cross-linking to other cell wall polymers (see below).

Rather more information is available concerning the associations formed by pectic polysaccharides. If we consider first homogalacturonan—the simplest case—it is well known that elements of these chains can pack together in association with calcium or other cations to form gels. The calcium fits into cavities between the chains which seem to be of the right size and such cavities are lined with carboxylate and other electronegative oxygen atoms which are favourable ligands (Fig. 4) (Grant *et al.*, 1973). Calcium binding is greatly enhanced by the presence of homogalacturonan sequences more than 20 residues in length.

Similar associations will occur between molecules of rhamnogalacturonan I but the behaviour is likely to be modified since packing is inhibited by the presence of the rhamnosyl residues. Equally, substitution of arabinogalactan, arabinan and galactan is likely to reduce regular packing (Rees

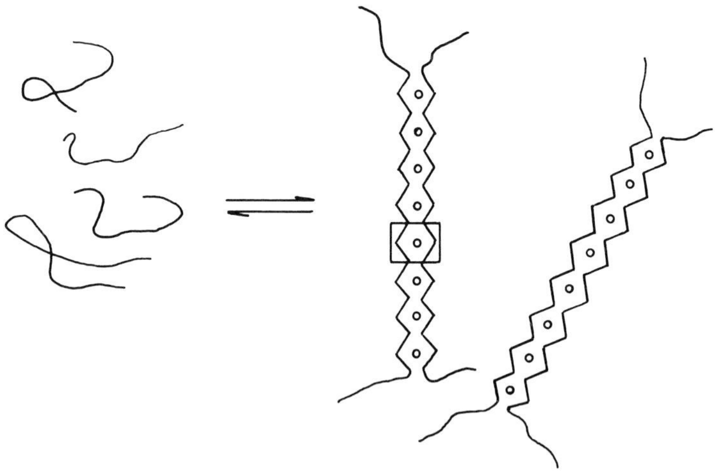

Fig. 4. Representation of association of polygalacturonate sequences by chelation of Ca^{2+} (Grant *et al.*, 1973). Circles represent calcium molecules, lines polygalacturonate sequences.

and Wight, 1969). Methyl esterification of the carboxyl groups of polyuronide residues results in much reduced interactions with cations and hence gel formation does not occur under the same conditions as with the polyelectrolyte. If, however, the water activity is lowered in the vicinity of the chains, gelling may occur and this process may be enhanced by low pH which supresses the ionization of those unesterified carboxyl groups which are present.

The strongest evidence that interactions occur between the pectic polysaccharides and hemicelluloses is that solubilization of the former by the use of endopolygalacturonase facilitates the removal of the latter from the wall. However, the nature of these interactions, and whether they occur between the hemicellulose and the neutral pectic polysaccharides, the acidic pectic polysaccharides, or both, is unclear.

It was at first thought that the hydroxyproline-rich protein of the wall was connected to polysaccharide components of the matrix through the galactosyl-serine linkage, but this is now open to doubt. However, it has been demonstrated that lectin-like protein fractions from cell walls of mung beans (*Phaseolus aureus*) bind specifically to galactosyl residues (Kauss and Bowles, 1976) and if such lectins are similar to those from potato tubers (Allen and Neuberger, 1973) which in turn closely resemble cell wall hydroxyproline-rich glycoprotein, it is possible that strong non-covalent glycoprotein/polysaccharide interactions may be a feature of cell walls (e.g. Kauss and Bowles, 1976).

THE CELL WALL COMPOSITE

For the cell walls examined so far, the picture emerges of cellulose fibres embedded in hemicellulose to which they are hydrogen-bonded. The cellulose/hemicellulose aggregates are interconnected—probably via non-covalent but possibly via covalent binding—by a network of pectic polysaccharide and/or protein which itself probably has a complex and varied tertiary structure. Fundamental differences appear to exist between the matrix compositions of mono- and dicotyledons and this is illustrated in Table I.

TABLE I

Percentage polymer composition of cell walls

Wall component	Suspension-cultured sycamore[a]	Pea epicotyl[b]	Maize coleoptile[c]
A. Pectic polysaccharides			
Rhamnogalacturonan I	7	} 11	—
Rhamnogalacturonan II	3		—
Homogalacturonan	6		6
Arabinan	9	8	—
Galactan/Arabinogalactan	9	8	7
B. Hemicelluloses			
Xyloglucan	19	23	—
Glucuronoarabinoxylan	5	—	38
Glucomannan	—	7	5
C. Cellulose	23	25	24
D. Hydroxyproline-rich glycoprotein	19	18	20

[a]McNeil *et al.*, 1979. [b]Gilkes and Hall, 1977. [c]Darvill *et al.*, 1977, 1978; Darvill, 1976.

TEXTURE

It is now appropriate to turn to the relationship between the data on cell walls and cell wall components outlined above and practical observations on the effects of various treatments on texture. Clearly, workers using particular processing techniques and particular plant material will need to consider the following proposals in the light of their own experience; nevertheless, it is possible to make certain generalizations about possible reasons why some of the more commonly used processing techniques have the effects they do and perhaps to suggest how less empirical approaches to the problem might be adopted.

It is perhaps appropriate to begin by dealing with a process where it is relatively easy to relate the effects of the treatment to cell wall structure; namely the process whereby calcium is included in the water used for blanching various vegetables in order to prevent sloughing and softening. This clearly relates to the role of calcium in gelation of pectic poiysaccharide (cf. p. 59). The increased gelation in itself results in firmer texture but because this also increases the stability of the interconnections between the cellulose/hemicellulose components and the rest of the matrix, the process is further enhanced. The activity of enzymes capable of de-esterifying the acidic pectic polysaccharides is also likely to be of importance here and low temperature pre-treatments may be used to enhance such de-esterification.

We can distinguish between effects which are mediated via processes in the middle lamella and hence relate to cell adhesion and those where the cell walls of individual cells are involved. In the above example, a major part of the effect is probably due to reactions in the middle lamella—which consists predominantly of polyuronide—and hence is related to cell adhesion.

Any simple pre-treatment is unlikely to affect only one component in the wall and the effect obtained is probably the result of several processes, not all of them necessarily acting in the same direction.

The significance of this statement is best seen in relation to the factors likely to enter into any processing procedure, namely pH and temperature. The possible effects of pH on polyuronide components has already been mentioned (p. 60), but there are further ramifications. Thus, the hydrogen bonding at the hemicellulose/cellulose interface may well be affected by a shift to acid pH. Equally, at moderate temperature and acid pH, a proportion of the arabinose and galactose side chains of the wall glycoprotein will be cleaved and as these side chains are probably important in determining interactions between glycoprotein and the rest of the matrix, a reduction in cohesion may occur.

Another aspect of the effect of pH relates to work on the mechanisms of cell extension. Thus, it appears that the increase in wall plasticity induced by growth regulators such as indolylacetic acid which in turn leads to increased growth, is mediated via an acceleration in the activity of an outwardly directed proton pump which causes a drop in the pH of the wall (Cleland, 1973). Such increases in plasticity can be induced by low pH treatment of tissue (Rayle and Cleland, 1970). There seem to exist in the wall, polysaccharide hydrolases and group transfer enzymes active at around pH 5. These enzymes catalyse the cleavage and possibly the reformation of linkages in hemicellulose and to a lesser extent in the pectic polysaccharide (Labavitch and Ray, 1974; Gilkes and Hall, 1977; Darvill *et al.*, 1977, 1978). The reduced interaction between cell wall components as a consequence of this rearrangement of linkages causes marked changes in the rheological properties of the wall resulting in an increased plastic modulus. Clearly, therefore, the pH of the wall at harvest and in the period between harvest and processing, quite apart from the pH during processing itself, is a subject which deserves close attention. Because growth regulators have marked effects upon the properties of cell walls, not only via the proton pump mechanism but also via the synthesis of wall constituents (Baker and Ray, 1965) and even via re-orientation of microfibrils (Ridge and Osborne, 1971), it might be worthwhile exploring the effects of such substances on textural properties.

To some extent the foregoing discussion raises more questions than it answers, but as noted in the introduction this is a consequence of the paucity of knowledge both on cell walls and on their relationship to textural quality. Nevertheless, progress in cell wall research has been rapid in the last ten years and if such progress continues and is linked in appropriate circumstances to textural quality, then the prospects for a rational rather than an empirical approach to processing for texture may well be feasible. Equally, such advances would make possible the rationalization of plant breeding programmes.

ACKNOWLEDGEMENTS

The author wishes to thank Professor P. Albersheim and Dr A.G. Darvill for making unpublished material available to him.

REFERENCES

Akiyama, Y. and Kato, K. (1977). *Agric. biol. Chem.* **41**, 79-81.
Allen, A.K. and Neuberger, A. (1973). *Biochem. J.* **135**, 307-314.
Baker, D.B. and Ray, P.M. (1965). *Pl. Physiol.* **40**, 345-352.
Bauer, W.D., Talmadge, K.W., Keegstra, K. and Albersheim, P. (1973). *Pl. Physiol.* **51**, 174-187.
Cleland, R. (1973). *Proc. natn. Acad. Sci. U.S.A.* **70**, 3092-3093.
Darvill, A.G. (1976). Studies on maize cell walls in relation to extension growth. Ph.D. thesis, University of Wales.
Darvill, A.G., Smith, C.J. and Hall, M.A. (1977). *In* "Regulation of Cell Membrane Activities in Plants" (E. Marrè and O. Ciferri, eds), pp. 275-282. North-Holland, Amsterdam and Oxford.
Darvill, A.G., Smith, C.J. and Hall, M.A. (1978). *New Phytol.* **80**, 503-516.
Dea, I.C.M., Rees, D.A., Beveridge, R.J. and Richards, G.N. (1973). *Carbohyd. Res.* **29**, 363-372.

Dea, I.C.M., Morris, E.R., Rees, D.A., Welsh, E.J., Barnes, H.A. and Price, J. (1977). *Carbohyd. Res.* **57**, 249-272.
Gilkes, N.R. and Hall, M.A. (1977). *New Phytol.* **78**, 1-15.
Grant, G.T., Morris, E.R., Rees, D.A., Smith, P.J.C. and Thom, D. (1973). *FEBS Lett.* **32**, 195-198.
Kauss, H. and Bowles, D.J. (1976). *Planta.* **130**, 169-174.
Keegstra, K., Talmadge, K.W., Bauer, W.D. and Albersheim, P. (1973). *Pl. Physiol.* **51**, 188-196.
Labavitch, J.M. and Ray, P.M. (1974). *Pl. Physiol.* **54**, 499-502.
Labavitch, J.M., Freeman, L.E. and Albersheim, P. (1976). *J. biol. Chem.* **251**, 5904-5910.
Lamport, D.T.A. (1970). *A. Rev. Pl. Physiol.* **21**, 235-270.
Lamport, D.T.A. and Miller, D.H. (1971). *Pl. Physiol.* **48**, 454-456.
Lamport, D.T.A., Katona, L. and Roerig, S. (1973). *Biochem. J.* **133**, 125-131.
McNeil, M., Darvill, A.G. and Albersheim, P. (1979). *Prog. Chem. org. nat. Prod.* **37**, 191-243.
Monro, J.A., Penny, D. and Bailey, R.W. (1976). *Phytochemistry* **15**, 1193-1198.
Northcote, D.H. (1972), *A. Rev. Pl. Physiol.* **23**, 113-132.
Rayle, D.L. and Cleland, R. (1970). *Pl. Physiol.* **46**, 250-253.
Rees, D.A. and Welsh, E.J. (1977). *Agnew. Chem. int.* (*Ed. Engl.*) **16**, 214-224.
Rees, D.A. and Wight, N.J. (1969). *Biochem. J.* **115**, 431-439.
Ridge, I. and Osborne, D.J. (1971). *Nature, Lond.* (*New Biol.*), **229**, 205-208.
Smith, C.J. (1977). Studies on cell wall structure and function in *Zea mays L.* Ph.D. thesis, University of Wales.
Talmadge, K.W., Keegstra, K., Bauer, W.D. and Albersheim, P. (1973). *Pl. Physiol.* **51**, 158-173.

Section II
The Foundations of Quality—Improving Quality by Breeding

II.1 The Role of Genetics in Breeding for Quality of Vegetables

G.A. MARX

New York State Agricultural Experiment Station,
Geneva N. Y. 14456, USA

INTRODUCTION

That heredity plays a role in determining the quality of fruits and vegetables is evident from the numerous varietal differences in quality known in virtually every crop species. For the most part, however, breeding for improved quality has been and remains unsystematic and empirical. There may be many reasons for this (Stevens, 1973) but essentially too little is known about what quality is, how it is measured, and how it is inherited. Without such information it is difficult to pursue purposeful, efficient breeding programmes.

Plant breeders rarely, if ever, set out to improve quality alone—that is, as the primary objective. They usually aim instead to improve the crop overall. Given the array of variables which command his attention at any one time, the plant breeder, understandably, tends to select for characters which he can readily observe and measure, and to ignore those which are difficult to evaluate. In the absence of specific selection criteria and without deliberate selection pressure, any improvement in quality may be rather a matter of chance. Moreover, the complexity of the situation is such that most breeders are resigned to making incremental improvements, progressively adding one or a few desired characters at a time. Breeders are often reluctant to use widely divergent material as parents in new crosses, preferring instead to work with stocks that approximate to the final type. This approach, though logical and successful, fails in some important aspects. It fails to provide information on genetic control of the attribute, and it leads to a narrowing of the genetic base. Typically, all but a few of the "best" plants in a segregating population are eliminated. The "best" plants are generally those which are superior in "overall merit". Such selected plants may not, however, be the best from the standpoint of a specific quality attribute, especially one which is not clearly defined. Thus, plants which are less attractive overall, and are therefore discarded, could hold the key to a valuable attribute.

GENERATING GENETIC INFORMATION

An alternative approach offers significantly more potential for quality improvement. It calls for an orderly acquisition of biophysical, biochemical, and genetical information as well as for the identification and preservation of genetically-defined stocks. Specific quality attributes need to be identified and characterized in as much detail as the existing technology will permit. This characterization may apply to any one of the criteria of quality, e.g. colour, texture, flavour, etc., at any level, from volatile flavour compounds to gross plant morphology. Once specific quality components have been identified and objective detection and assay methods developed, then the stage is set for genetic analysis.

For the genetic analysis to be maximally effective, it is desirable to include the greatest possible genetic diversity, including wild and exotic forms and related species. It may be necessary to include lines that are patently inferior in order to reveal the full extent of the existing heritable variation and to gain a full understanding of the genetic control of a given attribute. Such lines are also a source of potentially valuable but hidden genes.

Ideally, the information required to gain an understanding of the genetic control of quality will come from collaboration between biochemists/physiologists and plant geneticists. Indeed, some of the best known examples are products of such co-operation and come from species which are well-studied genetically, thus underscoring the value of genetics in bringing quality factors to light and in stimulating interdisciplinary research. Although some quality components can be easily discerned and genetically analysed without recourse to biochemical methods, the ultimate characterization of gene action invariably rests with the biochemist.

SOME EXAMPLES OF SIMPLE GENETIC CONTROL

Many attributes of quality are already known to be simply inherited. They are the most amenable to manipulation.

FRUIT COLOUR MUTANTS OF TOMATO

In tomato (*Lycopersicon esculentum* Miller), for example, a series of different fruit colours (Table I) are known to be controlled by different single genes (Rick, 1974).

TABLE I

Some of the mutants of tomato (*Lycopersicon esculentum* Miller) affecting fruit colour

Gene symbol	Gene name	Gene symbol	Gene name
at	*apricot*	*pd*	*pigment diluter*
B	*β-carotene*	*r*	*yellow flesh*
Del	*δ-carotene*	*rin*	*ripening inhibitor*
hp	*high pigment*	*sh*	*sherry*
nor	*non-ripening*	*t*	*tangerine*
Nr	*Never ripe*	*y*	colourless fruit epidermis
og^c	*old gold* (crimson)		

Whereas the colour differences can be discerned visually, the true nature of gene action is defined in specific biochemical or physical terms. The value of each mutant may vary according to how it contributes to different quality criteria. For example, *hp* (high pigment) (Thompson, 1955) and og^c (old gold) (Rick and Dempsey, 1961; Thompson *et al.*, 1967) have similar effects in that both intensify the red colour of the fruits and consequently both make the fruits visually more appealing than normal tomatoes (Robinson, 1974). From the standpoint of nutritional value, however, the action of the two genes is rather different. Whereas *hp* enhances total carotenes (and ascorbic acid), by 100%, og^c causes the fruit to be more red than normal by reducing β-carotene. Thus, *hp* increases vitamin C whereas og^c reduces vitamin A content. The fruits of two other mutants, *t* (tangerine) and *B* (*β-carotene*) also appear similar to the eye, both having orange-coloured fruits, but *B* gives ten times more β-carotene (pro-vitamin A) than normal (Robinson, 1974). In contrast, the fruits of *tt* plants accumulate ζ-carotene, which is not a vitamin A precursor. These and other genes of tomato have had a less direct, but by no means less important, part to play in an overall understanding of quality. They have been useful in formulating a scheme for carotenoid biosynthesis (Khudairi, 1972).

ENDOSPERM MUTANTS OF MAIZE

In another species, maize, the substitution of recessive alleles for the normal dominant allele at a single locus is largely responsible for converting ordinary field or dent corn to sweet corn. The substitution of su_1 (sugary-1) for *Su* not only increases the sugar content of the endosperm, but also increases the phytoglycogen content (Peat *et al.*, 1956; Ayers and Creech, 1969). It is the phytoglycogen, perhaps more than the sugar content, that gives sweet corn its distinctive edible qualities. Recently, sweet corn breeders have begun to avail themselves of a group of genes which are known to affect carbohydrate quality and quantity of the kernel, and which have long been a part of the maize geneticist's repertory. All of the following genes have been shown to influence kernel quality: *wx* (waxy), su_1 (sugary-1), su_2 (sugary-2), fl_1 (floury-1), fl_2 (floury-2), bt_1 (brittle-1), bt_2 (brittle-2), sh_1 (shrunken-1), sh_2 (shrunken-2), *du* (dull), *ae* (amylose extender), o_1 (opaque-1) and o_2 (opaque-2) (Neuffer and Coe, 1974). Six of these, *ae*, *du*, sh_2, su_1, su_2 and *wx* were studied alone and in combination by Creech (1965) and by Creech and McArdle (1966). Chemical analyses were made for kernel sugar, WSP (water soluble polysaccharides), starch, and dry matter. These and subsequent studies by Creech and his collaborators and by others, both before and afterward, not only provided useful insights into the specific genes that control carbohydrate metabolism, but also revealed mutants and combinations of mutants that might be useful in improving sweet corn quality. An example of this is given in Table II where the taste ratings for sweetness of five genotypes are compared with normal dent. The single mutant, sh_2, actually codes for considerably more sugar than su_1 but sh_2 kernels are virtually devoid of phytoglycogen and it was found that a high sugar content alone did not improve quality.

More recently still another gene of maize, designated *se* (sugary enhancer), has been discovered. It acts as a modifier of *su* and its effect is to increase the sugar content dramatically, comparable to sh_2, without loss of phytoglycogen (Ferguson *et al.*, 1978). Thus, there is a large and varied pool of genes available to the sweet corn breeder, giving him unusual latitude for quality improvement. Sweet corn varieties incorporating certain

TABLE II

Taste ratings for sweetness of six genotypes at 28 days after pollination, ranked 1-9, 9 being the sweetest (Creech, 1966)

Genotype	Taste index
Normal (Dent corn)	1·3
su_1 (Sweet corn)	3·3
ae wx	6·7
ae su_1 *wx*	6·7
sh_2	7·7
ae du wx	8·3

Least significant difference (5%) = 2·4, (1%) = 3·2.

of the genes and gene combinations cited above have begun to appear in seed trade catalogues.

BITTERNESS IN CUCUMBER

Bitterness versus non-bitterness is a quality attribute of cucumber (*Cucumis sativus* L.) which is determined by alternative alleles of the *bi* gene (Andeweg and De Bruyn, 1959). Most cucumbers possess a group of compounds, known as cucurbitacins, which are responsible for making them bitter to the taste of many but not all humans. After testing the cotyledons of 15 000 individual plants, Andeweg and De Bruyn (1959) succeeded in isolating a single non-bitter plant in the cultivar Improved Long Green. Non-bitterness is inherited as a monogenic recessive. The absence of bitterness later proved to be associated with an absence of cucurbitacins. It is widely believed, but not experimentally proved, that non-bitter cucumbers are those that humans find to be "burpless". Although non-bitter cucumbers are of interest mainly in connection with human taste preferences, some unexpected secondary issues have emerged. Cucumber varieties lacking cucurbitacins were found to be more resistant (i.e. non-preferred) than the normal types to spotted and to striped cucumber beetles (Da Costa and Jones, 1971a). Then it was shown that although non-bitter cucumbers (*bi/bi*) were more resistant to cucumber beetles than bitter (*Bi/Bi*) cucumbers, they were more susceptible to two-spotted mites (Da Costa and Jones, 1971b). Thus, cucurbitacins attract cucumber beetles and cause feeding, whereas the same compounds, regulated by the same gene, *Bi-bi*, have an adverse effect on two-spotted mites. Still another reminder that man is not the sole arbiter of quality is revealed by the fact that rabbits seem to prefer non-bitter varieties of cucumber (H. M. Munger, personal communication).

GENES CODING FOR SUBSTANTIAL AND SUBTLE DIFFERENCES IN QUALITY

Some genes of some species cause such fundamental changes in morphology and/or physiology as to alter the usage of the crop or even to regard the altered genotype as a separate crop. For example, the *o* gene in tomato is responsible for changing the normal spherical fruit shape (*O*) to the so-called plum- or pear-shape (*o/o*) (Butler, 1952). Concommitantly, the fruits are smaller and more crack resistant, have thicker juice (more pectins), smaller stem scars, and they "hold" better in the field. As a result, plum tomatoes are widely used for purée products whereas normal tomatoes are used fresh and for juice.

In the pea, the genes *Pa-pa* determine the difference between pale green and dark green edible seeds (Winge, 1936). Varieties of processing peas with pale green seeds are primarily used for canning whereas those with dark green seeds are used for freezing. Even more categorical are the single genes (either *p* or *v*) that account for the difference between normal peas and edible-podded peas (sugar peas, snow peas) (White, 1917). Each of these recessive genes removes most but not all of the parchment layer normally present on the interior pod wall (endocarp). The two recessive genes together (*p v*) remove all the parchment. Another gene, *n* (Wellensiek, 1925), combined with *p* or *v* or with *p v*, produces edible-podded peas with thick pod walls somewhat resembling bean pods. "Snap pea" is the name given this type of pea in recent times. Still another single gene difference in pea, *R* versus *r* (White, 1917), is mainly responsible for two major uses. Pea varieties with round or smooth seeds (*R/R*) are used as dry peas for soups and other preparations. Dry peas are often considered as a separate crop from the green peas, used fresh or for canning or freezing.

At the molecular level, volatile compounds considered to be those mainly responsible for distinct flavour differences in snap beans (*Phaseolus vulgaris* L.) have been identified by Stevens and Frazier (1967), and by Stevens (1970a). Differences in the concentration of oct-1-en-3-ol and linalool were evident among Pole Blue Lake, Gallatin 50, and Romano. Pyridine was found to be unique to Romano and was believed to be important in its harsh flavour. Oct-1-en-3-ol and linalool concentration proved to be simply inherited. Simply inherited volatile flavour compounds were also isolated in tomatoes (Stevens, 1970b).

USE OF WILD AND EXOTIC FORMS

Among the examples given above, the tomato genes giving redder fruit and higher β-carotene came from interspecific crosses, *Lycopersicon esculentum* × *L. chilense* and *L. esculentum* × *L. hirsutum* (Robinson, 1974). Curiously, fruits of *L. chilense*, the species which is the source of og^c, are green, not red. The gene *B* which gives ten times more β-carotene than normal came from another green-fruited species, *L. hirsutum. Lycopersicon minutum, L. pimpinellifolium,* and *L. chmielewskii* have been successfully exploited for genes which increase the sugar and solids content of the fruit (Rick, 1979). Even after the genes which are responsible for these increases had been transferred to large, red-fruited lines, a 40% gain in sugar content was realized. Such increases are important not only for processing tomatoes, but for fresh tomatoes as well. Evidence obtained by Stevens *et al.* (1979) indicates that increased sugar and acid content enhances the flavour of fresh tomatoes.

Among the many different mutants of maize which are known to affect carbohydrate synthesis and which were cited earlier, nearly all were isolated from field corn which is commonly considered to be a different crop from sweet corn. Nevertheless, the two crops share a common gene pool and now sweet corn breeders have begun to take advantage of the genes which were previously only a part of the geneticists' domain. The non-bitter cucumber, also cited earlier, was found as a mutant of an old, obsolete cultivar, illustrating that old varieties represent another valuable source of genes which might influence quality.

These few examples illustrate that maximal use should be made of all material which can be successfully crossed, including wild and exotic forms and related species. The results of such crosses may often be unpredictable, even surprising, because of unknown interactions among different genes and cytoplasms, but the potential benefits seem to outweigh the risks by a considerable margin. Moreover, the use of diverse sources and types of lines in breeding and inheritance studies gives an indication of the full range of heritable variation available for exploitation. Inheritance studies which, in some cases, may depend on sophisticated biochemical techniques to ascertain the nature of the genetic differences involved, may then furnish the genetical insight required to proceed in an orderly, systematic manner. This process simultaneously permits the isolation of lines with specific genotypes which can be characterized and retained for future use. Thus, both the information and the material is available as a foundation for further advances.

GENES AFFECTING RATE OF MATURATION

Some tomato mutants, which might seem to be mere curiosities, have added significantly to an understanding of the biochemistry of ripening. Plants carrying *Nr*, *nor*, or *rin* resist or prevent normal fruit ripening (Tigchelaar *et al.*, 1978), presumably because they cause blocks in the biosynthetic pathway leading to ripening. Such mutants therefore play a part in identifying the individual steps in the process. But there is also an immediate practical result: the fruits obtained by crossing normal with non-ripening tomatoes complete all the steps in the ripening process but they do so more slowly than normal (Herner and Sink, 1973; Brecht *et al.*, 1974; Tigchelaar *et al.*, 1978). This, then, is a mechanism to extend the shelf-life of harvested tomatoes, and it has important implications in storing, shipping, and marketing.

Several of the mutants of maize, already mentioned, affect the carbohydrate balance of the endosperm both qualitatively and quantitatively. Ordinary sweet corn (i.e. *su*) loses its quality rapidly after harvest because there is a rapid conversion of sugar to starch coupled with a loss of moisture and an increase in water soluble polysaccharides. These deteriorative changes can be slowed and quality retention enhanced by some of the genotypes which are now being evaluated.

The *R-r* locus of peas is responsible for the well known difference between round (smooth) and wrinkled pea seeds. But, as can be seen from the data in Table III (Kooistra, 1962), the gene has profound and manifold effects, many of which are related to the maturation process. Smooth-seeded varieties of peas mature more rapidly and remain in an acceptable quality range for a shorter period than do varieties of wrinkled-seeded peas (Fig. 1).

A second gene, *Rb-rb*, discovered by Kooistra (1962) and independently by Marshall (1963), also influences the sugar, starch, and amylose content of pea seeds as is shown in Table IV. It remains to be seen if *rb*, when combined with *r*, will serve as a means of extending the duration of prime quality.

At the level of gross morphology, a single gene, the self-pruning gene (*sp*), changes the reproductive behaviour of the tomato plant from indeterminate to determinate. The presence of *sp* is a practical necessity in

TABLE III

Characteristics associated with differences at the *R-r* locus of peas (Kooistra, 1962)

	Smooth peas (RR)	Wrinkled peas (rr)
Seed surface	Smooth	Wrinkled
Shape of starch grains	Oval	Round
Nature of starch grain	Without radial splits	With radial splits
Sugar content	Relatively low	Relatively high
Starch content	High	Low
Amylose content of starch	Low	High
Water content of unripe seed	Relatively low	Relatively high
Water loss during maturation	Relatively low	Relatively high
Water absorption capacity	Relatively low	Relatively high

TABLE IV

Differences in sugar, starch, and amylose content, as % of dry matter, in dry pea seeds associated with different combinations of the *R-r* and *Rb-rb* genes (Kooistra, 1962)

Genotype	Sugar	Starch	Amylose[a]
R- Rb-	6·3	49·9	37·9
rr Rb	10·6	31·5	65·4
R- rb rb	9·2	32·5	24·5
rr rb rb	9·5	27·4	42·9

[a] as % of the starch.

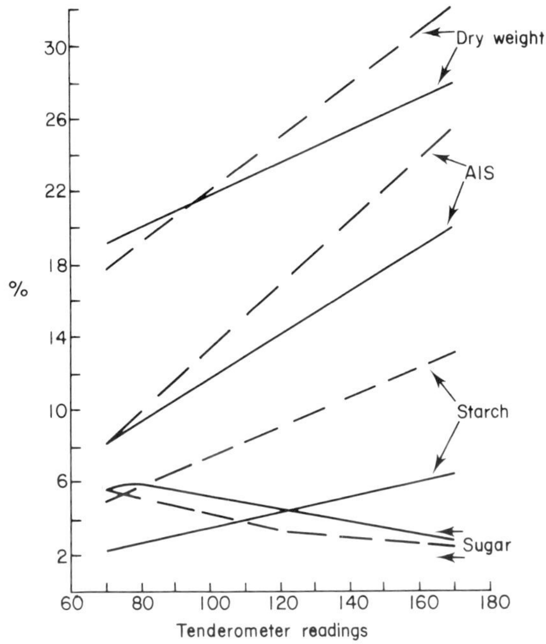

Fig. 1. Average relationships of dry weight, AIS (alcohol insoluble solids), sugar and starch content and tenderometer reading of wrinkled (————) and smooth (-------------) peas (Ottoson, 1958).

varieties intended for mechanical harvesting, because determinate lines exhibit concentrated fruit-set and ripening. It has been observed, however, that a reduction of the sugar content of the fruit accompanies the introduction of *sp* (Emery and Munger, 1970; Stevens, 1973). This and other changes associated with *sp* are interpreted to be an indirect rather than a direct effect of the gene. The concentrated fruit load places an inordinate demand on the supply of photosynthate in a relatively short period of time (Augustine *et al.*, 1976).

Peas also possess genes which cause a concentrated fruit-set. Perhaps a situation similar to that in determinate tomatoes also holds for fasciated and simultaneously flowering peas, but, if so, it has not yet been demonstrated. The increasing interest in the so-called "leafless" pea (*afila*) (Goldenberg, 1965) inspires the same question, but in this case because of the apparent reduction in leaf surface.

MATURITY, YIELD AND QUALITY IN PEAS

Even if all quality attributes were simply inherited—an unlikely prospect—

the complexity of quality might be only slightly diminished. The large number of quality attributes, the interaction among them and the sequence in which they become manifest conspire to assure a high degree of complexity. This can be illustrated by examining the reproductive behaviour of the garden pea. The focus is placed on individual components, which when combined in increasing levels of complexity lead ultimately to the dynamic state that exists in field populations.

MATURATION OF A SINGLE SEED

A solitary plant, separated in space, bearing a single pod and containing a single seed, represents the simplest possible situation and thus serves as a baseline for a consideration of the successive orders of complexity which follow. Even reduced to this simplest form, there is a dynamic relationship between time, maturity, yield, and quality (Fig. 2). About 15 days after anthesis, the fertilized ovule or "berry" has enlarged to a point where it begins to approach the lower limits of acceptable horticultural quality. The physical and chemical changes that ensue (Fig. 3), if allowed to proceed to completion, lead to a mature, dry seed. Within the limits of edible quality, the increase in fresh and dry weight, water, and cell volume are nearly linear. To the extent that these changes occur, they represent changes in maturity, yield, and quality. Over much of the maturation curve, quality is a function of maturity, the latter being subject to various physical and

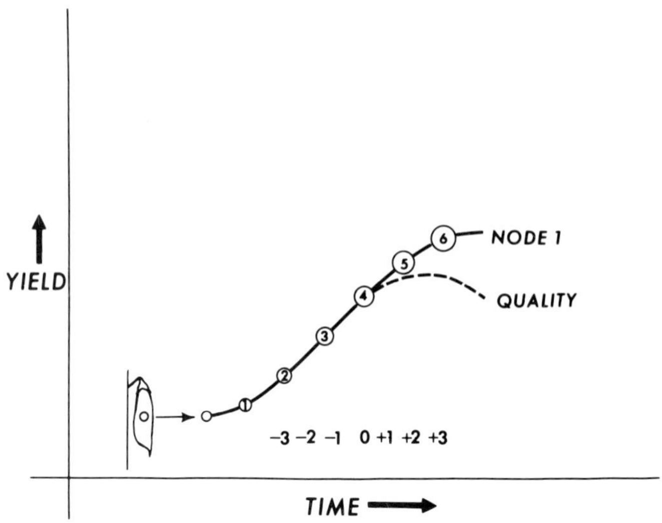

Fig. 2. Diagrammatic representation of the maturation of a single pea showing some of the changes in size, "yield" and quality with time. Numbers below curves represent days before (−) and days after (+) the hypothetical time of peak quality (0).

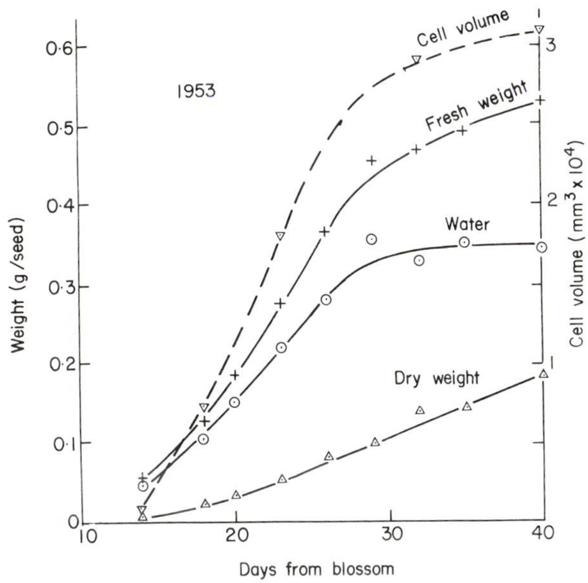

Fig. 3. Increase in fresh weight, dry weight, water, and mean cell volume in pea seeds with time (McKee *et al.*, 1955).

chemical methods of determination. Peak quality, however, occurs within a narrow range of maturity and it persists only briefly, literally one day or less. At first, maturity, yield, and quality are all positively correlated, but beyond a given point, quality begins to decline and this is followed later by a decline in yield (the result of a loss of seed moisture). Hence, optimal quality anticipates maximal yield and maximal yield anticipates maximal maturity. Since quality, rather than yield, dictates the time of harvest, some yield must be sacrificed to gain the desired edible quality, even at the level of a single pea.

Several genetically determined differences exert an influence in the train of events just outlined. The *R-r* genes, coding for smooth versus wrinkled seed, will, as already shown, determine the rate of maturity and the quality of the berry. Berry size is another genetically determined attribute that affects both yield and quality. The influence of berry size on yield is obvious since weight is a function of size (Table V), but the influence on quality is less direct. Small or large size, in and of itself, can influence preference, and, beyond that, differences in size affect quality by altering the volume/surface ratio which in turn affects "chewiness".

Thus, the maturation of a single berry in many respects resembles the maturation sequence which characterizes an entire population of field-grown

TABLE V

Mean weight (g) per pea of different sieve sizes at each of three harvests (cv. Perfected Freezer 60) (Vittum, M.T., unpublished data)

Harvest date, July 1963	Sieve size 2	3	4	5	6+	Ungraded
5	0·224	0·309	0·400	0·512	0·600	0·348
8	0·232	0·308	0·406	0·524	0·631	0·422
11	0·241	0·312	0·417	0·536	0·641	0·517

plants. The addition of more seeds/pods, pods/peduncle, and pods/plant merely compounds an already dynamic process.

VARIATION IN OVULES AND SEEDS PER POD

The normal complement of ovules per pod is 8-9. Perhaps as few as three genes regulate the known variation of this trait which ranges from a low of 5-6 to a high of 10-12. Ovule number only sets the limits for seed number; evidently another gene system governs the capacity to set seed (Marx and Mishanec, 1967). Seeds per pod, like seed size, is a component affecting maturity, yield, and quality. An additional one or two seeds per pod translates into a large increase when multiplied by the number of plants involved on a field scale. Moreover, the seeds within a pod tend to be more alike in maturity than seeds from other pods on the plant. Consequently, this character has an effect on uniformity of maturity.

VARIATION IN PODS PER PEDUNCLE

The number of pods per peduncle may range from one to 5-6 or more, but most varieties have two pods at each node. Two major genes control most of the variation in this character. Generally, but not always, there is greater similarity of maturity between pods on the same peduncle than between pods at different nodes. Hence, increasing the number of pods per peduncle has been an attractive challenge to the plant breeder for many years because it offers a means of increasing yield and uniformity of maturity.

VARIATION IN POD-BEARING NODES

The number of reproductive nodes per plant represents the final level of intra-plant complexity. Genetic studies of flowering and photoperiodism have led to an understanding of how this character is regulated (Murfet, 1977). Again, a relatively few major genes are involved. This genetic system is crucially important in the partitioning of photosynthate and in determining pod fill, pod set, and pod concentration, all of which bear importantly on maturity, yield, and quality.

All these variables, superimposed and interacting, determine the composition of the harvested product. Indeed, they determine how many peas of what size and what maturity are present in the harvested sample, but the specific contribution that each component makes is a function of time and therefore each varies with the time of harvest. However, it would be misleading to imply that the components act additively. On the contrary, because of component compensation, the components commonly are negatively correlated (Adams, 1967), and the interactions may be complex and variable.

Figure 4 graphically depicts some of the points considered above, and Fig. 5 and Table V illustrate a phenomenon which is frequently overlooked.

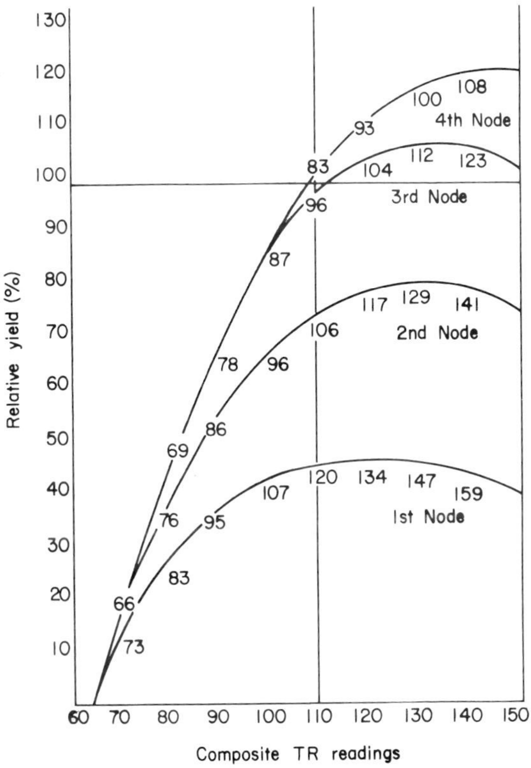

Fig. 4. The "normal" yield and maturity curves of peas, showing the relative distribution of peas at successive pod-bearing nodes. Peas from the various nodes will have the tenderometer readings given in the diagram as the average tenderometer readings change by an interval of 10 units. The data are typical of wrinkled-seeded varieties but the actual curves may vary widely depending on prevailing conditions (Ottoson, 1958).

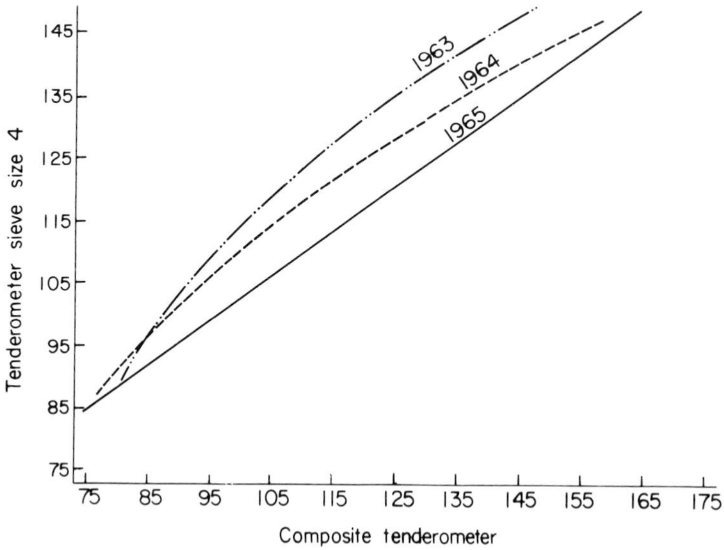

Fig. 5. Maturity of sieve size 4 in relation to composite tenderometer values for cv. Perfected Freezer 60 peas in 1963, 1964, and 1965 (from unpublished data of M.T. Vittum).

As harvest time is delayed there is not only an increase in the overall maturity of the plants sampled, but also a parallel increase in the weight (Table V) and maturity (Fig. 5) of the peas in any one size grade. Apparently this is a reflection of an accelerating maturation rate at the reproductive nodes nearer the apex.

The intra-plant variation resulting from the normal reproductive behaviour of the plant constitutes only a part of the overall variation in maturity, yield, and quality which is encountered under ordinary field conditions. Inter-plant variation, which is largely environmental in origin, adds significantly to the complexity.

It would seem justified to conclude from the few selected examples provided that genetic knowledge, especially knowledge of simply inherited attributes, is no less important to breeding for quality improvement than for any other economic character. Whereas polygenic control may complicate the process, it does not invalidate the arguments presented.

REFERENCES

Adams, M.W. (1967). *Crop Sci.* **7**, 505-510.

Andeweg, J.M. and De Bruyn, J.W. (1959). *Euphytica* **8**, 13-20.

Augustine, J.J., Stevens, M.A. and Breidenbach, R.W. (1976). *J. Am. Soc. hort. Sci.* **101**, 456-460.

Ayers, J.E. and Creech, R.G. (1969). *Crop Sci.* **9**, 739-741.

Brecht, P.E., Johnson, L. and Gonzalez, A. (1974). *Cornell Plantns* **30**, 40-41.
Butler, L. (1952). *J. Hered.* **43**, 25-35.
Creech, R.G. (1965). *Genetics* **52**, 1175-1186.
Creech, R.G. (1966). *Sci. Fmr* **13**, (3), 8-9.
Creech, R.G. and McArdle, F.J. (1966). *Crop Sci.* **6**, 192-194.
Da Costa, C.P. and Jones, C.M. (1971a). *Hort Science* **6**, 340-342.
Da Costa, C.P. and Jones, C.M. (1971b). *Science, N.Y.* **172**, 1145-1146.
Emery, G.C. and Munger, H.M. (1970). *J. Am. Soc. hort. Sci.* **95**, 410-412.
Ferguson, J.E., Rhodes, A.M. and Dickinson, D.B. (1978). *J. Hered.* **69**, 377-380.
Goldenberg, J.B. (1965). *Bol. genet.* **1**, 27-28.
Herner, R.C. and Sink, K.C. Jr. (1973). *Pl. Physiol.* **52**, 38-42.
Khudaira, K.A. (1972). *Am. Scient.* **60**, 696-707.
Kooistra, E. (1962). *Euphytica* **11**, 357-373.
Marshall, H.H. (1963). *Can. J. Pl. Sci.* **43**, 508-512.
Marx, G.A. and Mishanec, W. (1967). *Crop Sci.* **7**, 236-239.
McCollum, J.P. (1970). *Hort Science* **5**, 99.
McKee, H.S., Robertson, R.N. and Lee, J.B. (1955). *Aust. J. biol. Sci.* **8**, 137-163.
Murfet, I.C. (1977). *In* "The Physiology of the Garden Pea" (J.S. Pate and J.F. Sutcliffe, eds), pp. 385-430. Academic Press, New York and London.
Neuffer, M.G. and Coe, E.H. Jr. (1974). *In* "Handbook of Genetics. Plants, Plant Viruses, and Protists" (R.C. King, ed.) Vol.II, pp. 3-30. Plenum Press, New York and London.
Ottoson, L. (1958). *Växtodling* **9**, 1-112.
Peat, S., Whelan, W.J. and Turvey, J.R. (1956). *J. chem. Soc.* 2317-2322.
Rick, C.M. (1974). *In* "Handbook of Genetics. Plants, Plant Viruses, and Protists" (R.C. King, ed.), Vol.II, pp. 247-280. Plenum Press, New York and London.
Rick, C.M. (1979). *Am. Veg. Grow.* **27**, (1), 63-64,66.
Rick, C.M. and Dempsey, W.H. (1961). *Tomato genetic Coop. Newsl.* **11**, 21.
Robinson, R.W. (1974). *Cornell Plantns* **30**, 37-39.
Stevens, M.A. (1970a). *HortScience* **5**, 95-98.
Stevens, M.A. (1970b). *J. Am. Soc. hort. Sci.* **95**, 9-13.
Stevens, M.A. (1973). *HortScience* **8**, 110-112.
Stevens, M.A. and Frazier, W.A. (1967). *Proc. Am. Soc. hort. Sci.* **91**, 274-285.
Stevens, M.A., Kader, A.A. and Albright, M. (1979). *J. Am. Soc. hort. Sci.* **104**, 40-42.
Thompson, A.E. (1955). *Science, N.Y.* **121**, 896-897.
Thompson, A.E., Tomes, M.L., Erickson, H.T., Wann, E.V. and Armstrong, R.J. (1967). *J. Am. Soc. hort. Sci.* **91**, 495-504.
Tigchelaar, E.C., McGlasson, W.B. and Buescher, R.W. (1978). *HortScience* **13**, 508-513.
Wellensiek, S.J. (1925). *Biblphia. genet.* **2**, 343-476.
White, O.E. (1917). *Proc. Am. phil. Soc.* **56**, 487-588.
Winge, O. (1936). *C.r. Trav. Lab. Carlsberg* (*Sér. Physiol*). **21**, 271-393.

II.2 Breeding Vegetables for Freedom of Blemish

J.K.A. BLEASDALE

National Vegetable Research Station,
Wellesbourne, Warwick, CV35 9EF, England

INTRODUCTION

In the context of stored and processed vegetables a major factor affecting quality and yield is freedom from blemish. I therefore propose to review some contributions that plant breeders can, and are making, by providing cultivars that are less likely to be blemished either during growth or during storage. My examples are intended to illustrate principles and to highlight successes and problems, rather than to be comprehensive.

UNDESIRABLE CHARACTERISTICS

The plant breeder needs to establish which characteristics of a crop are largely under genetic control as opposed to mainly under environmental control. The breeder can, of course, do little about the latter but he can expect to remove a lot of the genetically controlled blemishes in a suitably designed breeding programme.

Quantitative techniques are now well established for partitioning the influence of environment and genes on the expression of a character. In essence, these involve growing a range of genetically different material, such as a collection of existing cultivars, in a range of environments. These different environments can be produced at one site by using a range of sowing dates, soil fertilities, plant populations and so on. The questions to be answered are "Do varieties have different degrees of expression of the blemish in question in the different environments and is this difference between varieties consistent over the various environments?" If they do differ, then there is a genetic basis for the blemish expression which may be strong or weak according to the range found within the material tested. If the varietal differences are not consistent over the different environments, the situation (e.g., genotype × environment interaction) leads to complications for selection for reduced expression of blemishes. Of course, a robust conclusion can only be reached if a wide range of cultivars is tested in a series of challenging environments.

CARROTS

In work with carrots (Chantenay and Autumn King types) similar to that described previously Dowker and his associates (Dowker, 1971; Dowker *et al.*, 1974, 1976a,b, 1978; Dowker and Jackson, 1975a,b, 1977) showed that root shape, shoulder conformation and internal colour were mainly under genetic control; splitting and purple-topped roots were about half genetically controlled and half environmentally controlled. Fanging, green-topped and cored roots, and bolting were rather more under environmental control, and mis-shapen roots, skin smoothness, external colour and mean root size were largely under environmental control. I have referred here to 12 quality characteristics of carrot roots, five of which could be improved by breeding. Their sheer number makes the task difficult because whilst improving one character no other should deteriorate.

The breeding programme must rely on being able to exert strong selection pressure on segregating material for the characteristic in question. The preliminary work done to establish the degree of genetic control is useful here because it also reveals the environments likely to give maximum expression to genetic differences. For example, the splitting of carrot roots was most prevalent in widely spaced crops grown for a long time, so breeding lines are screened accordingly. On the other hand, differences in internal colour, which are of considerable concern to canners, were best expressed in roots grown on certain peat soils at high plant populations. Thus, to make progress, selection of lines has to be in multiple environments.

CAULIFLOWER

The environments involved in breeding for reduction of blemishes in the carrot crop are relatively easily created in the field. However, this is not so with cauliflowers as curd quality is very much affected by the weather at the time of curding and this cannot be adequately controlled or predicted to ensure blemish expression and hence to get strong selection pressure. My colleague, Dr. P. Crisp, has developed a way of using tissue culture of the curd to reinforce field selection for freedom from bracts and freedom from purple discolouration.

Selection of segregating populations of plants in the field can commonly leave as many as 30% and sometimes more of the plants apparently free of either defect. However, if small pieces of curd are removed from these plants at the stage of optimum marketable maturity they can be aseptically cultured (Crisp and Walkey, 1974). If the curd is likely to have a purple discolouration, then the leaves on stems produced in five weeks from curds in aseptic culture will have purple colouration along their margins and on their abaxial surfaces (Crisp *et al.*, 1975a,b). Similarly, curd cultures from plants prone to bractiness produce much larger bracts than cultures of plants free from this defect. Using this technique as a second tier to field selection increases the effectiveness of selection and inbred lines of the Flora Blanca type have been selected using this technique. In a trial at the National Vegetable Research Station (NVRS), they were virtually free of these defects while Flora Blanca varieties from other sources were severely affected. These lines have been entered in National Institute of Agricultural Botany (NIAB) trials as potential varieties.

It is well established that the expression of many physiological disorders is under some genetic control. In the NIAB vegetable trials, the levels of the main disorders are recorded and over the years of trial there can often emerge some useful indication of the relative susceptibility of cultivars to disorders.

The disorders which still plague us in stored and processed vegetables have one feature in common—they cannot be produced to order. Thus, it is difficult to maintain selection pressure and so with some disorders it is virtually impossible to make any progress in breeding for resistance. One solution is to seek clearly visible and reliable characteristics that are correlated with the incidence of the disorder and use these characters as the basis for selection.

BRUSSELS SPROUTS

Such an approach was taken by Faulkner and Johnson (1963) in their work on internal browning (IB) of Brussels sprouts. They tested the correlation between the incidence of IB and either the amount of cover afforded to the sprout buttons by the large leaves on the main stem, or the degree of marginal browning on the leaves of the terminal rosette of the plant, or the internal density of the sprouts. They also demonstrated that susceptibility to IB can be inherited by showing that the progenies of plants with IB had a higher proportion of sprouts with IB than characterized the cultivar from which the parents were selected.

They found no correlation between the incidence of IB and either leaf cover or rosette scorch, but they did find that the proportion of low density sprouts in their breeding lines was well correlated with the incidence of IB which was also related to the amount of frost in the four weeks prior to picking (Fig. 1). It was concluded that sprout plants free from IB and having buttons of high internal density should be selected for breeding

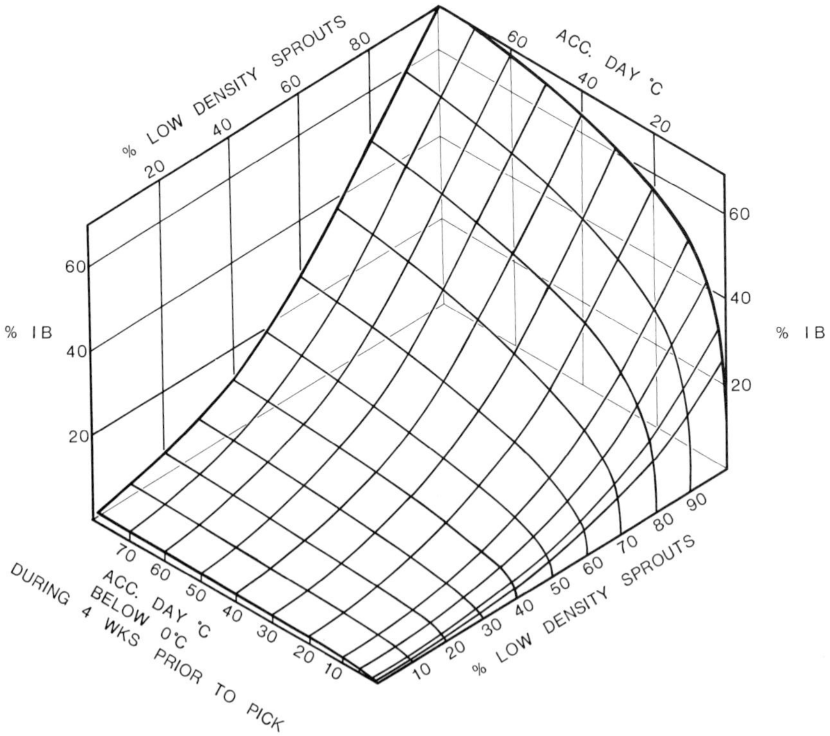

Fig. 1. Relationship between percentage low density sprouts, accumulated day-degrees below 0°C, and percentage internal browning (IB).

purposes, as this was the most likely way of obtaining cultivars with reduced susceptibility to IB.

This policy has been practised in the sprout breeding programme at Wellesbourne and hence we are confident that we have done as much as it is possible to do to ensure that the hybrids we have produced jointly with commercial breeders are not prone to IB.

Our soil scientists at Wellesbourne are now working to further our understanding of the causes of IB so as to improve our selection procedures and so ultimately to eliminate this disorder. This clearly illustrates the value of an interdisciplinary approach in breeding and this is nowhere more apparent than in breeding for a reduction in those blemishes caused by diseases and pests.

DISEASE BLEMISHES

FRENCH BEANS

I suspect that many processors are not aware that bean common mosaic virus (BCMV) can reduce the yield of crops of French beans and cause blemishes on the pods. This should not alarm them unduly as the cultivars grown in the UK are mostly resistant to the strains of BCMV known to occur here. However, BCMV is a seed-borne disease and two new strains (NL3 and NL4) have appeared on the world scene and it is likely to be only a matter of time before they are introduced on seed. Fortunately, breeders have already been active and two cultivars (Valja and 1970-73) have been bred in Holland with resistance to the old and new races of BCMV. This serves to remind us that breeding for disease resistance is often a race in which the breeder aims to keep ahead of the disease organism. At Wellesbourne we have found resistance to the new virus strains in cv. Turkish Brown and there are indications that the basis for this resistance may be usefully different from that which was used to produce the Dutch cultivars (Fig. 2).

Halo-blight can cause severe pod blemish in French bean crops but as it is a seed-borne disease it is kept in check by strict precautions during seed production. This adds to the cost of seed so it is attractive to consider the possibilities of having varieties with resistance to the two known races of halo-blight. The traditional source of resistance is PI 150414 and breeders are increasingly using this line to breed resistant French beans. Our breeders

Fig. 2. Resistance to the newer strains of bean common mosaic virus has been found in cv. Turkish Brown (*right*) which when crossed with a susceptible cultivar gives a resistant F_1 (*centre*).

and pathologists at Wellesbourne have examined the sources and inheritance of resistance to halo-blight (Taylor *et al.*, 1978) and have recently found that OSU 10183 seems likely to be a good source of polygenic resistance.

This could be particularly valuable in our breeding programme directed at producing a navy bean that can be grown competitively in the UK. Given more stable resistance to halo-blight we should be able to produce crops for canning as baked beans as well as seed for these crops.

WHITE CABBAGE

In the last decade the UK production of white cabbage has trebled and now stands at about 110 000 tonnes per annum. Much of this output is processed to be sold as coleslaw, the heads being stored during winter and spring for this processing and for market sales. In store, several internal necrotic disorders are liable to develop and to cause costly losses. One of these disorders, characterized by dark necrotic lesions of 5-10 mm diameter on the internal leaves of the hearts, has now been shown (Walkey and Webb, 1978) to be caused by turnip mosaic virus (TuMV). This large lesion necrosis (LLN) should not be confused with "pepper spotting" in which numerous small lesions occur at random on the leaves.

Walkey and Webb (1978) showed that at harvest the heads from cabbage visibly infected with TuMV weighed about 20% less than apparently healthy heads. During storage (Table I) this difference was increased, by the extra trimming necessary, to about 30%. They have subsequently grown some 50 cultivars of white cabbage and inoculated them with TuMV to screen them for resistance/tolerance. The first year's results have revealed that some cultivars are tolerant and this useful finding will be checked in subsequent work.

It has also been noted in variety trials that some cultivars either very rarely or never develop pepper spotting. Thus it seems likely that within existing cultivars of white cabbage we should be able to find some cultivars that are not prone to LLN or pepper spotting. If this is not so or such cultivars prove to be unsatisfactory for other reasons, then it should clearly be possible to breed for freedom from these necrotic disorders.

PEST BLEMISHES

Amongst processed crops the most common blemishes attributable to pests are the mines on carrots caused by the larvae of carrot fly and the presence of cabbage fly larvae in Brussels sprout buttons. Insecticides are available which will produce >90% control of these pests but a number of factors make it desirable to see how much plant resistance to these pests might be

TABLE I

Occurrence of internal LLN and "head" weight from healthy and TuMV-infected cabbage, recorded at monthly intervals during cold storage[a]

Virus assessment at harvest[b]	Date of examination	"Heads" with LLN	Necrosis severity[c]		Mean "head" wt (g) Untrimmed	Trimmed
Healthy	9 Dec. 75	1/15	1·0		2260	2114
	7 Jan. 76	1/14	1·0		1998	1853
	4 Feb. 76	0/15	0		2080	1959
	2 Mar. 76	0/15	0		2163	2041
	7 April 76	0/15	0		2101	1949
	4 May 76	1/13	1·0		1872	1692
				Mean	2079[d]	1935[d]
TuMV-infected	9 Dec. 75	7/15	3·1		1611	1221
	7 Jan. 76	12/15	1·7		1903	1608
	4 Feb. 76	9/15	1·9		1836	1459
	2 Mar. 76	12/15	1·3		1673	1393
	7 April 76	13/15	1·9		1649	1407
	4 May 76	16/22	1·9		1532	1343
				Mean	1700[d] (18·2)[e]	1405[d] (27·4)[e]

[a]Plants harvested and placed in cold store 11 November 1975. [b]Based on symptoms. [c]Mean necrosis severity for individual "heads" based on a 0 (no necrosis) to 4 (severe) scale. [d]"Head" weight difference highly significant at the 0·1% level. [e]% reduction in "head" weight compared with healthy "heads".

obtained by breeding. At Wellesbourne our work with cabbage root fly has been done using cauliflower and radish as test crops. As the processing and storage of these crops is not much affected by cabbage root fly, I direct attention here to our work on carrot fly.

CARROTS

By exposing a wide range of carrot cultivars in an NVRS field, where carrot fly have been nurtured for many years, we have been able to find some cultivars that are consistently less damaged than others. Subsequently this behaviour has been tested at other sites in the UK in co-operation with ADAS, and more recently in seven different sites in Europe under the auspices of the International Organization for Biological Control. Clause's cv. Sytan Original represents the extreme of "resistance" so far found and Danvers 126 the extreme of susceptibility (Fig. 3). Breeding work is well under way and some selections at the F_3 have been tested. It is evident that complete resistance is unlikely to be achieved but useful partial resistance is almost certainly there. We are convinced that this partial resistance is needed to play a part in an integrated control system.

Integration of all the control measures open to us is the only sure way to improve the control of some pests. Many factors make it unlikely that new insecticides will emerge to give us the almost complete freedom from blemish so desired by many processors. Further, the minimal use of insecticides is now seen as socially desirable and also the surest way of preserving the effectiveness of any given insecticide. Thus partial control by plant resistance is needed to facilitate reduced insecticide usage. Partial plant resistance could produce its effect in several ways. For example, the resistant plants may be less attractive to the pest. The reduced invasion may result in less damage even if the invading insects can then thrive on the

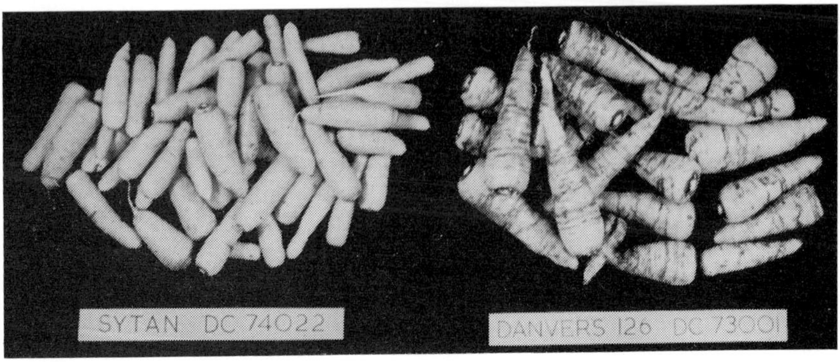

Fig. 3. Roots of cv. Sytan have been shown to be far less susceptible to carrot fly attack than roots of cv. Danvers 126. The difference in the degree of larval damage is clearly seen here in roots exposed to similar initial attacks.

plants. It is, however, more likely that "resistant" plants will reduce the survival or fecundity of invading pests. We do not yet know the mechanism of the resistance to carrot fly. Indeed, we hope that some of our resistant lines may have different resistance mechanisms which can be combined in some new cultivar with accumulated resistance of a higher order than any so far found. However, we do know a lot about the mechanism of insecticides and they all have one important feature in common.

Insecticides have a consistent proportional effect on pest numbers. Thus a given insecticide, applied in a controlled way at a consistent dose, will consistently kill, say, 90% of the insects present. Thus the higher the target population the higher will be the number of surviving insects after spraying and the greater will be the level of damage. This simple summary of the mechanics of insecticide action reveals the fundamental importance of having low insect populations to control. If five larvae per metre of a carrot crop row cause an acceptable level of blemish, then a 95% effective insecticide will be satisfactory if the target population is 100 larvae/m. But it will be no good at all if the target population is any higher. The more we understand about the epidemiology of the carrot fly the more will we be able usefully to use crop rotation and isolation to reduce pest levels. Partial plant resistance to the pest will further reduce the population making it possible to get a greater freedom from blemish with existing insecticides and perhaps making it viable to use less insecticide and so encourage natural predation of the pest.

CROSS PROTECTION TO CAUSE LESS BLEMISH

Finally, I want to turn to a somewhat novel approach to protecting vegetables from the ravages of virus disease. This consists of breeding a weak and symptomless strain of the virus, deliberately infecting the crop with this strain and, by doing so, protecting it from infection by the natural and more virulent strains of that virus.

This cross-protection was first suggested, as a commercial method of controlling virus disease, by Broadbent (1964). Eventually, Rast (1975) showed that, when tomato plants were infected with a weak mutant of tobacco mosaic virus (TMV), they were protected from infection by severe strains of TMV and optimum yields could be obtained.

My colleague, Dr. J. Tomlinson, working with Professor R.J. Shepherd of the University of California, has recently shown that a weakened strain of cauliflower mosaic virus (CaMV) could cross-protect Brussels sprout plants of cv. Fasolt (Tomlinson and Shepherd, 1978). Although this result encourages us to believe that such cross-protection could be used with considerable benefit on a wide range of brassica crops, it is ironic that the weakened strain was too weak to infect cauliflower from which its parent was obtained.

There is something very appealing about breeding our pests to protect our crops against their attack. The introduction of lethal genes into wild populations of insects has had some success.

One thing we can conclude from all present attempts to control the blemishes on our stored and processed crops, is that it is essential to know thine enemy. The host/pest or disease relationship is a finely tuned one and the more we understand how the fine tuning is controlled the more will breeding be able to contribute to control.

REFERENCES

Broadbent, L. (1964). *Ann. appl. Biol.* **54**, 209-224.
Crisp, P. and Walkey, D.G.A. (1974). *Euphytica* **24**, 305-313.
Crisp, P., Gray, A.R. and Jewell, Patricia A. (1975a). *Euphytica* **24**, 177-180.
Crisp, P., Jewell, Patricia A. and Gray, A.R. (1975b). *Euphytica* **24**, 459-465.
Dowker, B.D. (1971). *J. hort. Sci.* **46**, 485-497.
Dowker, B.D. and Jackson, J.C. (1975). *Ann. appl. Biol.* **79**, 361-365.
Dowker, B.D. and Jackson, J.C. (1977). *J. hort. Sci.* **52**, 299-307.
Dowker, B.D., Fennell, J.F.M. and Jackson, J.C. (1974). *J. hort. Sci.* **49**, 311-321.
Dowker, B.D., Fennell, J.F.M., Jackson, J.C. and Arthey, V.D. (1975). *Ann. appl. Biol.* **81**, 377-383.
Dowker, B.D., Fennell, J.F.M. and Jackson, J.C. (1976a). *J. hort. Sci.* **51**, 235-244.
Dowker, B.D., Jackson, J.C. and Fennell, J.F.M. (1976b). *J. hort. Sci.* **51**, 245-252.
Dowker, B.D., Jackson, J.C. and Phelps, K. (1978). *J. hort. Sci.* **53**, 131-137.
Faulkner, G.J. and Johnson, A.G. (1963). *Euphytica* **12**, 299-310.
Rast, A. Th. B. (1975). *Publtiës Proefstn Groenten-en Fruitteelt Glas Naaldwijk*, 192, pp. 76.
Taylor, J.D., Innes, N.L., Dudley, C.L. and Griffiths, W.A. (1978). *Ann. appl. Biol.* **90**, 101-110.
Tomlinson, J.A. and Shepherd, R.J. (1978). *Ann. appl. Biol.* **90**, 223-231.
Walkey, D.G.A. and Webb, M.J.W. (1978). *Ann. appl. Biol.* **89**, 435-441.

II.3 Breeding High Quality High Yielding Apples

F.H. ALSTON

East Malling Research Station, Maidstone, Kent, ME19 6BJ, England

INTRODUCTION

The prime aim of the breeder is to produce new apple varieties with high fruit quality. There is little merit in increasing yield unless such advances can be accompanied by high quality.

Fruit quality falls into four broad categories; eating quality, appearance, storage quality and processing quality. Eating quality covers all aspects of flavour, flesh colour and texture, while appearance concerns not only colour and freedom from russeting and cracking but also size and shape. Storage quality and shelf-life also depend on a number of factors including resistance to shrivelling, bitter pit, low temperature breakdown, senescent breakdown and to diseases such as *Gloeosporium* and *Phytophthora.* The processing qualities which most concern breeders are canning and cooking performance, ease of peeling, and juice quality.

Apple breeding is still considered to be a relatively long-term project although new varieties can be released only 12 years from the original cross (Watkins and Alston, 1973). The highly heterozygous nature of the crop necessitates the production of large progenies since the genetic basis of some quality and yield components is complex. Breeding prospects have, however, been greatly improved since some basic quality factors have been found to be simply inherited (Alston and Watkins, 1974).

EATING QUALITY

A combination of acids, sugars and aromatic substances forms the basis of apple flavour. The texture and colour of the flesh and the form of the core also affect the appreciation of an apple.

ACIDITY

The important acidic component of flavour, mainly malic acid, is controlled by a single dominant gene (Nybom, 1959). Most varieties are heterozygous for the gene, hence it is common to find that at least 25% of seedlings have an unacceptably low acidity and have to be discarded on that basis alone. Some parents also produce seedlings with fruit of too high an acidity. In some crosses over 30% of the seedlings have to be discarded because the fruit is too acid (Visser and Verhaegh, 1978). When breeding processing apples the converse applies, nearly 70% have to be discarded because the fruit acidity is too low.

SUGARS

Fructose, sucrose and glucose are the principal sugars in apples. In most cases 50% of the sugar is fructose. The amount of sugar is polygenically controlled (Brown and Harvey, 1971).

Differences in acidity within the range determined by the gene for acidity can be related to variations in sugar content, to provide an important basis for quality selection (Brown, 1971; Brown and Harvey, 1971). High quality dessert apples are in the medium-acid category (pH 3·2-3·5) and have a relatively high sugar content (14-16%). Most good cooking apples have high acidity (pH 2·8-3·2) and a moderate sugar content (11-13%). Successful North American apples such as Red Delicious and Spartan are in the low acid (pH 3·5-3·7), low sugar (9-11%) category.

AROMATICS

Gas chromatography was used to study the aromatic components of flavour in apple seedlings derived from Cox (Meigh, 1968). Differences in maturity made a precise interpretation of the results difficult. Present techniques of chemical analysis are too laborious and time consuming for use as routine selection methods. However, these techniques are useful for the detailed analysis of the aroma of advanced selections. A detailed study of the aroma components of "Malling" Suntan showed that it carries some of the important flavour components of Cox, one of its parents (Lewis and Williams, 1978).

Confronted by hundreds of seedlings the breeder is compelled to base the initial selection for good aroma on taste at maturity. In this way it is possible to classify Cox derivatives into two distinct groups; those with an aroma similar to Cox and those without such a desirable flavour (Table I). "Malling" Suntan was initially selected in this way. Cox appears heterozygous for its distinctive aromatic component which in some crosses segregates as if it were determined by a single dominant gene (Alston and Watkins, 1974). Surveys for additional sources of good flavour show that few varieties equal Cox as a parent for high eating quality.

TABLE I

Segregation for presence or absence of "Cox aroma" (Alston and Watkins, 1974)

	Present	Absent
Cox × Red Delicious	14	12
× Crawley Beauty	12	16
× Madoue Rouge	8	8
× Cravert Rouge	14	25

FRUIT FLESH

Little is known of the inheritance of texture components although Golden Delicious transmits fine texture and melting juicy flesh to a high proportion of its progeny.

A penetrometer is used to screen large populations for texture. Seedlings are discarded which have fruit flesh that is punctured at harvest by a 11 mm diam. probe with pressures below 8 kg. Such fruits are too soft particularly after 3 months' storage.

The desirable yellow-cream flesh found in Cox appears to be dominant (Crane and Lawrence, 1933) and indeed a number of Cox derivatives breed true for that character.

Undesirable core types include those with wide core lines and open core sections.

APPEARANCE

The aim is to produce new varieties which are attractive and distinctive with good colour and a skin finish free from grease and russet except where a completely russeted variety is the objective.

COLOUR

The presence of anthocyanin stripes in apple skin is controlled by a single dominant gene (Crane and Lawrence, 1933; Wilcox and Angelo, 1936), a point of considerable practical importance (Table II). Care is taken to choose parents which are either homozygous recessive for the gene like Golden Delicious or heterozygous like Cox if green or yellow apples are required. Conversely, the use of homozygous dominant parents such as Worcester Pearmain and Red Delicious results in progenies with a high proportion of red fruits. There is a considerable variation both in the proportion of coloured skin and the way it is laid down within these basic genotypes.

TABLE II

Segregation for presence or absence of anthocyanin in fruit skin (Alston and Watkins, 1974)

Red × Yellow	Present	Absent
Cox × 895	12	13
Cox × 956	3	3
James Grieve × Golden Delicious	12	10
× Starkspur Golden Delicious	15	10
Worcester × Golden Delicious	11	0
× Starkspur Golden Delicious	35	0

RUSSET

The unpleasant partial russeting and cracking which occurs in Cox and the network russeting of Golden Delicious are accentuated by low temperatures in spring and early summer and by some fungicides. Observations in Cox progenies suggest that Cox carries a dominant gene for russeting (Alston and Watkins, 1974) the effects of which are modified by minor genes.

There are prospects of improving skin finish in Cox progenies by the careful choice of russet-free parents. Experience suggests that more than one cross will be necessary to combine the superb skin finish of varieties like Idared and "Malling" Jester with all the good eating qualities we associate with Cox.

GREASY SKIN

A further deleterious skin feature, greasy skin, is also under simple genetic control (Alston and Watkins, 1974). Lord Lambourne carries a dominant gene for this character; it is essential that seedlings with a dry skin are selected from progenies of that variety.

SHAPE

Although preferences vary, in general the almost spherical shape of Cox is preferred in the United Kingdom. Consequently, seedlings with flat, ribbed, conical or angular fruits are usually discarded.

Some people consider that shape is of secondary importance in dessert apples. There is, for example, much interest in an East Malling selection (A133/8) despite its tendency to produce slightly asymmetrical fruits; its good skin finish, superb texture and flavour and high yield apparently outweigh its shape disadvantages.

Regularly shaped fruits are of paramount importance in a new variety for the processing industry. It must be possible to core and peel the fruit completely on a mechanical peeler without any hand finishing as is necessary with angular asymmetrical varieties such as Bramley's Seedling.

SIZE

Size is one of the most important selection criteria, and many seedlings are discarded on this basis. The aim is for uniform fruit size, preferably in the 65-75 mm range for dessert apples and in the 75-85 mm range for processing apples. In general, large fruit size is recessive, progeny means being below the mid-parent value (Crane and Lawrence, 1933; Brown, 1960).

Although some new diploid varieties with improved fruit size have been produced the most promising approach to this problem is through polyploidy. Triploids appear very successful, particularly "natural" triploids (Knight, 1966), which occur by chance in a proportion of about 1 in 300 of seedlings from diploid × diploid crosses (Einset, 1947). These "natural" triploids normally arise from the fertilization of an unreduced diploid egg cell by a normal haploid pollen grain. Thus, two-thirds of the genetic constitution of such a triploid can be derived almost unchanged from the female parent. Crispin and "Malling" Suntan, examples of such triploids, produce a much higher proportion of top grade fruit than their maternal parents Golden Delicious and Cox while at the same time retaining many of their characteristic eating quality, shape and colour features.

Natural triploids can be pre-selected in the nursery by retaining the most vigorous seedlings with large leaves.

STORAGE

Large samples of fruit are required to assess storage potential precisely. Therefore, full-scale storage trials are precluded until fruit is available from advanced trials.

Preliminary selection is made on the first few fruits of young seedlings by inspecting samples stored in air at 3·5°C at monthly intervals during the storage season. At this stage selections which show severe bitter pit, core flush, shrivelling, breakdown and storage rots are discarded.

Disorders associated with a low calcium concentration in the fruit (bitter pit, water core and senescent breakdown) present the greatest challenge to the breeder. All the large fruited, high quality derivatives of Cox produced so far appear to suffer from these disorders to the same extent as Cox, damage being accentuated in large fruits. The very late maturing variety "Malling" Kent, like Spartan, has fruit with a high calcium concentration (Perring, 1978). Both these varieties are relatively free from bitter pit and breakdown during storage; neither, however, has the fine eating quality associated with Cox. An attempt is being made to solve this problem in another way by breeding new rootstocks which are more efficient in calcium uptake (Watkins and Werts, 1977). In this way it might be possible to improve calcium uptake while maintaining good eating quality.

PROCESSING

The main objective here is to select heavy and precocious cropping varieties which need the minimum of attention and will provide a regular source of low priced fruit. To achieve this, desirable quality features of varieties like Bramley (high acidity, moderate sugar content, pale cream flesh, large fruit, short stalk) are being combined with the precocity, regular cropping and regular fruit shape of new varieties like "Malling" Greensleeves.

Small-scale cooking tests are made at East Malling but we send large samples of promising selections to Long Ashton Research Station for authoritative processing assessments.

PEST AND DISEASE RESISTANCE

There is increasing concern about the extensive use of sprays to control pests and diseases in agriculture and horticulture, and prospects of increasingly stringent regulations to control the sale of produce from sprayed areas. Any reduction in spray regimes brought about by the introduction of resistant varieties will be considered to be an improvement in eating quality by some consumers.

Although sources of resistance have been found to all the main pests and diseases of the apple (Alston, 1971), work is concentrated on those which need an extensive spray programme for satisfactory control, such as mildew (*Podosphaera leucotricha*), or those which without control would severely affect fruit quality such as scab (*Venturia inaequalis*) and rosy apple aphid (*Dysaphis plantaginea*). The incorporation of mildew resistance will also have an indirect effect through the elimination of sprays that cause russeting.

YIELD

Early selection for yield is made only four years from germination among seedlings budded on to the dwarfing rootstock M 27 by selecting trees that produce more than 1·5 kg of fruit in that year. This heavy cropping potential has been confirmed in yield trials on M 7, M 26 and MM 106. Such high yielding varieties have a high fruit/shoot ratio and promise to be valuable in intensive orchard systems. Some of the seedlings selected in this way have produced two to three times more fruit than Cox in the National Fruit Trials.

Extensive use is being made of these selections as parents together with spur-type and compact sports to produce new types of tree well suited to intensive production systems (Watkins and Alston, 1973). Dwarfing rootstocks like M 27 and the development of heavy cropping precocious varieties such as "Malling" Greensleeves (Alston *et al.*, 1978) permit the early and rapid testing of new varieties.

INTEGRATION OF CHARACTERS

The rapid combination of resistance, habit and yield characters with all the important quality factors is hampered by the highly heterozygous nature of the apple and through the severe inbreeding depression that results from the use of a common recurrent parent in a backcrossing programme. Some degree of inbreeding is unavoidable if quality is to be maintained since most high quality cultivars are derivatives of the same variety, Cox. Gradually all the important characters are combined in selections which are then crossed with Cox. We hope to select from such progenies triploids which will combine most of the high quality of Cox with new features. In this way "Malling" Suntan combines Cox quality with late-flowering transferred from the old variety Court Pendu Plat, whilst other Cox types combine Cox quality and disease resistance.

There now is sufficient information to define selection objectives at the first fruiting stage. Suggested optimum values for selection criteria are shown in Table III.

The very good performance of some recent East Malling selections in variety trials should be sufficient to allay any doubts about the possibility of combining high yield and high quality. "Malling" Suntan released in 1975 (Alston and Watkins, 1975) gave an accumulated crop over six years which was over 60% better than Cox. When assessed in terms of fruit 65 mm and over in Classes 1 and 2 it was nearly six times better than Cox (Table IV). Another high quality selection, A133/8, gave an accumulated crop three times that of Cox in the first five years. In terms of fruit 65 mm and above in Classes 1 and 2 it gave nearly 17 times the crop of Cox (Table V).

The consumer should have a choice between two or three distinct dessert cultivars. Therefore we have selected two varieties which are quite different from Cox in both flavour and appearance, "Malling" Greensleeves and "Malling" Jester, for the early and mid-season market. Both appear very good pollinators for Cox and A133/8 and could thereby contribute substantially to the production of those high quality varieties.

High yield, good fruit size and excellent eating quality are generally

TABLE III

Optimum criteria[a] for quality and yield selection in apple (adapted from Alston and Watkins, 1974)

Fruit type	Colour grade (0-4)	Russet grade (0-6)	pH	Sugar (%)	Size (mm diam.)	Shape (ratio height/diam.)	Yield (kg) 4 years from germination
"Cox-type"	3	0-2					
Green	0	0-1	3·4	15	75	·85	1·5
Red	4	0-1					
Russet	0-1	5-6					
Processing	0	0-1	2·9	12	85	·75	2·0

[a]Colour grade: 0 (no red pigment) -4 (completely red). Russet grade: 0 (no russet) -6 (complete russet).

TABLE IV

Six years (1968-73) accumulated crops (kg/tree) of Suntan and Cox on M 7 at the National Fruit Trials

	Total	65 mm and above	Class 1 and Class 2	
			Total	65 mm and above
Suntan	95·2	90·4	88·0	83·5
Cox	58·2	16·8	51·6	14·9

TABLE V

Five years (1973-77) accumulated crop (kg/tree) of A133/8 and Cox on M 26 at the National Fruit Trials

	Total	65 mm and above	Class 1 and Class 2	
			Total	65 mm and above
A133/8	41·7	40·3	38·4	36·7
Cox	12·8	2·9	9·6	2·2

compatible. However, the associations between Cox-type flavour and poor calcium uptake and Cox-type flavour and russeting create breeding problems which need to be solved.

Preliminary results suggest that high yield, good fruit size and shape and excellent cooking quality are compatible. Thus, the prospect for establishing a reliable source of fruit for the processing industry is good.

In all breeding work priorities have to be recognized and compromises made. New characters are incorporated progressively, the best combinations being released successively for trial.

ACKNOWLEDGEMENT

The collaboration of the National Fruit Trials in supplying data from variety trials is gratefully acknowledged.

REFERENCES

Alston, F.H. (1971). *Proc. Angers Fruit Breed. Symp. 1970*, 231-248.

Alston, F.H. and Watkins, R. (1974). *Proc. Eucarpia Fruit Sect. Symp. V Fruit Breed. 1973*, 14-29.

Alston, F.H. and Watkins, R. (1975). *Rep. E. Malling Res. Stn for 1974*, 143-146.

Alston, F.H., Watkins, R. and Bates, J.W. (1978). *Rep. E. Malling Res. Stn for 1977*, 205-206.

Brown, A.G. (1960). *Euphytica* **9**, 327-337.
Brown, A.G. (1971). *Proc. Angers Fruit Breed. Symp. 1970*, 185-192.
Brown, A.G. and Harvey, D.M. (1971). *Euphytica* **20**, 68-80.
Crane, M.B. and Lawrence, W.J.C. (1933). *J. Genet.* **28**, 265-296.
Einset, J. (1947). *Fm Res.* **13**, 5.
Nybom, N. (1959). *Hereditas* **45**, 332-350.
Knight, R.L. (1966). *Proc. Balsgard Fruit Breed. Symp., Fjälkestad 1964*, 129-146.
Lewis, M.J. and Williams, A.A. (1978). *Rep. Long Ashton Res. Stn for 1977*, 130.
Meigh, D.F. (1968). *Rep. Ditton Lab. 1967-68*, 53.
Perring, M.A. (1978). *Rep. E. Malling Res. Stn for 1977*, 142.
Visser, T. and Verhaegh, J.J. (1978). *Euphytica* **27**, 753-760.
Watkins, R. and Alston, F.H. (1973). *In* "Fruit Present and Future" Vol. II, 65-74. Royal Horticultural Society, London.
Watkins, R. and Werts, J.M. (1977). *Rep. E. Malling Res. Stn for 1976*, 117.
Wilcox, A.N. and Angelo, E. (1936). *Proc. Am. Soc. hort. Sci.* **33**, 108-113.

Section II
The Foundations of Quality—Modifying Cultural Techniques to Develop Quality

II.4 Field Emergence and Seedling Uniformity

S. MATTHEWS

Wolfson Unit of Seed Technology,
School of Agriculture, University of Aberdeen,
Aberdeen, Scotland

INTRODUCTION

The requirements of modern marketing and processing have been one of the most influential forces for change in the growing of vegetables in this country in the last 20 years. The prime concern has been to present to the consumer an acceptable, unvarying and convenient product. The agronomic changes in sowing date and density, and in plant distribution, needed to achieve this have had a marked influence on the growers' expectations of field emergence and seedling uniformity in the many vegetable crops that are sown directly on a field scale.

The need for greater control of some aspects of production has arisen when the outlet for the production has required specific standards of, say, maturity or size. Peas for freezing have to be harvested at a critical stage of maturity, and the timing of the harvests, so as to spread the work at the freezing plant, is an increasing requirement associated with a specific market outlet. Similarly, the control of the size and shape of carrot roots becomes an important objective for the grower when the outlet is whole roots for canning. Thus, quality in the product, as it leaves the grower, could be interpreted as the achievement of the required standard as specified by the purchaser. If the purchaser is a food distributor operating on a large scale these standards could be influenced more by the requirements of mass production and marketing than by consumer tastes.

This paper examines how the quality expected in the product—quality being interpreted broadly as the achievement of a required standard—can be affected by sowing date and sowing density and the implications this has for field emergence, seedling uniformity and eventually seed quality. An attempt will be made to illustrate how the control of aspects of emergence can aid the production of the required standards of both maturity and size in vegetables.

FIELD EMERGENCE IN PEAS

Field emergence has long been regarded as a problem in the growing of garden peas (*Pisum sativum* L.) (Eastham, 1925). Both the condition of the soil and of the seed have been seen to influence the proportion of the sown seed that produces emerged seedlings. Although low soil temperatures prolong the time taken to emerge and tend to lower emergence, high soil moistures appear to be the more important adverse influence (Hull, 1937). Reduced emergence has been commonly attributed to soil-borne infection by fungi of *Pythium* spp. (Baylis, 1941; Matthews and Whitbread, 1968) and some improvement in emergence can be achieved by fungicidal seed treatments especially in early sowings (Perry, 1967) when the levels of emergence are generally lower compared with later sowings. This effect of early sowing has increased in significance because of the influence of the processing industry on the growing of the crop.

In 1976, 53 000 hectares of peas were grown in the UK for canning and quick-freezing. This represented 60% of the garden pea crop in area and £19 m, 63% of the total value. The rest of the crop (31 000 hectares) was harvested dry for human consumption. Both canning and freezing involve harvesting at a critical stage of maturity to obtain an economic yield of peas containing the desired level of sugars. In order to organize the harvests around industrial processing plants, a spread of the harvesting time is required. This has led to the planning of sowings by food processors so as to provide a steady supply of produce over a period of weeks rather than days. As a consequence, consideration about the suitability of soil conditions and weather for emergence have had less influence on the timing of sowings than was the case when less of the crop was processed. This, in turn, has drawn attention to the importance of seed condition and the problem of so-called low vigour seed.

THE EMERGENCE PROBLEM AND SEED VIGOUR

All seed lots, the production units in which crop seeds are produced and tested, must have a minimum percentage germination before they can be sold—for peas this is 80%. Despite the widespread use of seed lots of uniformly high laboratory germination a considerable range in percentage field emergence can be found among commercially available seed lots (Fig. 1; Matthews and Bradnock, 1967). The seed lots with acceptable levels of germination but poor emergence ability in the field have become known as

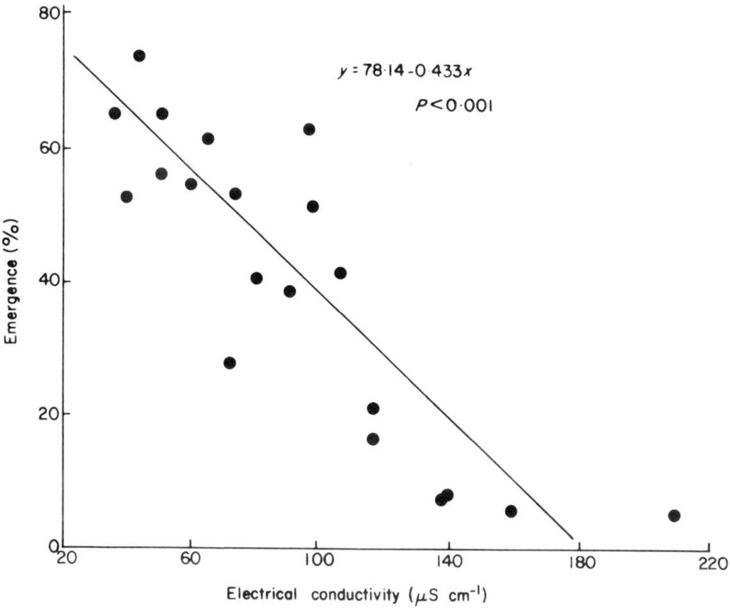

Fig. 1. Relation between field emergence and the electrical conductivity of seed soak water for peas.

low vigour seed and much effort has gone into finding convenient laboratory methods of detecting such seed lots before sowing.

The characteristic of low vigour seed that has proved the most useful as the basis for a routine test method is the high level leakage of solutes, including electrolytes, that occurs when such seeds are soaked in water. The inverse relation between the electrical conductivity of soak water and field emergence (Fig. 1) has been confirmed several times and has given rise to the electrical conductivity test, which has been in routine use in the UK for the past eight years. The test gives a conductivity reading which processors and growers can use as a guide to their seed sowing decisions. Seed lots giving high conductivity readings ($>30\ \mu S\ cm^{-1}\ g^{-1}$ seed weight) are either not sown at all or only used in later sowings. Thus, the emergence problems that accompany early sowings can be lessened by the sowing of only high vigour seed lots that give low conductivity readings.

THE PRODUCTION OF HIGH VIGOUR SEED

The detection of low vigour seed lots to guide sowing decisions is not the most satisfactory solution. It would be more useful to give guidance on how to avoid producing such seed in the first place. Much of our work over the last five years has sought to improve our understanding of the production problem to do just this.

We have found that seed lots that leak solutes readily into soak water show considerable areas of dead tissue (as revealed by the vital stain tetrazolium chloride) on the outer surfaces of their cotyledons after imbibition (Matthews and Rogerson, 1976). As long as the embryo axis remains alive, pea seeds, like other grain legumes, can tolerate considerable areas of dead tissue on their cotyledons before they fail to germinate. It is this dead tissue that contributes large amounts of solutes to the water in which low vigour seeds are soaked. A characteristic of dead cells is their inability to retain cell contents.

Two causes of the production of dead areas on the outer areas of the cotyledons have been identified. First, deterioration, either in the field in wet ripening and harvesting conditions (Matthews, 1973) or in prolonged storage in commercial warehouses (Powell and Matthews, 1978b), can produce seeds that have high levels of germination but low vigour. Low vigour seed can also be produced very readily in warm, moist storage conditions in the laboratory in weeks or even days (Powell and Matthews, 1977) and laboratory manipulations of this kind can be used to predict which seed lots will deteriorate least when stored for years in commercial warehouses (Powell and Matthews, 1978b).

Second, the rapid uptake of water into dry seeds through inconspicuous cracks in the seed coat has been found to kill areas of tissue on the outside surfaces of the cotyledons (Powell and Matthews, 1979). This so-called imbibition damage (Powell and Matthews, 1978a) can be eliminated if dry seeds or embryos (seeds without seed coats) are imbibed slowly in solutions of low water potential (e.g. polyethylene glycol solutions).

Cracks in seed coats have been produced in very dry seeds in simulations of the kind of treatment seeds receive in mechanical handling during seed processing. When very dry seeds were dropped 3 m on to a metal surface, seed coats were cracked and the seed showed imbibition damage which was avoided when the seeds were allowed to imbibe only slowly in solutions of polyethylene glycol (Matthews *et al.*, 1980).

We have recently observed imbibition damage in seeds that had been imbibed in wet soil in the field as well as in water (Table I). Imbibition damage appears to be a general feature of grain legumes since we have seen the phenomenon in dwarf French beans, soya beans, navy beans and field beans. This could have significance to the production of high vigour seeds in these crops.

Although many of these observations need to be extended to cover actual seed crops, it seems that the production of low vigour pea seeds might be avoided by greater care in the harvesting and post-harvest handling, and

TABLE I

Laboratory germination and field emergence of seed lots of peas, and the vital staining of the outer surfaces of the cotyledons with tetrazolium chloride after imbibition in different conditions

Seed lot	Laboratory germination (%)	Field emergence—wet soil (%)	Percentage of cotyledons with complete staining		
			Polyethylene glycol	Water	Wet field soil
A	96	86	98	85	93
B	97	73	98	85	91
C	92	61	88	59	75
D	87	37	85	61	55
E	95	27	93	68	63

perhaps the earlier harvesting of some seed crops in the dry summer climates of seed-producing areas such as New Zealand and Hungary. In this way the production of seed coat cracks and consequently seeds prone to imbibition damage could be avoided. The situation would also be improved by more informed decisions about seed holding so that seed lots with poor storage potential are used first. Improvements in the physiological quality of the pea seeds in trade in this country would help to avoid emergence failures even when the quality requirements of the food processor at harvest lead to sowings being made in unsuitable soil conditions.

EMERGENCE, SPACING AND PRODUCE SIZE

The growing of vegetable produce to size specifications for mass marketing and processing is most effectively achieved by controlling the spacing of plants. When crops are direct-drilled, reliable or predictable field emergence is essential if the required target densities are to be produced. The way in which field emergence is of significance differs between crops. This can be illustrated by reference to a few examples.

Spacing studies on carrots for canning whole have led to recommendations of the population densities that growers should aim for so as to produce a large proportion of roots of a specific size for the processor. It is not economically disastrous if the crop is not uniform with respect to root size since the required sizes can be graded out. However, some control over the plant population is useful in order to avoid producing too many large-sized roots through wide-spacing. Growers are advised what populations to aim for and the weight of seed to sow per unit area to produce these populations. The sowing rates suggested take into account seed weight, laboratory germination and seed bed conditions in an attempt to allow for variations in emergence because of seed and field factors (Bleasdale, 1963).

Most of the broccoli spears (calabrese) and an increasing proportion of the cauliflower heads are reaching the consumer in the form of frozen packs. Within the packs the produce is often divided into small pieces for the convenience of the consumer. The processors require broccoli spears within the diameter range 2·5-7·0 cm with a length of about 15 cm (Thompson and Taylor, 1976). These sizes are achieved either by sowing at high densities (40 plants m^{-2}) and using whole spears or, alternatively, by growing at low densities and longitudinally sectioning larger spears grown from wide-spaced crops. Wide-spacing is now the more usual system used. The growing of cauliflower for freezing has shown a similar change in emphasis from high densities for mini-heads to low densities for large heads that are sectioned (Tyler, personal communication). When crops are direct-drilled, reliable or predictable emergence is important for both the high and the low density systems. Poor emergence of calabrese at high densities will produce over-sized spears that are too small to be sectioned economically and at low densities, gaps in wide-spaced crops, which would lead to reduced yields.

Another example where density is used to give a particular size of produce can be seen in the production of white cabbage for sale from store from November to June. This crop, although relatively new, has reached a national tonnage of around 100 000 and is still increasing. Different outlets which tend to be large, either chain stores for pre-packs or processors, have different size requirements: small heads (up to 1 kg) for the pre-packs and large heads (2-3 kg) for economic prolonged storage (7-8 months) in cool (0-1°C), humid (95% r.h.) conditions for processing. The production of these size grades to order by controlling the plant population imposes additional demands on the emergence ability of the seed.

LABORATORY GERMINATION AND FIELD EMERGENCE

The requirements of a minimum standard laboratory germination have done much to improve the field emergence of vegetable seeds. Even so, our own experience in two years of field trials in several locations has revealed how wide a percentage of field emergence can be obtained from commercial seed lots, differing only slightly in their percentage laboratory germination (Table II).

Thus even among lots with similar, high and acceptable levels of germination poor emerging lots can be found. Apparently, in some lots the germinating seeds have poor emergence ability or, as it is sometimes referred

TABLE II

Range of laboratory germinations and mean field emergences (number of sowings shown in parentheses) for seed lots of vegetable crops

Crop	Year	No. of lots	Range of lot means (%)	
			Laboratory germination	Field emergence
Cabbage	1977	31	80-98	7-78 (4)
	1978	14	75-96	38-63 (3)
Cauliflower	1978	12	85-96	10-64 (3)
Sprouts	1978	15	71-98	25-72 (3)
Carrots	1977	31	73-90	33-52 (3)
	1978	14	68-91	29-59 (6)
Onions	1977	17	73-97	25-70 (4)
	1978	12	77-92	46-72 (6)

to, have low vigour. Our experience, and that of others (e.g. Hegarty, 1976) has also shown that soil conditions can reduce emergence. However, in our trials there was a marked tendency for some lots to emerge well in all sowings whilst others show consistently poor emergence (Matthews, 1980).

Although under many circumstances laboratory germination seems unable to detect clear differences in the relative proportions of low vigour seeds in a lot, which is indicated by the percentage of viable seeds that emerge (relative field emergence), laboratory germination can sometimes be correlated with relative field emergence (Matthews, 1980). It is not surprising that the percentage laboratory germination is indicative of the vigour of the viable seeds, because of the way populations of seeds, and indeed most organisms, die.

The hypothetical seed deterioration curve depicted in Fig. 2 shows that there is an initial slow decline in the percentage of germinable seeds over a relatively long period of time before a rapid fall and then the final stage when the last few members of the population lose the ability to germinate. Vigour differences, that is, differences in the emergence ability of seed lots with similar high levels of germination, come about because of the prolonged period of the initial decline. Some lots, e.g. A in Fig. 2 are at the start of the slow initial decline and are of high vigour, others, e.g. B, are well down the decline and are of low vigour. Since there is a small decline in germination during the initial period, differences in laboratory germination will be indicative of differences in vigour. However, the magnitude of some of these differences is so small that the sample size used in routine germination tests (400 seeds) would be insufficient to detect them but at the same time it would be impracticable to increase the sample size.

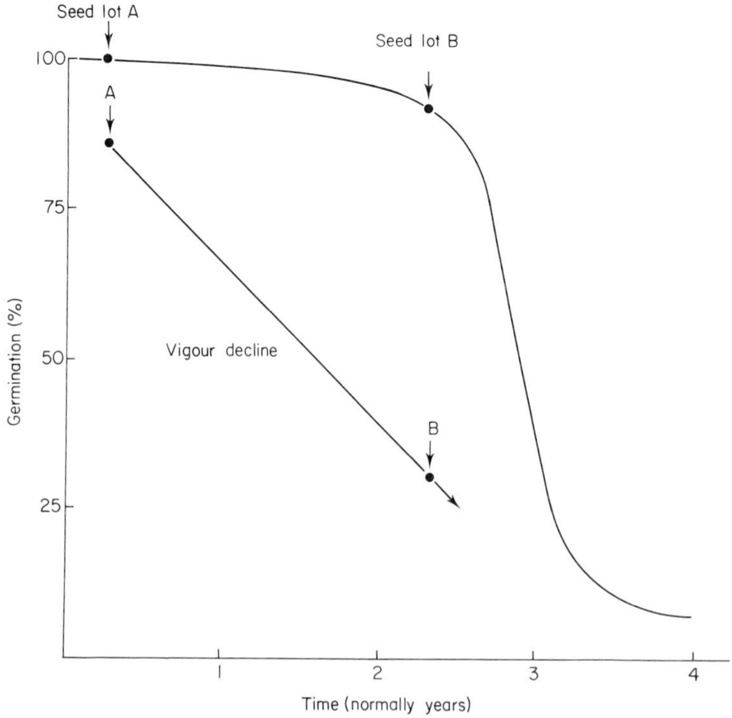

Fig. 2. The deterioration curve.

An alternative approach is to develop an indirect means of determining the relative positions of seed lots on the slow initial decline. We have recently devised such a vigour test which we have termed controlled deterioration.

CONTROLLED DETERIORATION

In essence, seed lots are deteriorated rapidly by holding seeds at elevated moisture contents (around 20% of fresh weight) at high temperatures (45°C) in aluminium foil packets in a water bath. The germination after this deterioration treatment is determined and it is this percentage which is used as a measure of vigour. Seed lot A, that is, the high vigour lot (Fig. 2), would still germinate well, whereas B would have been pushed down the steep part of the deterioration curve and would therefore give a germination percentage close to zero.

Closer correlations between germination after deterioration and field emergence have been found for onions ($r = 0{\cdot}77^{***}$), lettuce ($r = 0{\cdot}71^{***}$) and for sprouts ($r = 0{\cdot}79^{***}$) than between normal laboratory germination

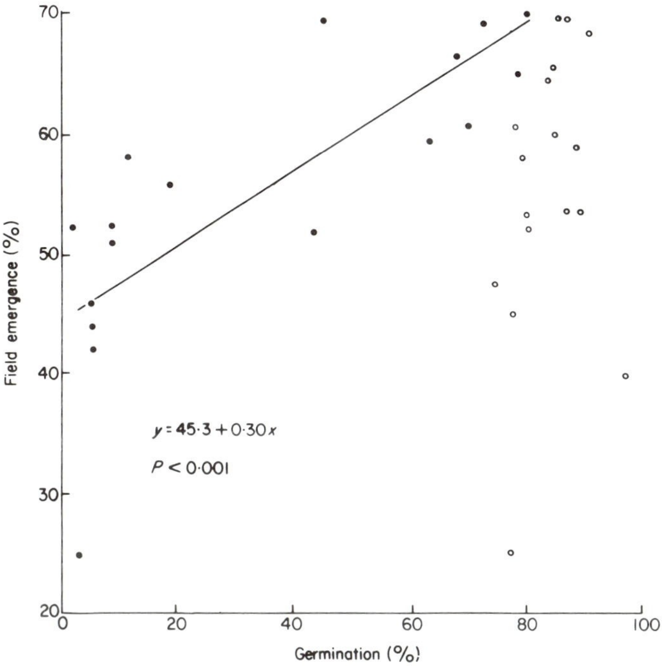

Fig. 3. The relation of normal laboratory germination (∘) and germination after controlled deterioration (•) to percentage field emergence of onions.

and emergence (Matthews, 1980; Fig. 3). In addition weak lots are seen to be clearly different, the range of germinations after deterioration being considerable in the case of all species (lettuce: 1-89%; onion: 1-77%; sprouts: 25-72%; cabbage: 6-86%; cauliflower: 0-88%). The test is also proving to be highly repeatable which is often a difficulty with tests of vigour involving comparisons of seedling growth (Smith *et al.*, 1973) or germination rates because of the sensitivity of both these growth characteristics to small differences in temperature.

THE USE OF VIGOUR TESTS

The use of vigour tests is not straightforward since there are both commercial as well as technical considerations involved. The complete rejection of low vigour lots would be wasteful and economically damaging to seed suppliers if they were unaware of how to avoid producing such material. However, vigour tests could be used as research tools in the search for improved methods of production and as an aid to seed companies in their decisions on seed holding. If only high vigour material was retained for prolonged storage much of the low vigour seed that is presently being used by growers would be eliminated. More care by growers in the handling and storage of seed between receipt from suppliers and actual sowing would also improve the vigour of the seeds that are sown.

Vigour tests might also assist the grower to produce a required plant population but expectations of the value of vigour tests in this rôle should not be raised too high. Although relative emergences between lots can often be clearly indicated by a vigour test, the absolute percentage emergence that a vigour level will produce can vary enormously because of the unpredictability of soil conditions. Even so, some improvements in the advice to growers could be made by using vigour tests. For high density populations, particularly high sowing rates could be used for low vigour lots. If vigour tests were used the number of seeds sown per station for wide-spacing could be determined more precisely so as to ensure at least one emerged seedling at each station without at the same time wasting seed. This would be particularly valuable for costly F_1 hybrid sprout seed which can cost over £300 per kg. In sprouts, where seed can account for as much as 30-40% of the production costs, there has already been a move to reduce the number of seeds sown per station for lots with high laboratory germinations. The justification for doing this has been the finding of a significant positive relation between laboratory germination and field emergence at the Scottish Horticultural Research Institute (Tyler, personal communication).

SEEDLING UNIFORMITY

Recent work on lettuce (Gray, 1978) and earlier work on cauliflower (Salter and Fradgley, 1969) have shown that uniformity in seedling size arising from variations in the times of emergence can persist. As a result there is sometimes a lack of uniformity in the maturation of the crop which is important when one aspect of quality is uniformity of size on a large scale for mechanized handling and mass marketing. Gray (1978) has also pointed out that the variation in plant weight at harvest in carrots was the same as at emergence which is in turn influenced by the size of the embryo within the indehiscent fruit (Gray and Steckel, 1980). Although specific carrot root sizes can be graded out there would be economic advantage in a more uniform crop of the required size, if that was an aspect of produce quality specified by the purchaser.

ACKNOWLEDGEMENTS

I would like to thank the Wolfson Foundation for their financial support and my colleagues in the Wolfson Unit of Seed Technology for their help and co-operation. I would also like to thank Mr F. Tyler of the Humberside region of ADAS and Mr R. Thompson of the Scottish Horticultural Research Institute for much helpful discussion.

REFERENCES

Baylis, G.T.S. (1941). *Ann. appl. Biol.* **28**, 210-218.

Bleasdale, J.K.A. (1963). *Short Term Leafl. Minist. Agric. Fish. Fd* No. 27, pp. 19.

Eastham, A. (1925). Rep. Fourth int. Seed Test. Congr. pp. 227. HMSO, London.

Gray, D. (1978). *Acta Hort.* **83**, 309-315.

Gray, D. and Steckel, J.R.A. (1980). *In* "Seed Production" (P. Hebblethwaite, ed.), pp. 475-484. Butterworths, London.

Hegarty, T.W. (1976). *J. hort. Sci.* **51**, 133-146.

Hull, R. (1937). *Ann. appl. Biol.* **24**, 681-689.

Matthews, S. (1973). *Ann. appl. Biol.* **73**, 211-219.

Matthews, S. (1980). *In* "Seed Production" (P. Hebblethwaite, ed.), pp. 647-660. Butterworths, London.

Matthews, S. and Bradnock, T. (1967). *Proc. int. Seed Test. Ass.* **32**, 553-563.

Matthews, S. and Rogerson, N.E. (1976). *J. exp. Bot.* **27**, 961-968.

Matthews, S. and Whitbread, R. (1968). *Pl. Path.* **17**, 11-17.

Matthews, S., Powell, A.A. and Rogerson, N.E. (1980). *In* "Seed Production" (P. Hebblethwaite, ed.), pp. 513-526. Butterworths, London.

Perry, D.A. (1967). *Proc. int. Seed Test. Ass.* **32**, 3-12.

Powell, A.A. and Matthews, S. (1977). *J. exp. Bot.* **28**, 225-234.

Powell, A.A. and Matthews, S. (1978a). *J. exp. Bot.* **29**, 1215-1229.

Powell, A.A. and Matthews, S. (1978b). *Acta Hort.* **83**, 133-140.

Powell, A.A. and Matthews, S. (1979). *J. exp. Bot.* **30**, 193-197.

Salter, P.J. and Fradgley, J.R.A. (1969). *J. hort. Sci.* **44**, 141-154.

Smith, O.E., Welch, N.C. and McCoy, O.D. (1973). *J. Am. Soc. hort. Sci.* **98**, 552-556.

Thompson, R. and Taylor, W. (1976). *J. hort. Sci.* **51**, 147-157.

II.5 Timing and Supply of Water in Relation to Quality

M.K.V. CARR

National College of Agricultural Engineering, Silsoe, Bedford MK45 4HS, England

INTRODUCTION

The economic significance of water in relation to quality is well recognized for many crop species. However, systematic field studies of the effect of timing and supply of water on the development of quality and, with a few exceptions, the associated biochemistry are rare. This is perhaps understandable as only in the last decade has it become possible to describe and quantify some of the physiological and metabolic changes which occur in plants subjected to water stress.

Water deficits influence crop yields in dynamic and complex ways. The effects are dependent on the severity, duration and timing of the stress relative to the developmental stages of the plant whilst interactions with other environmental and cultural variables modify the responses still further (Hsiao *et al.*, 1976).

Crop yield is, however, relatively easy to understand: quality is a far more nebulous concept. It involves many subjective as well as real criteria, the development of which not only depend on the interplay of genetic and environmental factors during the growth of the crop, but also on the methods of harvesting, processing and storage which follow. It is not therefore surprising that many reports in the literature merely describe empirically observed relationships between water availability and some easily observed or measured aspects of quality. With a few notable exceptions, the underlying processes, or the interrelationships between the influence of water and other environmental factors, are rarely understood or explained.

In this review, examples of some of the effects of timing and supply of water on the development of quality in temperate vegetable and fruit crops are summarized under the following headings: leafy crops, root and tuber crops, and seed and fruit crops.

LEAFY CROPS

Crops in this group lose water rapidly after harvesting, if kept under ambient conditions, and become unsaleable if there is a weight loss of only 5-6% (Burton, 1972). Appearance, size, and flavour as well as water content are therefore important quality parameters.

Since leaf expansion is particularly sensitive to water stress, irrigation plays an important role in production. In the case of cabbage (leaves), cauliflowers (abortive flowers on thick hypertrophied branches), and lettuce (leaves), a single irrigation shortly before harvest is particularly beneficial although more frequent irrigation will result in higher yields (Salter, 1961; Drew, 1966; Sale, 1966a,b). Irrigated cabbages mature a few days earlier and are more uniform than those that are unirrigated, but Drew (1966) found that the ratio of marketable weight to total weight of individual heads was similar in all treatments. On the basis of higher moisture contents and improved appearance Sale (1966a,b) judged irrigated lettuce to be of better quality than those of similar fresh weight that were not irrigated.

In the case of celery (swollen petioles) frequent irrigations of closely spaced plants intended for early harvest is considered essential for high yields and good blanching quality. Provided there is irrigation, planting density can also be used to control head size (Rickard, 1970).

PHYSIOLOGICAL DISORDERS

In a number of vegetable and fruit crops various physiological disorders occur which are associated with local deficiences within the plant of certain elements, particularly calcium. The water status of the plant and the pattern of water movement to the organ concerned appear to play an important role in the severity of these disorders (Wiersum, 1966).

For example, "buttoning" in cauliflowers, which results in small unmarketable plants, is associated with water stress during the early seedling stages and immediately after transplanting into the field (Salter and Goode, 1967). Similarly, water stress when the curds are growing rapidly can increase the incidence of another physiological disorder known as "browning" which causes curd discolouration (Winter *et al.*, 1968).

In celery a disorder known as "blackheart" can be controlled with calcium sprays and by irrigation (Pieters, 1971) or, better, a combination of both. Thus, Cox and Dearman (1978), showed that, although blackheart is only important in England in areas of low summer rainfall, the use of calcium sprays in 1975 reduced its incidence in an unirrigated crop from 13 to 2% and from 5 to 0% in a crop which had been irrigated.

"Tipburn" is an important physiological disorder of lettuce which, in field-grown crops occurs in warm summer months as plants reach maturity. High relative growth rates induced by a number of environmental and cultural practices, including water supply, appear to increase the incidence of tipburn (Cox *et al.*, 1976).

FLAVOUR

Although, flavour is difficult to define, it is now generally recognized that abundant water may reduce the flavour intensity of vegetables (Arthey, 1975). There is some evidence to support this view. For example, Salter (1956) found that irrigation reduced the flavour of cooked cauliflower curds, and Freeman and Mossadeghi (1973) have shown that restricted water supplies can enhance the flavour of cabbages as well as glasshouse-grown watercress and onions. From the results of sensory and gas chromatographic tests they postulated that this may have been caused by either (i) a mild form of sulphur deficiency or, more likely, (ii) the accumulation of low molecular weight metabolites, such as organic acids, sugars and amino acids (flavour precursors) in plants subjected to water stress. During normal growth these would be rapidly transformed into macromolecular compounds like cellulose or proteins.

In contrast, Massey *et al.* (1962) found that water stress had no effect on the flavour of broccoli although succulence was reduced following a 4% increase in the dry matter content. According to Arthey (1975) drought may also be responsible for bitterness in brussel sprouts.

ROOT AND TUBER CROPS

In these crops the useful products are carbohydrate-rich storage organs sold on the basis of their fresh weight. The maximum permissible water loss, before the products become unsaleable, is 8-10% with rates of water loss after harvest being less than those from leafy crops (Burton, 1972).

Appearance, size, dry matter content, flavour and internal structure and composition, which effect storage properties, are important parameters of quality in root crops, although the relative importance of each varies with the product.

APPEARANCE AND SIZE

The availability of water plays an important role in the development of these two quality parameters although there is conflicting evidence about its influence on the incidence of split and fanged roots in carrots. For example, Fradgley (1971) found that alternating wet and dry periods had no effect on the number of misshapen roots. In contrast, Stanhill (1957) and Salter and Williams (1964) had previously observed that, although timing was important, a plentiful supply of water following a dry period usually increased the proportion of deformed roots especially at high plant densities. Regardless of planting density, dry conditions before harvest always reduced the proportion of split and fanged roots. Carrots, particularly those of the Chantenay group, may crack longitudinally in sandy soils when autumn rains follow a hot dry summer (Arthey, 1975).

In the case of potatoes irregular water supplies during the period of tuber expansion can induce secondary lateral growth (knobby tubers) or secondary longitudinal growth (off-types) although the prevailing temperatures also seem to be important (Bodlaender *et al.*, 1964; Ruf, 1964; Salter and Goode, 1967). A period of water stress followed by adequate water supplies can also induce tuber cracking (Robins and Domingo, 1956).

Irrigation before tuber initiation can be used to control tuber number, although there is evidence of differences in response between cultivars (North, 1960), and tuber size. Irrigation during the period of tuber expansion increases the weight of tubers in the large size grades and reduces the weight of unmarketable produce (Wellings, 1973).

Similarly frequent irrigation (at a soil moisture deficit of only 15-20 mm) for four to six weeks after tuber initiation will control the incidence of common scab (*Streptomyces scabies*), a disease associated with dry conditions which spoils the appearance of tubers. In 1970 the degree of infection of susceptible cultivars was reduced from 20-30% to 1-3% with a resultant increase in the weight of premium grade potatoes of 30-40 tonnes ha^{-1} (Lapwood *et al.*, 1971).

PHYSIOLOGICAL DISORDERS

An important physiological disorder of carrots known as "cavity spot" is associated by some workers with wet summers and irrigation although other environmental factors, including calcium nutrition, may also be involved.

Similarly a disorder of potatoes known as "hollow heart" is thought to result from a period of water stress in the tuber followed by a rapid influx of water which causes excessive rates of tuber expansion (Crumbley *et al.*, 1975).

FLAVOUR

Dry conditions again appear to improve flavour. For example, Stanhill (1957) found that carrots which were not watered during growth were sweeter, had a stronger flavour, smoother texture and deeper colour than roots of plants which were irrigated daily. Similarly, Stanhill (1958a,b) found that irrigation, particularly during the four-week-period before harvest, reduced the flavour of cooked turnips. Roots from plots kept dry at this time were described as being "more strongly flavoured", having a "tougher texture" with less "watery flesh" than those of unirrigated plants.

In one study reported in England the flavour of early potatoes irrigated once, when the tubers were small, or not at all, was considered to be "better" than that of tubers of plants irrigated more frequently during the period of tuber expansion. The descriptions used by the tasting panel were "well flavoured" and "moderate" respectively (Harrison, 1963).

DRY MATTER CONTENT AND SPECIFIC GRAVITY

These two attributes of quality are important for the potato processing industry but the evidence for the influence of water supplies on both parameters is conflicting. In the UK it appears that irrigation reduces the dry matter content although the degree of reduction varies with the cultivar and the season (Wellings, 1973). Similarly Gooding (1966) reported that the dry matter content was highest in years when the rainfall in August was low. By contrast the results of a survey by Metzger *et al.* (1938) in Colorado showed that the dry matter content was always increased by irrigation.

Sanders *et al.* (1972) associated the reduction in dry matter content and to a lesser extent, specific gravity, which followed the use of mist irrigation, with delays in maturity. Smith (1961) also found that irrigation reduced the specific gravity and in both instances the effect was to darken the colour of crisps or chips relative to those made from unirrigated potatoes. In contrast, Drewitt (1970) in New Zealand found that although the specific gravity generally increased with increasing frequency of irrigation, there was no relationship between crisp colour and specific gravity. Irrigation, however, regardless of frequency, caused a deterioration in colour standard although cultivars differed slightly in their responses.

As Hughes (1974) has pointed out, varying soil and climatic conditions which effect small changes in the balance between dry matter accumulation in the tuber and water content, as well as differences in maturity date, could explain some of these anomalous results.

STORAGE PROPERTIES

Hughes (1974) has also emphasized the complexity of trying to identify the effect of individual environmental factors on the many morphological and physiological aspects of quality in potatoes. The effect of the environment is not over at harvest; it continues into storage and there are some indications that the water status of tubers influences their susceptibility to internal bruising (blackspot) and to the more serious "shatter bruise". Although the importance of tuber turgidity relative to other parameters is not yet fully understood (Hughes, 1974), it does appear to interact with the temperatures at the time of impact (Smittle *et al.*, 1974).

Improved water availability during the middle of the growing season can lead to an increase in the incidence of "enzymic browning". This is a condition caused by the oxidation of phenolic compounds particularly tyrosine, by phenolase, which is sometimes observed when potatoes are cut and peeled (Mapson *et al.*, 1963).

The results of a three-year commercial scale project by the Potato Marketing Board in Britain indicated that, in general, irrigation did not affect potato quality or suitability for storage. Losses in store were generally lower in tubers from plants that had received irrigation than in those from unirrigated plants, whilst the cooking quality of tubers from both groups was similar.

SEED AND FRUIT CROPS

Water availability plays an important role in the development of quality in seed and fruit crops, with timing relative to the three major phases of development, flowering, fruit expansion and ripening or maturation being particularly important (Kaufmann, 1972).

SIZE AND APPEARANCE

The effect of water stress on seed or fruit size depends on two interacting factors, namely the effect of early stress on fruit number and of a later stress on the rate of fruit expansion. Size, on a fresh weight or volume basis, is usually reduced by late stress. However, if the number of fruits formed is reduced as a result of stress during the early stages of reproductive growth the size of individual fruits may be increased because there is then less inter-fruit competition.

The pea crop is particularly sensitive to water stress at certain growth stages and correct timing of irrigation can influence such quality parameters as the number of marketable pods, the mean weight and number of peas in a pod and the size of individual peas (Salter, 1963). In the case of tomatoes frequent irrigation during fruit expansion increases the proportion of large fruits (Salter and Goode, 1967; Waister and Hudson, 1970). Although there is often a delayed cropping response to irrigation of apples (Goode, 1970), irrigating after fruit-set (from July to mid-September) increases fruit size. Thus Goode *et al.* (1978a) found that so-called "late" irrigation increased the weight of the two highest grades of apples by 67%. By comparison when irrigation was applied throughout the growing season the effects on fruit size were less consistent (Goode and Hyrycz, 1964; Goode and Ingram, 1971).

Most of the evidence reviewed by Salter and Goode (1967) and Kaufmann (1972) indicates the importance of maintaining a supply of easily available water during the period of fruit expansion in deciduous and evergreen fruit crops. This is true also of raspberries (Goode and Hyrycz, 1968) and black-currants (Goode and Hyrycz, 1970); similarly recent studies with apples in Israel have shown that there is a linear relationship between final fruit volume and the number of days in which the available soil water content in the top 0·6 m exceeds 30-40% (Assaf *et al.*, 1974, 1975; Shalhevet *et al.*, 1976).

Irregular watering during fruit expansion can cause splitting or cracking. This has been reported for peas and tomatoes (Arthey, 1975) as well as for apples and a range of other fruits including prunes, cherries, citrus and grapes (Kaufmann, 1972). This phenomenon occurs when heavy rain follows a period of dry weather and is associated with the increased availability of soil water when accompanied by high relative humidities and, frequently, low temperatures. This combination leads to high internal pressures which cause the fruits to split when the tensile forces in the outer tissues exceed the breaking point (Walter, 1967a; Kaufmann, 1972).

The splits do not have to be obvious. For example, Goode *et al.* (1975) demonstrated how uneven water supplies during fruit swelling can induce minute cracks to form in the skin of apples, mainly on the uncoloured side of the fruit. These are sometimes only discernible after the fruit has been stored but, as a result of excessive wilting and shrivelling during storage, a considerable loss of marketable fruit can result. Mist irrigation as well as controlled conventional irrigation reduces the amount of fine cracking (Goode *et al.*, 1979).

Colour is an important criterion of quality in many fruit crops but the influence of water alone on colour development is complicated by the interaction with other environmental factors which effect colour, such as light intensity and temperature. Thus, one report from Israel suggests that irrigation reduces the colour content of tomatoes (Rudich *et al.*, 1977), whilst another indicates that it has no effect (Shalhevet *et al.*, 1976). Similarly some of the reports reviewed by Walter (1967b) showed that irrigation could improve fruit colour whereas Goode and Hyrycz (1964) and Goode *et al.* (1978a) found little effect of irrigation on colour development in apples in England.

In the USA, evaporative cooling by sprinkling intermittently during the day time can considerably enhance the development of colour in Red Delicious apples as well as other quality criteria (Griffin *et al.*, 1974; Unrath and Sneed, 1974). The enzyme which triggers production of the anthocyanin pigment is apparently broken down prematurely at high temperatures.

PHYSIOLOGICAL DISORDERS

As with crops referred to earlier there is some evidence that water supplies affect the incidence of certain physiological disorders associated with the availability of calcium. Thus, Waister and Hudson (1970) found that frequent irrigation during the period of fruit expansion reduced the incidence of a disorder in tomatoes known as ''blossom end wilt''.

Similarly ''bitter pit'' in apples is associated with an imbalance in the fruit of potassium and magnesium ions relative to calcium (Sharples, 1975) but the effect of water on the incidence of this disorder is complicated. It appears that if irrigation aids calcium uptake and retention in the fruit then bitter pit is reduced, but if irrigation dilutes the calcium content by increasing fruit size, then the incidence of bitter pit is increased. Mist irrigation has been shown to reduce the extent of the disorder compared with conventional irrigation (Goode *et al.*, 1979).

OTHER QUALITY CRITERIA

A tasting panel could discern no difference in the flavour of either raw or cooked peas from plants grown under wet or dry conditions (Freeman, 1970). However, the fibre content of bush bean pods is increased by water stress after flowering (Maurer *et al.*, 1969; Kemp *et al.*, 1974) and there is also evidence that the crude protein and fat contents, when expressed as a fresh weight basis, are increased (Maurer *et al.*, 1969). Water stress can also lead to irregular ripening in beans (Dubetz and Mahalle, 1969), an increase in the number of malformed pods (Vittum and Flocker, 1967) and a higher dry matter content in the pod (Maurer *et al.*, 1969), all of which are undesirable characteristics for processors.

Rudich *et al.* (1977) found that drip irrigation of outdoor tomatoes grown for processing reduced such quality criteria as the total soluble solids content, the Brix index and viscosity whilst increasing acidity and the number of rotten fruit. These results confirmed those of earlier work in the USA (Martin *et al.*, 1966; Vittum and Flocker, 1967) and they all emphasize how important the last irrigation is relative to harvest if a realistic balance between yield and quality is to be obtained.

In Israel, Gulfatreich *et al.* (1974) found that irrigation of apples reduced most criteria of fruit quality other than fruit size which, incidentally, was the one of overriding importance. These included firmness, total soluble solids, acid content and flavour but it was only in the later years of this six-year experiment that the differences became marked. However, it was also observed that fruits from the irrigated plots were less mature at harvest than those that had not been irrigated.

Much work has been done on the effect of water on the quality of citrus fruits. The results tend to confirm those for other crops with fruits subjected to water stress, having a reduced juice content and increased citric acid, sugar percentage, total soluble solids and thicker peels. Nevertheless, the results are often inconsistent from year to year which means that other climatic factors must also be involved (Hilgeman and Reuther, 1967; Shalhevet *et al.*, 1976).

CONCLUSIONS

In 1967 Salter and Goode concluded from their review of the literature that the effects of moisture stress on crop quality factors were still largely unknown and that a better understanding of these effects would make it easier to assess the need for a compromise between quantity and quality. They also stated that little was known about the interrelationship between moisture stress in plants and biochemical relationships. Only a little progress has perhaps been made since then.

One of the major problems is to separate the effects of water on crop quality from those of other environmental variables and the complex interactions which can occur are highlighted by the studies on potatoes (Hughes, 1974).

Nevertheless certain broad generalizations can be made. For example, in most vegetables and fruits, a degree of water stress, and this does not appear to have been quantified, can improve the flavour of the product as well as several other biochemical components of quality. In contrast the effect on appearance, succulence and fibre content may be detrimental. Where size is important water stress during the period of rapid expansion, whether of roots or fruits, will reduce fresh weight and volume. If moisture supplies are irregular at this time misshapen or split roots, tubers or fruits can result, the extent also depending on the ambient relative humidity and temperature. Even minute cracks in apples caused by water stress can result in large losses in store. Evidence for the direct effect of water supplies on colour development in fruit and on dry matter content and specific gravity in potatoes is less consistent because of the interaction with other environmental factors. Plant water status is clearly implicated in the incidence of a number of calcium-related physiological disorders in several vegetable and fruit crops, although with tipburn and bitter pit the mechanisms appear to be particularly complicated.

However sparse the data base might be there is enough empirical and subjective evdience to show that carefully controlled and well-timed applications of water can play a large part in maintaining high standards of quality. In addition the irrigation water itself must be of good quality because of the increasing danger of the contamination of supplies by industrial and domestic sewage (Nichols *et al.*, 1971; Williams, 1972). However, water is the one environmental variable which can be controlled on a field scale and it seems appropriate to consider how this can best be achieved in practice.

The results of the mist irrigation experiments with apples by Goode *et al.* (1979) lend support to the hypothesis that, with fruit crops in particular, environmental stress (induced by a dry soil and/or dry air) and physiological stress (imposed by the cropping level) are additive. Misting and irrigation together can minimize environmental stress and thereby permit the tree to withstand a higher cropping stress. As Goode *et al.* (1979) pointed out the use of mist in a commercial fruit orchard is worth consideration with large potential benefits in terms of yield and quality.

Perhaps the concept could be extended to allow an appropriately designed solid-set irrigation system to be used for:

(i) Frost avoidance by bloom delay through bud cooling in the early spring (Griffin, 1976).
(ii) Frost protection (perhaps by misting).
(iii) Irrigation.
(iv) Misting in mid-summer.

The development and testing of an integrated system of this kind would clearly require the close co-operation of people from many disciplines but is surely worth investigating.

Certainly, if irrigation is going to be used as a means of maintaining quality, improved methods of scheduling irrigation and of applying water less wastefully and more uniformly than most systems now used in field crops must be found. Trickle irrigation has many advantages for top and soft fruit providing the water is of a high quality and the benefits from regular liquid nutrient feeds, that this system of irrigation allows has recently been confirmed by Goode *et al.* (1978b).

There is much commercial interest in the development of trickle irrigation systems suitable for use in widely-spaced field-grown row crops. These, if successful, would allow precise quantities of water to be applied at the right time and in the right place and, if necessary, the systems could be automated. It is only through precise controls of water supplies to crops in ways like this that the full benefits, in terms of improved yield and quality, will be obtained.

REFERENCES

Arthey, V.D. (1975). "Quality of Horticultural Products". Butterworths, London.
Assaf, R., Bravdo, B. and Levin, I. (1974). *J. hort. Sci.* **49**, 53-64.
Assaf, R., Levin, I. and Bravdo, B. (1975). *J. hort. Sci.* **50**, 481-493.
Bodlaender, K.B.A., Lugt, C. and Marinus, J. (1964). *Eur. Potato J.* **7**, 57-71.
Burton, W.G. (1972). *Agric. Engng. Lond.* **53**, 43-52.
Cox, E.F. and Dearman, A.S. (1978). *Expl Hort.* **30**, 1-6.
Cox, E.F., McKee, J.M.T. and Dearman, A.S. (1976). *J. hort. Sci.* **51**, 297-309.
Crumbly, I.J., Nelson, D.C. and Duysen, M.E. (1975). *Am. Potato J.* **52**, 266-274.
Drew, D.H. (1966). *J. hort. Sci.* **41**, 103-114.
Drewitt, E.G. (1970). *Tech. Rep. Winchmore Irrig. Res. St.* No. 3, pp. 29.
Dubetz, S. and Mahalle, P.S. (1969). *J. Am. Soc. hort. Sci.* **94**, 479-481.
Fradgley, J.R.A. (1971). *Rep. natn. Veg. Res. Stn for 1970*, 86.
Freeman, G.G. (1970). *Rep. natn. Veg. Res. Stn for 1969*, 60-62.
Freeman, G.G. and Mossadeghi, N. (1973). *J. hort. Sci.* **48**, 365-378.
Goode, J.E. (1970). *In* "The Role of Water in Agriculture" (J.A. Taylor, ed.), pp. 161-170. Pergamon Press, Oxford and New York.
Goode, J.E. and Hyrycz, K.J. (1964). *J. hort. Sci.* **39**, 254-276.
Goode, J.E. and Hyrycz, K.J. (1968). *J. hort. Sci.* **43**, 215-230.
Goode, J.E. and Hyrycz, K.J. (1970). *J. hort. Sci.* **45**, 379-391.
Goode, J.E. and Ingram, J. (1971). *J. hort. Sci.* **46**, 195-208.
Goode, J.E., Fuller, M.M. and Hyrycz, K.J. (1975). *J. hort. Sci.* **50**, 265-269.
Goode, J.E., Higgs, K.H. and Hyrycz, K.J. (1978a). *J. hort. Sci.* **53**, 295-306.
Goode, J.E., Higgs, K.H. and Hyrycz, K.J. (1978b). *J. hort. Sci.* **53**, 307-316.
Goode, J.E., Higgs, K.H. and Hyrycz, K.J. (1979). *J. hort. Sci.* **54**, 1-11.
Gooding, E.G.B. (1966). *Expl Agric.* **2**, 129-137.
Griffin, R.E. (1976). *Am. Soc. agric. Engrs.* Pap. No. 76-1040, pp. 15.
Griffin, R.E., Alfaro, J.F., Keller, J., Hanson, G.R., Anderson, J.L., Ashcroft, G.L. and Richardson, E.A. (1974). *Am. Soc. agric. Engrs.* Pap. No. 74-5033.
Guelfatreich, S., Assaf, R., Bravdo, B.A. and Levin, I. (1974). *J. hort. Sci.* **49**, 217-225.
Harrison, D.J. (1963). *Rep. Efford exp. Hort. Stn for 1962*, 51-52.
Hilgeman, R.H. and Reuther, W. (1967). *In* "Irrigation of Agricultural Lands" (R.M. Hagan, H.R. Haise, T.W. Edminster and R.C. Dinauer, eds), Vol.II, pp. 704-718. American Society of Agronomy, Madison, Wisconsin.
Hsiao, T.C., Fereres, E., Acevedo, E. and Henderson, D.W. (1976). *In* "Water and Plant Life" (O.L. Lange, L. Kappen and E.D. Schulze eds), pp. 281-305. Springer-Verlag, Berlin and New York.
Hughes, J.C. (1974). *Potato Res.* **17**, 512-547.
Kaufmann, M.R. (1972). *In* "Water Deficits and Plant Growth" (T.T. Kozlowski, ed.), Vol. III, pp. 91-124. Academic Press, New York and London.
Kemp, G.A., Krogman, K.K. and Hobbs, E.H. (1974). *Can. J. Pl. Sci.* **54**, 521-528.
Lapwood, D.H., Wellings, L.W. and Hawkins, J.H. (1971). *Pl. Path.* **20**, 157-163.
Mapson, L.W., Swain, T. and Tomalin, A.W. (1963). *J. Sci. Fd Agric.* **14**, 673-684.
Martin, P.E., Lingle, J.C., Hagan, R.M. and Flocker, W.J. (1966). *Calif. Agric.* **20**(6), 12-14.
Massey, P.H.Jr., Eheart, J.F., Young, R.W. and Mattus, G.E. (1962). *Proc. Am. Soc. hort. Sci.* **81**, 316-323.
Maurer, A.R., Ormrod, D.P. and Scott, N.J. (1969). *Can. J. Pl. Sci.* **49**, 271-278.

Metzger, C.H., Tobiska, J.W., Douglass, E. and Vail, C.E. (1938). *Proc. Am. Soc. hort. Sci.* **35**, 635-643.
Nichols, A.A., Davies, P.A., King, K.P., Winter, E.J. and Blackwall, F.L.C. (1971). *J. hort. Sci.* **46**, 425-433.
North, J.J. (1960). *J. R. agric. Soc.* **121**, 7-20.
Pieters, J.H. (1971). *Meded. Proefstn Groenteteelt volle Grond.* No. 55, pp. 27.
Rickard, P.C. (1970). *Rep. Arthur Rickwood exp. Husb. Fm for 1970,* 36-39.
Robins, J.S. and Domingo, C.E. (1956). *Agron. J.* **48**, 488-492.
Rudich, J., Kalmar, D., Geizenberg, C. and Harel, S. (1977). *J. hort. Sci.* **52**, 391-399.
Ruf, R.H. (1964). *Am. Potato J.* **41**, 377-381.
Sale, P.J.M. (1966a). *J. hort. Sci.* **41**, 31-42.
Sale, P.J.M. (1966b). *J. hort. Sci.* **41**, 43-52.
Salter, P.J. (1956). *Rep. natn. Veg. Res. Stn for 1955*, 37-38.
Salter, P.J. (1961). *J. hort. Sci.* **36**, 241-253.
Salter, P.J. (1963). *J. hort. Sci.* **38**, 321-334.
Salter, P.J. and Goode, J.E. (1967). "Crop Responses to Water at Different Stages of Growth". *Res. Rev. Commonw. Bur. Hort. Plantn Crops* No. 2, pp. 246.
Salter, P.J. and Williams, J.B. (1964). *Rep. natn. Veg. Res. Stn for 1963*, 48.
Sanders, D.C., Nylund, R.E., Quisumbing, E.C. and Shetty, K.V.P. (1972). *Am. Potato J.* **49**, 243-254.
Shalhevet, J., Mantell, A., Bielorai, H. and Shimshi, D. (1976). "Irrigation of Field and Orchard Crops under Semi-arid Conditions". International Irrigation Information Center, Bet Dagan and Ottawa.
Sharples, R.O. (1975). *In* "Climate and the Orchard" (H.C. Pereira, ed.) pp. 113-124. *Res. Rev. Commonw. Bur. Hort. Plantn. Crops* No. 5.
Smith, O. (1961). *Am. Potato J.* **38**, 265-271.
Smittle, D.A. and Thornton, R.E., Peterson, C.L. and Dean, B.B. (1974). *Am. Potato J.* **51**, 152-164.
Stanhill, G. (1957). *Rep. natn. Veg. Res. Stn for 1956,* 47.
Stanhill, G. (1958a). *J. hort. Sci.* **33**, 108-118.
Stanhill, G. (1958b). *J. hort. Sci.* **33**, 264-274.
Unrath, C.R. and Sneed, R.E. (1974). *J. Am. Soc. hort. Sci.* **99**, 372-375.
Vittum, M.T. and Flocker, W.J. (1967). *In* "Irrigation of Agricultural Lands" (R.M. Hagan, H.R. Haise, T.W. Edminster and R.C. Dinauer, eds), Vol. II, pp. 674-685. American Society of Agronomy, Madison, Wisconsin.
Waister, P.D. and Hudson, J.P. (1970). *J. hort. Sci.* **45**, 359-370.
Walter, T.E. (1967a). *Rep. E. Malling Res. Stn for 1966*, 83-95.
Walter, T.E. (1967b). *Rep. E. Malling Res. Stn for 1966*, 70-82.
Wellings, L.W. (1973). *Expl Husb.* **24**, 54-69.
Wiersum, L.K. (1966). *Acta bot. neerl.* **15**, 406-418.
Williams, J.H. (1972). *A.D.A.S. q.Rev.* **6**, 106-122.
Winter, E.J., Laflin, T., Craze, S.P. and Carter, A.R. (1968). *Expl Hort.* **18**, 52-59.

II.6 Diseases in Relation to Quality and Storage Life, with Special Reference to Root-Crops

D.H. LAPWOOD

Rothamsted Experimental Station, Harpenden, Herts. AL5 2JQ, England

INTRODUCTION

Diseases may affect the quality of root-crops in several ways. They may, (i), decrease plant stand or vigour and thereby the yield, size, shape and eating quality, aspects not considered here or, (ii), by blemishing the surface affect presentation at the retail outlet, increase water loss and therefore decrease keeping quality in store and increase wastage from deep peeling when processed, and (iii), directly destroy the tissues by rotting them.

Diseases may be controlled by use of chemicals or in non-chemical ways such as breeding or modifying cultural methods; the latter will be emphasized whenever possible.

The main root vegetables used for human consumption (Table I) are parsnip, turnip and swede, which are only of minor importance and will be considered together and red beet (beetroot), carrot and potato, considered separately.

PARSNIP, SWEDE AND TURNIP

Parsnips, like swedes and turnips, are usually left in the ground over winter and harvested as required. Some, preferably lifted by hand, may be stored in clamps, barns or temperature-controlled stores as a reserve, to ensure continuity of supply during severe weather. Two types of parsnip are in demand for processing, viz. small parsnips for canning and prepackaging and a limited quantity of large ones for soup. Parsnip canker is the main disease problem and occurs on the shoulder or crown of the roots especially in wet seasons. Various fungi, including *Phoma* sp., *Itersonilia pastinacae*,

Channon, and *Mycocentrospora acerina* (Hartig) Deighton, can cause the symptom (Channon, 1965). Control can be achieved by using a canker resistant cultivar like Avon Resister, by late planting in mid April/early May to produce smaller roots by harvest and by earthing-up to decrease the chance of spores reaching the crowns.

TABLE I

Root vegetables grown for human consumption in England and Wales, 5 year average, 1970-74

	Cropped or production area ($10^3 \times$ hectares)	Yield per hectare (tonnes)	Production ($10^3 \times$ tonnes)
Parsnip	2·4	21·8	51
Red beet	3·4	30·9	105
Turnip/Swede	3·6	30·9	110
Carrot	13·3	37·9	504
Potato (maincrop)	165	31·4	5170

Data from Ministry of Agriculture, Fisheries and Food, Agricultural Statistics, England and Wales, 1975. HMSO.

Swedes may be used for human consumption or stock feed, but some, as with turnips, may be used in processing and in a "stew pack" mix with carrots and onions. Turnips and swedes are very susceptible to boron deficiency which causes brown rotting (heart rot) a condition that can be prevented by the application of borax. Club root, *Plasmodiophora brassicae* Wor. is a most serious disease causing severe root distortion and stunted growth. The fungus which affects most crucifers, can remain viable in soil for several years and care should be taken to provide sufficient intervals between brassica crops in the rotation. Liming and improved drainage may decrease the disease and a few turnip and swede cultivars are resistant. Dry rot, *Leptosphaeria maculans* (Desm.) Ces and de Not, causes a root rot which may spread during storage. The fungus may be seed-borne, for which a thiram-soak treatment (Maude, 1967) provides an effective control, or soil-borne. In the latter case a minimum of four years between susceptible crops is required if the disease is to be controlled.

RED BEET (BEETROOT)

Red beet is frost-susceptible and has to be lifted and stored in outdoor clamps, in indoor unventilated or fan ventilated stores using night air for cooling, when beet may be kept until the end of March; in refrigerated stores they can be kept until July (Davies *et al.*, 1976). Most maincrop red beet are grown under contract for processing. Hand-lifted beet suffer less injury and keep much better than machine-lifted and damage to the crown should be avoided and therefore it is unwise to grade before storage. Red beet, like turnip and swede, are liable to heart rot from boron deficiency which may lead to an attack by *Pleospora betae* (Bjorling); this fungus also causes a damping-off of seedlings (black leg) for which treatment of seed with thiram has proved effective (Maude, 1967). It can survive in surface debris and because its longevity in soil is unknown, long rotations are advised. Sugar beet (and mangolds and seakale beet) are affected and delaying planting until soils are warm, to ensure quick emergence and early development, decreases incidence of the disease in these crops.

Scab, caused by *Streptomyces* spp., probably *Streptomyces scabies* (Thaxt.) Waksm. and Henrici, consists of deep corky pits or raised corky projections on the beet surface which are troublesome to the processor. The disease is only locally important, especially in the Fenland areas of East Anglia where the situation is made worse because scab-susceptible sugar beet and potato occur in the rotation. Recent work has shown that infection occurs at the decortication of the hypocotyl when tissues remain susceptible for only about two weeks (Adams and Lapwood, 1978). Other than avoiding scab-infested sites, no proven control measures are available.

CARROT

Carrot production is second only to the potato, far exceeding that of all other root vegetables (Table I). Most carrots are left in the ground until required, perhaps with earthing-up of soil or straw covering to protect them from the severest winter weather. Some contract crops may be lifted into soil-covered trenches. Carrots stored in barns or in slatted-floor stores with night air cooling usually represent buffer stocks against bad weather, when field supplies are not available. More recently carrots have been put into polythene-lined bins or boxes (usually of 0·5 tonne capacity) for refrigerated storage which extends post-harvest life to 30-40 weeks (Robinson *et al.*, 1975). The processors' demand for continuity of supply of a high quality product, plus increasing labour costs and unwillingness to work outside in severe weather, is enforcing changes towards mechanization and longer-term storage. Tucker (1974) reported 10-20% more damage to mechanically-harvested carrots, and Apeland (1974) a 30% decrease in storage potential over a 6-7 month storage period at 0°C, compared with hand-lifted crops. Losses were in part due to increased damage and attack by *Mycocentrospora acerina*, which may infect harvest wounds (Davies, 1977). A wound healing (curing) method has been devised (Davies, 1977) in which roots are kept at 24-26°C at 98% r.h. for two days but it has not been adopted commercially.

The three most important pathogens limiting the storage of carrots in East Anglia are considered to be *Botrytis cinerea* Fr., *Mycocentrospora acerina* and *Sclerotinia sclerotiorum* (Lib.) de Bary (Derbyshire, 1971). Lesions of *Botrytis cinerea* may be found in the growing crop at the bases of petioles near the crown (Goodliffe and Heale, 1975), but roots are more often infected during storage by air-borne spores and especially if they become sub-turgid (Goodliffe and Heale, 1977) when the disease may spread from root to root by contact. Lining bins or boxes with polythene to prevent dehydration has largely overcome this problem (Derbyshire and Crisp, 1971).

The rot caused by *Mycocentrospora acerina* is seldom seen at harvest although chlamydospores and thick-walled mycelium, which can survive the winter (Wall and Lewis, 1978), may be found on senescing petioles near the crown and in adhering soil (Davies, 1977). Infection appears to be associated with damage and first rots may not be found until after some 12 weeks of storage. There is no evidence of root to root spread in store.

Sclerotinia sclerotiorum attacks a wide range of cultivated plants and the disease is favoured by wet and cool conditions. Although symptoms may be found in the field, infested areas are difficult to avoid when machine harvesting. Roots carry sclerotia into store and infections most probably occur through wounds. Diseased carrots develop a mycelium bearing sclerotia characteristically with droplets of moisture, and the fungus spreads to infect other roots so that pockets of infection develop. Rotation with non-susceptible crops is essential as the sclerotia can survive in soil for several years, although some may germinate the following summer to produce disc- or cup-shaped fruiting bodies which can discharge large numbers of spores into the air over several weeks. A minimum break of three years between susceptible crops is recommended.

To control these pathogens a broad-spectrum fungicide is required and a post-harvest pre-storage dip in benomyl proved effective (Derbyshire and Crisp, 1978). This achieved an average of 96% marketable yield after 5 months' storage over the three storage seasons, 1971-73. As a result of this work the trade has shown a greater interest in carrot storage.

Some diseases have not responded to chemical treatment. Young carrots lifted warm in early autumn and left over a weekend in bulk without cooling may develop a surface blackening caused by *Chalaropsis thielavioides* Peyron. making them unsuitable for prepacks. Cooling is the only satisfactory treatment (Derbyshire and Shipway, 1978). Crater rot, caused by *Rhizoctonia carotae*, Rader, has been considered a serious threat to carrot storage in Sweden (Jensen, 1971) and was first recognized in Eastern

England in 1968. Sclerotia either taken into store with the crop or contaminated boxes used to store a previously affected crop, germinate and, in conditions of high humidity inside a polythene-lined box, mycelium soon covers the roots. In one trial (Derbyshire and Shipway, 1978), 21% of machine-harvested carrots were infected despite benomyl treatment.

Bacterial soft rots caused by *Erwinia carotovora* or *Pseudomonas* spp. may develop at a low level of incidence during prolonged cool storage. However, with the rise in temperature during pre-packing and transport, the incidence of rot may be sufficiently increased to make the produce unsaleable. Washing water may also contain large numbers of *E. carotovora* and washing prior to long-term storage is not recommended. However, maintaining 25 ppm chlorine in the water used to wash finger carrots for prepacking has proved effective (Segall and Dow, 1973).

POTATO

Potatoes are the most widely grown of all the root vegetables with an annual maincrop production of around 5 million tonnes (Table I). The potato is very susceptible to frost and so all crops are lifted and stored. As with other roots, stores vary greatly from the outside "Dickie Pie", the barn floor, stores with night air cooling, to the sophisticated refrigerated store (Eddowes, 1978). Still by far the greater proportion of potatoes is sold loose (Table II), and these are farm-stored for varying periods depending on market demands. The remainder are used by processors and prepackers.

TABLE II

Estimates of potato usage in Great Britain 1977/78

Use	$10^3 \times$ tonnes	%	Use	$10^3 \times$ tonnes	%
Raw loose	3547	72	Raw prepacked	390	7·8
Crisped	403	8	Dehydrated	188	3·8
Frozen or chilled	402	8	Canned	23	0·5
Total 4 953 000 tonnes					

Data from the Potato Marketing Board, London.

Processors at present account for some 20% of the ware tonnage and this could be 30% if imports were stopped. This is an expanding market and over the next five years may be expected to take 35-40% (it has already reached 50% in the USA), which if the 0·5 million tonnes for the fish and chip trade is included is nearly 2 million tonnes. Processors and prepackers with their production lines require continuity of supply of a high quality product, and some are attempting to store potatoes from one harvest to the next. The requirements of different processors differ in that the dehydrator can take advantage of low storage temperature (*c.* 7°C) but the crisper has to store around 10°C to avoid low temperature sweetening which affects crisp quality. When this occurs potatoes have to be warmed (reconditioned) at 18°C for several weeks, which sometimes results in heavy losses from disease. Thus, loss of water through damage, sprouting and skin diseases, plus loss of crop through rots become increasingly important factors in long-term storage.

DISEASES AFFECTING STORAGE LIFE

The diseases which affect the storage life and quality of potatoes are of two main types; those that blemish and those that rot (Table III). The rots may be "dry" or "wet" and the latter may be subdivided into those in which the tissues remain intact, although watery, and those that disintegrate the tissues completely. The main problems to the pre-packers are ones of (1) presentation and hence the importance of skin spot and common scab

TABLE III

Diseases affecting the storage life and quality of potatoes

Blemishes	Inoculum source	
	Soil	Seed tuber
Common scab (*Streptomyces scabies*)	+[a]	+
Skin spot (*Polyscytalum pustulans*)	−	+
Silver scurf (*Helminthosporium solani*)	−	+
Rots—dry		
Dry rot (*Fusarium solani* var. *coeruleum*)	+	+[a]
Gangrene (*Phoma exigua* var. *foveata*)	−	+
Late blight (*Phytophthora infestans*)	−	+
Rots—wet (intact)		
Pink rot (*Phytophthora erythroseptica*)	+[a]	+
Watery wound rot (*Pythium ultimum*)	+	−
-wet (disintegrating)		
Bacterial soft rot (*Erwinia carotovora* vars. *carotovora* and *atroseptica*)	−	+

[a] Probably the more important inoculum source.

and, with pigmented cultivars like King Edward, silver scurf, and (2) rotting by the bacterial soft rots. For those who store potatoes at *c.* 7°C, skinspot and gangrene are the main problems, the former adding to peeling losses, while at *c.* 10°C the main problems are Fusarium dry rot and the bacterial soft rots. Pink rot, late blight, watery wound rot and dry rot will not only cause direct loss of stored tubers, but under poor storage conditions, they may be centres from which rapidly spreading bacterial degeneration starts.

Pathogens may be seed- or soil-borne, some are both but generally one of the sources is more important than the other (Table III), and therefore control measures have to be directed at the main field sources of inoculum.

SEED-BORNE DISEASES

If seed tubers were healthy, or if "health" could be improved, then there would be little or no seed-borne disease (Hirst *et al.*, 1970). Unlike the other root vegetables, potatoes are propagated vegetatively from relatively bulky "seed" tubers and not from small "true" seeds. There is an extensive seed producing industry which has become located in those areas of the UK with low aphid populations, in order to produce virus-free stocks. Certification schemes have traditionally concentrated on freedom from virus diseases and until relatively recently little attention was paid to the fungus diseases although crops for certification have always been "rogued" for "blackleg" (*Erwinia carotovora* var. *atroseptica* Van Hall, Dye). Now the seed of all potato cultivars is produced from virus-tested stem cuttings (VTSC) and these are multiplied first by specialist VT growers, who then pass the produce on to the normal seed multipliers, who sell to the trade. The use of stem cuttings was an attempt to free tubers from seed-borne diseases (Graham and Hardie, 1971), but unfortunately during the years of multiplication both fungal (Hide, 1978) and bacterial (Graham *et al.*, 1976) pathogens rapidly reinfect the stocks. In the past, organo-mercury dips were used to free tubers from disease but these often created more problems than they solved and fortunately more effective and safer chemicals are now to hand. Two of these, thiabendazole and 2-aminobutane, are now widely used on seed. Thiabendazole will control silver scurf, skin spot, gangrene and dry rot and the benefit of treatment prior to storage of the seed crop can be maintained through to ware with silver scurf and skin spot but not gangrene (Hide and Bell, 1978). 2-Aminobutane has to be applied in a specially constructed gas chamber (Graham *et al.*, 1973); it will not control gangrene and the benefits of a pre-storage treatment are not felt in the ware crop.

There is no effective bacteriocide and therefore the grower of high quality ware would be advised always to buy Foundation Seed (FS), certified seed which is guaranteed not to contain more than 0·25% plants with blackleg, whereas Stock Seed (SS) and AA certificate may have 1% and "CC" 4% (Ebbels, 1979).

Late blight overwinters in tubers and may be brought on to the farm with seed, or it may already be present locally in the residue of the previous ware crop dumped in dell holes or left in piles to rot, or in groundkeepers. Any one of these sources could be sufficient, given suitable weather, to initiate a blight attack and therefore all should be eliminated as far as practicable.

SOIL-BORNE DISEASES

These are more difficult to control than seed-borne diseases and therefore there is greater reliance on cultural methods.

Common scab (*S. scabies*) occurs on the seed tuber but this inoculum is probably of less importance than that in soil. Some control can be achieved by avoiding known scab-infested sites and land recently limed or ploughed from long-term ley, or by choosing a resistant cultivar. Ploughing in a green manure crop, treating soil with chemical prior to planting and acidifying soil with sulphur have all been tried with varying success (Lapwood, 1973). As it is a disease which affects the crop only when dry weather coincides with the early stages of tuber development, irrigation during this period has proved very effective (Lapwood *et al.*, 1973), although not as popular as would be a fungicide spray. Research into foliar-applied downward translocated systemic chemicals has shown that scab might be controlled in this way (McIntosh, 1975) but safer and more effective chemicals are required before it becomes a practical or commercial proposition.

Pink rot (*P. erythroseptica*) occurs widely but sporadically and has been troublesome during recent warm summers especially where crops have been irrigated, causing a wilt as well as a rot of progeny tubers. Affected tubers taken into store, rot and may initiate bacterial soft rot decay. The fungus produces short-lived sporangia and long-lived oospores, the latter may be carried by the seed tuber but soil-borne inoculum is probably the more important and a break of at least four years between potato crops is recommended.

Watery wound rot (*Pythium ultimum*) is only of importance in exceptionally mild autumns when tubers are lifted warm. However, when it does occur it can cause severe loss and the rotting of small numbers of tubers in store may be followed by extensive bacterial decay. The only control seems to be to avoid lifting under warm soil conditions especially when tubers are immature and to attempt to minimize damage.

PRE-HARVEST AND POST-HARVEST DISEASE

Some diseases will be evident before harvest, such as late blight (foliage lesions and tuber blight), common scab (tuber surface lesions), pink rot (wilting foliage and tuber rot), blackleg (wilting stems with black lesion at the base and under exceptionally wet conditions, tuber rot). With some diseases infection may occur in the growing crop but symptoms only appear later during storage. These include silver scurf, skin spot and latent lenticel infection of the *Erwinia* soft rot bacteria. In others, infection takes place at harvest through wounds, from inoculum that has built up in the growing crop. These include dry rot, gangrene, watery wound rot and also late blight, skin spot and the bacterial soft rots.

There are several measures that can be taken to avoid tuber blight, such as an adequate fungicide spray programme, well-earthed and pointed ridges, burning-off the haulm before blight develops and leaving at least two weeks between death of haulm and lifting to allow any build-up of soil-borne inoculum to decrease.

Susceptibility of harvest wounds may be decreased indirectly by using the right balance of NPK fertilizer to reduce proneness to wounding, by lifting when the crop is most resistant or by keeping the level of available inoculum low, and decreased directly by reducing the amount of damage done at harvest by machinery.

Susceptibility to dry rot is decreased by NPK fertilizer application (Boyd, 1967) but susceptibility to gangrene is increased by low NPK and also by high rates because of increased susceptibility of the tissues to damage. High rates increase growth cracks and susceptibility to bacterial soft rot (Graham and Harper, 1966).

Susceptibility of tubers to gangrene is said to be high during the early stages of the growing season but gradually decreases towards harvest, and a similar pattern is shown by dry rot. However, in practice it is difficult to separate susceptibility from changes in amount of inoculum available for infection. With gangrene there appear to be two peaks, one soon after planting infected seed and then again after burning-off. This latter may be related to the formation of pycnidia on dying stems shortly after burning-off, and the washing down of spores into the soil (Logan *et al.*, 1976). The amount of storage disease seems to be related to the date chosen for burning-off and harvest (Fox and Dashwood, 1975), on the chemical used and on the time interval between burning-off and harvest (Logan *et al.*, 1976; Adams and Griffith, 1978).

The most important factor affecting the amount of post-harvest disease is the amount and type of damage at lifting, and there is urgent need to devise machines and methods to decrease it. Shotton (personal communication) has shown that by the intelligent use of existing machinery, adjusting the forward speed of the tractor, the digger web speed, and the use of agitation, damage can be drastically reduced.

Recently thiabendazole and 2-aminobutane have been cleared for use on the ware crop. With the former, care is required not to overwet the tubers (correct rate is 2 l per tonne) when misting the tubers (Logan and Copeland, 1975) as this will encourage bacterial soft rotting in store, whereas the use of the latter is said to increase the incidence of Fusarium dry rot. None of the commercially-available bacteriocides can be considered effective and the only method of controlling bacterial soft rot is through good store management. Unless soft rots are already present on loading, a curing (wound healing) period during which the bulk is held at a temperature of 15°C for 14 days before decreasing to holding temperature (*c.* 7°C), will prevent infection by the bacteria and by other wound parasites such as skin spot, gangrene and dry rot.

CONCLUSIONS

(1) The long-term cool storage of root vegetables, other than potatoes, is a relatively new venture and still not widely accepted or practised and so the disease problems, unlike those of the potato, remain largely undefined and little work has been done on etiology and control.

(2) Cultural methods may be used with success to decrease the incidence of some diseases, but are usually more effective when combined with a chemical treatment. Chemicals, however, should not be used as a substitute for good farming practice.

(3) Soil-borne pathogens are more difficult to control than seed-borne pathogens, but the new generation of systemic fungicides may help. There is as yet no effective bacteriocide and therefore the soft rot bacteria remain the greatest threat to the quality and long-term storage of root vegetables.

(4) Increased mechanization has increased the level of damage to the harvested crop with consequent loss of storage potential from increased water loss and wound parasites. Commercially acceptable curing methods are available for potatoes but not carrots.

(5) The incidence of disease can vary widely in different stocks and in different seasons and therefore some way of predicting storage potential from in growth or immediately pre-harvest assessments would be of great benefit and could result in a dramatic decrease in storage losses.

REFERENCES

Adams, M.J. and Griffith, R.L. (1978). *Ann. appl. Biol.* **88**, 51-55.

Adams, M.J. and Lapwood, D.H. (1978). *Pl. Path.* **27**, 97-98.

Apeland, J. (1974). *Acta Hort.* **38**, 353-357.

Boyd, A.E.W. (1967). *Ann. appl. Biol.* **60**, 231-240.

Channon, A.G. (1965). *Ann. appl. Biol.* **56**, 119-128.

Davies, W.P. (1977). *Ann. appl. Biol.* **85**, 163-164.

Davies, A.C.W., Oswin, P., Rutherford, P.P., Tucker, W.G. and Phelps, K. (1976). *Expl Hort.* **28**, 15-30.

Derbyshire, D.M. (1971). *Comml Grow.* 16 April 1971, 703.

Derbyshire, D.M. and Crisp, A.F. (1971). *Proc. 6th Br. Insectic. Fungic. Conf.* **1**, 167-172.

Derbyshire, D.M. and Crisp, A.F. (1978). *Expl Hort.* **30**, 23-28.

Derbyshire, D.M. and Shipway, M.R. (1978). *Outlook Agric.* **9**, 246-252.

Ebbels, D.L. (1979). *ADAS q. Rev.* **32**, 21-58.

Eddowes, M. (1978). *Outlook Agric.* **9**, 253-259.

Fox, R.A. and Dashwood, E.P. (1975). *Proc. 6th trienn. Conf. Eur. Ass. Potato Res.* 88-89.

Goodliffe, J.P. and Heale, J.B. (1975). *Ann. appl. Biol.* **80**, 243-246.

Goodliffe, J.P. and Heale, J.B. (1977). *Ann. appl. Biol.* **85**, 163.

Graham, D.C. and Hardie, J.L. (1971). *Proc. 6th Br. Insectic. Fungic. Conf.* **1**, 219-224.

Graham, D.C. and Harper, P.C. (1966). *Eur. Potato J.* **9**, 141-145.

Graham, D.C., Hamilton, G.A., Quinn, C.E. and Ruthven, A.D. (1973). *Potato Res.* **16**, 109-125.

Graham, D.C., Quinn, C.E. and Harrison, M.D. (1976). *Potato Res.* **19**, 3-20.

Hide, G.A. (1978). *Potato Res.* **21**, 277-289.

Hide, G.A. and Bell, F. (1978). *Ann. appl. Biol.* **90**, 427-436.

Hirst, J.M., Hide, G.A., Griffith, R.L. and Stedman, O.J. (1970). *J. R. agric. Soc.* **131**, 87-106.

Jensen, A. (1971). *Acta Hort.* **20**, 125-129.

Lapwood, D.H. (1973). *In* "Actinomycetales" (G. Sykes and F.A. Skinner, eds), pp. 253-260. Academic Press, New York and London.

Lapwood, D.H., Wellings, L.W. and Hawkins, J.H. (1973). *Pl. Path.* **22**, 35-41.

Logan, C. and Copeland, R.B. (1975). *Proc. 8th Br. Insectic. Fungic. Conf.* **2**, 589-595.

Logan, C., Copeland, R.B. and Little, G. (1976). *Ann. appl. Biol.* **84**, 221-229.

McIntosh, A.H. (1975). *Proc. 8th Br. Insectic. Fungic. Conf.* **2**, 609-617.

Maude, R.B. (1967). *Proc. 4th Br. Insectic. Fungic. Conf.* **1**, 259-264.

Robinson, J.E., Browne, K.M. and Burton, W.G. (1975). *Ann. appl. Biol.* **81**, 399-408.

Segall, R.H. and Dow, A.T. (1973). *Pl. Dis. Reptr.* **57**, 896-899.

Tucker, W.G. (1974). *Acta Hort.* **38**, 359-372.

Wall, C.J. and Lewis, B.G. (1978). *Trans. Br. mycol. Soc.* **71**, 143-146.

II.7 Modifying Cultural Techniques to Develop Quality—the Effect of Weed Control and Herbicides

J.M. KING

Processors and Growers Research Organisation, Thornhaugh, Peterborough, PE8 6HJ, England

INTRODUCTION

Quality in processed vegetables is a combination of many factors. Flavour and texture in certain vegetables are directly linked to maturity when harvested, while appearance and shape and size are often affected by cultural techniques. Individual facets of quality are of little value alone and for produce to be of top quality all must be combined. Standards for produce destined for processing have always been very high, and producers have had to find ways of growing crops to meet them. Research has enabled many of the targets to be achieved. In recent years we have seen a "spin-off" of this technology in the production of crops grown for pre-packing and fresh sale.

DIRECT EFFECTS OF WEEDS ON THE QUALITY OF VEGETABLES

Competition from weeds can affect the growth and quality for processing of vegetables in several ways, and before discussing how herbicides and other methods of weed control can be used to modify cultural techniques it is important to appreciate the effects of weeds on quality.

In many vegetables the size of the harvested produce is a major aspect of its quality. Carrots are an excellent example, but it also applies to crops such as red beet, onions, cauliflowers, and Brussels sprouts. Weed competition can affect the growth and the size of the harvested product. Shadbolt and Holm (1956) studied the effects of weed competition on carrots, onions and red beet, and found that it caused progressive reductions in the diameter of the roots and bulbs. Work at the National Vegetable

Research Station (Hewson and Roberts, 1971), showed that weed competition increased the proportion of thick-necked onions. In brassica crops, weed competition can also affect the quality of the produce, for example, Lawson (1972) reported that spring cabbage plants subjected to weed competition had longer internodes, which resulted in loose, unmarketable heads.

Competition from weeds can also affect the maturity of crops, an essential element of quality in legumes such as peas, broad and dwarf beans. King and Handley (1972), found that vining peas matured more rapidly in the presence of a dense stand of chickweed (*Stellaria media*). Although published data is not available it is possible that crops maturing under stress from weed competition give poorer quality produce with inferior texture and flavour.

Weeds may also increase the incidence or severity of diseases which could affect quality. Rademacher (1967), found a higher incidence of *Ascochyta pisi*, a disease which can produce staining of pea seeds, in weed-infested pea plots. There are many records of weeds acting as hosts of pests and diseases of vegetables, and thus causing problems which could lead to poorer quality at harvest.

Contamination during harvesting is a very real problem in the processing pea and bean crops, which are mechanically harvested. Elliot (1972), states that the presence of weeds such as chickweed (*S. media*) and knotgrass (*Polygonum aviculare*), which are easily uprooted when cutting peas, lead to increased contamination of the vined product with soil and stones. A more serious contamination problem, however, occurs when buds, flowers, berries or broken fragments of weed plants are found in the vined produce entering the factory. Elliot lists the following as weeds which can create such problems—mayweeds (*Matricaria* spp.), creeping thistle (*Cirsium arvense*), black nightshade (*Solanum nigrum*) and poppy (*Papaver* spp.). Pea crops infested with mayweeds could also lead to contamination and flavour changes during vining due to their strong aromatic scent. Such occurrences though rare, are known and could result in crop rejection by the processor. Contamination of the final pack is extremely serious to the processor and as pointed out by Elliot (1972), the Pure Foods and Drug Act of 1955 states that the content of a packet must be declared on the outside and it must not contain any extaneous vegetable matter of any kind.

Finally, the presence of weeds can seriously slow down or delay the harvesting of vegetables (Gane, 1968). Where maturity is important for optimum quality this can have serious repercussions, for instance, if pea crops contain a high percentage of weed debris the speed of the beaters may have to be increased in order to thresh the material, and this could result in increased damage to the green peas (Elliot, 1972). Similarly, with mechanically harvested dwarf green beans, weed infestations of species such as fat hen (*Chenopodium album*), which is strongly branched, can lead to increased damage to the bean pods caused by the action of the picking times and the weed.

INDIRECT EFFECT OF WEEDS AND WEED CONTROL METHODS ON THE QUALITY OF VEGETABLES

ROW WIDTHS

When man first began to cultivate crops he probably worried little about weeds and sowing seeds at optimum plant densities or the effect of weeds on the produce. As farming became more organized and mechanical cultivation was introduced he began to sow his crops in rows, between which he could hoe or cultivate to remove weed competition. This system remained in use for many crops until relatively recently. Cereals were the first to be grown in narrow rows where cultivations were not possible and others such as peas soon followed. This change was made possible by the discovery of herbicides which could selectively remove weeds without damaging the crop. Not only does the use of more even plant distribution generally lead to higher yields, but in many crops where specific size grades are required by the customer or processor, it is often the only way size specifications can be met. Obvious examples are root crops such as carrots (Bleasdale, 1963) and red beet. To produce small "baby" carrots for canning or finger Amsterdams for freezing whole, a bed system is required with high plant densities in rows often only 2-3 cm apart, and weeds must be wholly controlled by the use of herbicides. High plant densities control size of root crops, while regular distribution leads to more even development, maturity and size. Peas are already grown in rows of 18 cm apart or less, but in dwarf green beans the change to narrow rows has still not been fully achieved in spite of the availability of suitable harvesters. Improved herbicide programmes have now been developed and this has opened the way to changed growing systems, but there remains one major stumbling block, that of volunteer crop problems of which potatoes are by far the most serious (King, 1974). In some brassica crops the lack of really effective herbicides capable of dealing with a wide range of weed species still prevents full use being made of the growing systems which would give higher yields of top quality produce. In others such as Brussels sprouts the optimum row widths for quality and yields are still sufficiently wide to allow some mechanical cultivation. Work at the Scottish Horticultural Research Institute and the Processors and Growers Research Organisation has demonstrated that it is possible to produce evenly-maturing, high density broccoli (calabrese) in narrow rows which results in smaller primary heads. These could be processed whole without the splitting which is necessary for large heads grown at low plant density. One of the problems which has prevented this system from being widely used commercially is the difficulty of controlling weeds. Similar problems were also encountered when "mini" cauliflowers were grown commercially.

DRILLING OR TRANSPLANTING

In brassicas a more effective range of chemicals is available for use after transplanting than for pre-emergence use, and thus it is often the drilled brassicas which create the most weed problems. This has resulted in the majority of Brussels sprout crops being transplanted rather than drilled, although there is no evidence to suggest that the quality of the produce is affected by this system of establishing the crop.

SEEDBED PREPARATION

For herbicides to work efficiently it is necessary for weeds and crop plants to emerge evenly. This applies particularly to post-emergence treatments where growth stages are important. A well-prepared seedbed is therefore required and in the case of incorporated or pre-emergence treatments a relatively fine tilth is essential. In turn this should reflect in better and more even emergence of the crop plant which is important if final quality is to be high.

ADVERSE EFFECTS OF HERBICIDES ON QUALITY

Although herbicides are an essential part of many growing systems they are not infallible, sometimes failing to control specific weeds and occasionally damaging the crop. Pre-emergence herbicides which "thin" the plant stand can seriously affect both yield and quality because the value of the optimum plant arrangement pattern is lost. Materials which stunt or delay early growth of crops to be harvested at the green stage can affect crop quality. Lawson and Rubens (1970), working with peas in Scotland found that a pre-emergence treatment of a mixture containing chlorpropham caused irregular emergence, stunting of some plants and encouraged secondary tillering. This resulted in a marked unevenness in maturity of the peas at harvest and there was a higher proportion of haulm to peas which would reduce harvesting efficiency. Pre-treatment of the soil with TCA for wild oats or couch control may predispose peas to damage from post-emergence herbicides or disease by affecting the protective layer of wax on their leaves. Contact herbicides such as dinoseb-amine may scorch the leaves of affected plants, while Proctor and Armsby (1958) reported increased attack from downy mildew (*Peronospora viciae*) on peas grown in soil previously treated with TCA.

Changes have taken place in recent years in the way in which final plant stand is achieved. In the past a higher seed rate was sown in row crops, and the final population was achieved when the crop was hoed or "chopped out". Precision seeders are now widely used to sow the seed to a stand, and weeds are controlled by herbicides. Such a system relies on the use of high quality seed, seedbed conditions which enable good germination and emergence, and safe herbicides which do not adversely affect seedling numbers or growth. If any one of these factors fails it can lead to thin stands or unevenly spaced plants with a corresponding reduction in quality at harvest. A few chemicals may cause flavour changes in produce of treated plants (see Lyon, D. pp. 267-272, this Symposium).

CONCLUSIONS

To produce high quality produce of the correct maturity, flavour, texture and size, crops must be grown in systems which allow the required numbers of plants to develop rapidly and without competition either from neighbouring plants or weeds. In some vegetable crops we have almost achieved this situation, in others there is still some considerable way to go. Carrots and peas are examples where the standard of weed control is sufficiently high for them to be grown in narrow rows which provide conditions suitable for both high yields and quality. In broad beans, such conditions could also be provided, but their value is low as the crop does not readily respond to narrow rows or very high plant density. In dwarf green beans, weed control is probably effective enough to allow the "bed system" production likely to give higher yields and even maturity, but on many farms volunteer potatoes are preventing uptake of the new system. Volunteer potatoes are a particular problem in dwarf beans due to broken fragments of stem which contaminate the produce after mechanical harvesting. If these cannot be removed before slicing they may appear in the final pack as they are almost impossible to remove following this operation. Volunteer potatoes can also cause serious problems in peas where the potato berries formed by some cultivars can contaminate the vined produce. In brassicas their presence may prevent the grower utilizing the optimum row width and because they are seldom evenly distributed they may affect the growth of some of the crop plants more than others, possibly leading to differences in size or relative maturity at harvest. Research on volunteer potatoes has not yet provided a satisfactory method of control and growers may have to decide between growing potatoes or vegetable crops, but not both. In onions, herbicides are used extensively and under most conditions weeds can be controlled although sometimes at the expense of a check to the development of the crops. Similarly, in brassicas and red beet the present range of herbicides does not always provide safe and satisfactory weed control. This can restrict the use of growing systems which would provide better quality.

There is therefore a need for an increased range of herbicides particularly for those crops where inadequate weed control is still restricting the choice of growing systems. There is also the problem of volunteers, which in many crops may be more serious than weeds.

Finally, as weed control efficiency improves it is essential that the grower accepts changes in production methods so that higher yields and better quality may be obtained.

REFERENCES

Bleasdale, J.K.A. (1963). *Short Term Leaft Minist. Agric. Fish Fd.* No. 27, pp. 19.

Elliot, A.H.A. (1972). *Proc. 11th Br. Weed Control Conf.* **3**, 932-942.

Gane, A.J. (1968). *Proc. 9th Br. Weed Control Conf.* **3**, 1219-1222.

Hewson, R.T. and Roberts, H.A. (1971). *J. hort. Sci.* **46**, 471-483.

King, J.M. (1974). *Proc. 12th Br. Weed Control Conf.* **1**, 309-313.

King, J.M. and Handley, R.P. (1972). *Proc. 11th Br. Weed Control Conf.* **2**, 539-546.

Lawson, H.M. (1972). *Weed Res.* **12**, 254-267.

Lawson, H.M. and Rubens, T.G. (1970). *Proc. 10th Br. Weed Control Conf.* **2**, 638-645.

Proctor, J.M. and Armsby, W.A. (1958). *Proc. 4th Br. Weed Control Conf.* **1**, 156-161.

Rademacher, B. (1967). *Mitt. biol. BundAnst. Ld-u. Forstw.* **121**, 177-185.

Shadbolt, C.A. and Holm, L.G. (1956). *Weeds* **4**, 111-123.

II.8 Effect of Weed Control and Herbicides on Quality in Fruit

K.G. STOTT*, D.S. JOHNSON** and G.R. STINCHCOMBE*

**University of Bristol, Department of Agriculture and Horticulture, Long Ashton Research Station, Long Ashton, Bristol BS18 9AF, England*
***East Malling Research Station, East Malling, Maidstone, Kent ME19 6BJ, England*

INTRODUCTION

As far as we are aware nobody has set out to improve fruit quality by employing herbicides. Rather, herbicides have been used as a management tool to increase productivity and it has been the concern of the research and advisory services to check that in the pursuit of high yields, new management practices have not impaired quality. This paper presents aspects of work on herbicides that relate to fruit quality and was undertaken jointly by Long Ashton and East Malling Research Stations.

TRENDS IN ORCHARD WEED CONTROL AND MANAGEMENT

It is important to appreciate the changes in top fruit management that herbicides have made possible. Traditionally top fruit trees were large, widely spaced and weeds were controlled by continuous cultivations. Then in the 1940s, work at Long Ashton by Wallace (1953) and at East Malling by Rogers *et al.* (1948) showed that a mown grass sward under these big trees could be beneficial—mobilizing potassium and phosphorus and leading to redder, better quality fruit. The grass was not too competitive; the big trees shaded the grass and had deep extensive root systems. Alleyways were mown regularly and grass was cut round the trees by hand.

In the 1950s, labour became scarce and as modern herbicides became available, they were sprayed around the base of the tree as a cheaper alternative to scything. Later the whole tree row was sprayed, creating the strip and grass alley system. Latterly the herbicide strip has been widened and the grass alleyway narrowed until finally in some systems, the grass alley has disappeared altogether and the trees are grown in overall weed-free bare soil—the ultimate in soil management/herbicide systems.

Parallel with these developments there have been other changes in management. Growers can no longer afford ladder work to harvest fruit. To reduce tree size, dwarfing rootstocks have been introduced. With modern small trees, herbicides are essential for weed control under the branches down the tree rows.

These small trees do not shade the alley. In fact, the interception of light by these new row alley systems is very poor (Jackson, 1978). To intercept more light the modern trend is to increase the stocking rate by having double rows of trees or even beds of trees; the ultimate being the "Long Ashton Meadow Orchard system" with complete soil coverage at 70 000 trees/ha (29 000 trees/acre) at a spacing of 45 × 30 cm (Luckwill and Child, 1973). Light interception is greatly improved, but without herbicides these super-intensive systems would be impossible.

Herbicides have allowed such radical changes in orchard management that one could expect changes in the quality of the product. Two questions arise, firstly, have herbicides in themselves affected fruit quality, in the sense of producing taints or off-flavours. Secondly, by making new systems of management possible have they influenced mineral nutrition, soil water supply etc. to affect yield, fruit size and, in this way, fruit quality.

TAINT

Lyons (this Symposium, p. 268) defines taint as an unnatural flavour or odour not associated with food, whereas an off-flavour is a natural taste occurring in the wrong food. We studied the effect of herbicides on taint in the 1960s, as part of experiments to discover the tolerance of fruit trees to the then new residual herbicides.

Simazine and diuron were applied annually from 1960 in a replicated trial at Long Ashton at the relatively high doses of 3 and 6 lb active material/acre (3·4 and 6·7 kg/ha) beneath Cheddar Cross, Cox's Orange Pippin, Worcester and Jonathan apple and Conference pear until 1973, which was 14 consecutive years. These high rates gave excellent weed control and led to small but continuing increases in yield compared with the control plot that was kept weed-free manually (Stott, 1967).

To investigate fruit quality (acceptance and taint), the fruit were subjected to triangular tasting tests where testers were asked to discriminate the odd sample from two like samples and also whether they preferred the odd one. The results for the eight years 1965-72 showed that the panel was unable to discriminate between fruit from treated or untreated plots in any of the four apple cultivars or the one cultivar of pear. In only one year, Worcester fruit from the high simazine treatment were correctly identified, and were preferred. We concluded that annual rates of the residual herbicides simazine and diuron three times higher than those required for adequate weed control did not lead to taint or off-flavours.

To check if similar results have been reported, a literature search by computer back to 1964 produced only 12 references. In only three was there any suggestion of taint attributable to herbicide usage. Table I lists the range of 13 herbicides and seven crops investigated. Further, Lyons (see p. 269, this Symposium) reports that 1391 taint tests with 173 herbicides have succeeded in implicating only chloroxuron, MCPB and simazine + prometryne and even for these there were doubts as to whether the results were conclusive. In conclusion it appears that herbicides are unlikely to cause taint in top fruit.

YIELD

To examine the effect of herbicides and soil management on yield and fruit quality we started a one hectare experiment in 1970 at Long Ashton with virus-tested Cox and Golden Delicious (Stott, 1976). From 1972 we compared combinations of overall grass, or other cover crops, with the now accepted commercial system of alleyway grass and weed-free herbicide tree strip and the innovative overall herbicide weed-free system. Table II summarizes the main treatments and their effects during the initial phase of the experiment. From 1974 treatments were modified: Stott *et al.* (1976) describe the full range of cover crops, herbicides and fertilizer treatments, summarized here in Table V.

The most productive treatments were those in overall herbicide (Table II). In most years the weed-free plots gave significantly the largest yields and were no more expensive to maintain than the commercial treatment (Treatment 1). Table III which compares the mean annual yields for the period 1973-78 shows that the overall weed-free plots (Treatment 6) have significantly out-yielded those with standard commercial management (Treatment 1). Likewise Robinson and O'Kennedy (1978) in Ireland reported a 20% increase in yield for overall herbicides compared with overall grass, and at East Malling the change from a herbicide strip to overall herbicide in an established orchard produced a 32% increase in crop over the period 1972-75 (Atkinson and White, 1976). Many other workers have reported similar results (Atkinson *et al.*, 1980).

With such yield increases it is pertinent to enquire if quality has been sacrificed. In reviewing the effects of soil management on fruit quality

TABLE I

Effect of herbicides on taint in fruit 1964-75[a]

Herbicide	Apple	Pear	Strawberry	Raspberry	Loganberry	Gooseberry	Black currant
Atrazine	×	×		×			×
Bromacil				×			
Chloroxuron			×	×	×	×	
Chlorthamide				×		×	×
Dalapon			×	×	×	×	
Dichlorobenil						×	
Diphenamide			×	×	×	×	
Diuron	×	×					
Lenacil			×	×	×	×	
MCPA						×	
MCPB						×	
Trifluralin			×				
Simazine	×	×		×		×	×

1. × = No taint. 2. [a]Based on Lockheed literature search—12 references retrieved.

TABLE II

Growth and yield of Golden Delicious in 1973

Treatment	Stem diameter (mm)	Diameter increment (mm)	Fruit weight (kg/tree)	Leaf nutrients	
				%N	%P
3. Overall sward, grass growth retarded by MH + 2,4-D.[a] Pronounced grass competition.	39	7·5	3	1·76	0·34
1. Standard commercial management. Alley sward cut; tree row weed-free.	49	11·9	24	2·35	0·17
4. Alley and tree row overall white clover; free growing.	49	12·9	24	2·37	0·16
6. Overall weed-free simazine 2·2 kg a.i./ha/year.	53	13·9	34	2·60	0·15
Significant difference at:					
$P=0{\cdot}05$	5·1	3·4	4·8	0·27	0·05
$P=0{\cdot}01$	7·0	4·6	6·6	0·37	0·07
$P=0{\cdot}001$	9·4	6·2	8·9	0·51	0·09

[a]Maleic hydrazide + 2,4-D.

TABLE III

Yields and crop values for Cox's Orange Pippin

Treatment - herbicide and fertilizer	Yield (tonnes/ha) Mean annual yield, 1973-78	Yield, 1978	Crop value 1978 (£/ha based on market returns)
1. Alley sward cut; tree row weed-free. NPK.	14·1	20·1	3891
	**	*	
6. Overall weed-free simazine 2·2 kg a.i./ha/year. NPK.	18·2	31·4	6511
Gain Treatment 6		9·96	2620

*,** Difference significant at $P = 0{\cdot}05$ and $P = 0{\cdot}01$ respectively.

Proebsting (1970) suggests that the effects of soil management practices are exerted through their effects on mineral nutrient supply, principally nitrogen, soil-plant water relations and air temperatures. Overall weed-free systems using herbicides could be expected to influence quality in top fruit because compared with overall grass or even grass alleyway systems, the soil has been shown to contain more moisture (Atkinson and White, 1976; Stott *et al.*, 1977), the trees more leaf nitrogen (Atkinson and White, 1976; Stott, 1976) and air temperatures are likely to be higher with less risk of frost damage (Rogers *et al.*, 1948; Stott *et al.*, 1971). The main components of fruit quality that herbicide/soil management systems affect are: fruit size and colour; fruit juice quality; and storage quality.

FRUIT SIZE AND COLOUR

Since 1973 fruit from the Long Ashton trial has been assessed each year for size and colour. Mean fruit weight is recorded and fruit are sized into 5 mm diameter classes. Cox is colour graded into four classes: up to ¼, ¼-½, ½-¾ and over ¾ red. Golden Delicious is graded into three classes, pink, yellow and green.

SIZE

Noteworthy effects of treatments on size were recorded only in 1973. After three years in overall grass both Cox and Golden Delicious trees in Treatment 3 were suffering extreme nitrogen deficiency—poor growth, poor yields, low leaf and fruit nitrogen (Table II)—and produced fruits of a much smaller size (Fig. 1). These results showed that although the growth of grass was retarded by maleic hydrazide and 2,4-D, the overall grass sward was unacceptably competitive (Stott, 1976). In 1974/75 we killed off the sward in the tree row and applied additional nitrogen fertilizer (NNPK) in an attempt to improve tree vigour and yield. The result of the changed management was seen the next year in larger, greener fruit of Cox and to a lesser extent of Golden Delicious. Figure 1 also shows that apart from Treatment 3, all the other treatments produced a similar range of fruit sizes. The same general distribution of sizes has been recorded each year, including 1978, which gave record yields for both Cox and Golden Delicious. Minor differences have been found due to treatments, but no consistent effects have been noted. Thus, in 1978 overall weed-free treatments had significantly fewer fruit at 65 mm diameter than those with grass alleyways or overall clover ($P = 0{\cdot}05$), i.e. they had slightly smaller fruit. But they had the largest fruits in 1977 and average in 1975 and 1974. This illustrates the variability in response to herbicides that characterizes these and similar experiments (Perring, 1975; Johnson and Johnson, 1980).

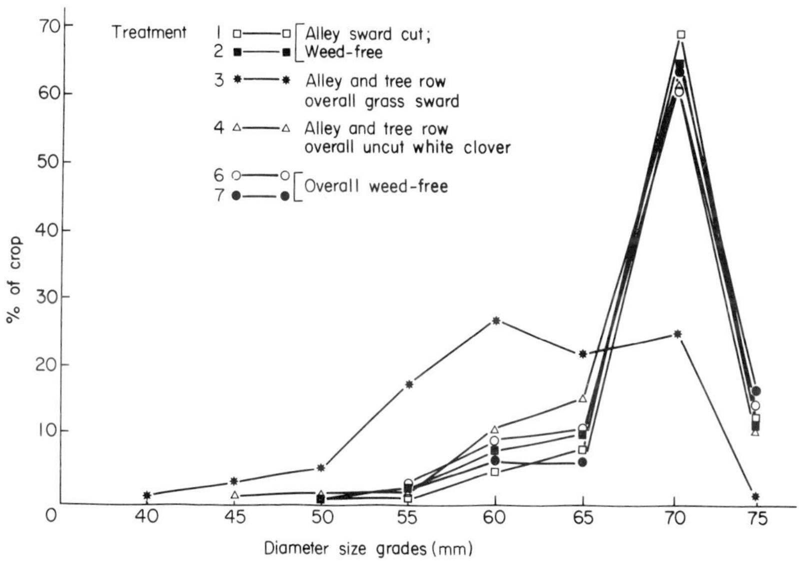

Fig. 1. Effects of herbicide/management treatments on Golden Delicious fruit size, 1973.

COLOUR

Table IV shows that the conventional treatment of mown grass alley and weed-free tree row has consistently produced redder Cox than the overall weed-free treatments, the difference averaging about 6% more fruit—say 35% in the ½-¾red category ($P=0{\cdot}01$). These results confirm results which were obtained over a six-year period with Cox and Worcester in an earlier experiment at Long Ashton (Stott *et al.*, 1974).

TABLE IV

Fruit colour—Cox's Orange Pippin (% of fruit in ½-¾ red class)

Treatment	1974	1975	1976	1978	Period mean
1. Alley sward cut; tree row weed-free.	35·8	36·8	44·0	23·0	34·9 **
6, 7 and 8. Overall weed-free (different rates of simazine or N fertilizer).	30·2	32·3	33·6	18·5	28·7

** Difference significant at $P=0{\cdot}01$.

We did not detect an effect of treatments on the colour of Golden Delicious. However, Gormley *et al.* (1973) in Ireland found that in two years out of three, Golden Delicious fruit from the grass plots was of higher colour and quality than from non-grass, whereas the weed-free overall herbicide plots produced the highest yield in all three years, and that the fruit, though greener was of acceptable quality.

However, the effect of herbicide treatments on fruit colour can be overridden. In 1978 the large crops of both Cox and Golden Delicious resulted in a much higher proportion of green fruit (44% over ¾-green) than treatments have usually produced: the average for all other years was 8%. All treatments responded in the same way. Hence, although herbicides acting through management can affect fruit colour, the effect is small compared with other factors which can exert major effects on yield, e.g. blossom density (high, 1978), pollination period (long and frost-free, 1978) and summer water balance (good in 1978).

It appears that herbicide/soil management systems do not produce unacceptable effects on fruit size and colour, and that these effects are more than balanced by increases in yield.

FRUIT JUICE QUALITY

For the past six years (1973-78) Dr L.F. Burroughs of the Food and Beverages Division at Long Ashton Research Station has examined the main components of the juice of fruit from the different herbicide treatments, measuring specific gravity (and Brix degrees), acidity in mequiv./litre (and % malic acid), pH, % tannin, soluble nitrogen (mg/100 ml juice) and total fruit nitrogen (mg/100 g).

Differences between treatments have only been slight, suggesting that different management systems do not greatly affect juice quality. Table V gives selected results for the first and last years. Correlation analysis showed that an increase in crop can lead to significantly lower juice specific gravity and acidity. No other correlations were significant in each year, and notably we did not find a significant correlation between colour, size and juice characteristics.

Of all the parameters assessed, juice nitrogen fluctuated most. Lea and Timberlake (1977) also have found juice nitrogen difficult to correlate with other parameters, such as pH, % tannin or specific gravity. The low juice nitrogen content of Treatment 3 in 1973 (Table V) reflects the low leaf nitrogen and poor growth of this treatment, caused by growing young apple trees for two years in overall grass, which was unduly competitive because sward vigour was not adequately controlled by MH + 2,4-D (Stott, 1976). However, juice nitrogen levels recovered when the competitive grass sward was killed and additional fertilizer applied.

To discover if nitrogen fertilizer is needed in overall weed-free systems of management, Treatment 8 is maintained weed-free, and differs only from Treatment 6 in that it has not received nitrogenous fertilizer for the last five years. Table VI suggests that after five years, the absence of nitrogen fertilizer application may be beginning to reduce Cox apple yields, and Table V shows that in 1978, Cox juice nitrogen was notably lower than in Treatment 6. But the only significant reduction yet found is that by 1978 leaf nitrogen in Cox (but not in Golden Delicious) was depressed ($P = 0{\cdot}05$) (Table VII). These results suggest that trees grown in overall weed-free systems of management maintained by herbicides need little fertilizer nitrogen. Conversely, when this system is combined with traditional grass-based fertilizer rates it can lead to the production of fruit with a high nitrogen content and hence a risk of processing and storage defects. Treatment 6, which receives the standard Long Ashton fertilizer dosage of 148 units N and K/ha and 47 units P/ha, (N and K 75 and P 24 kg/ha) has maintained high levels of leaf N (3·0% in 1978) and in some years high levels of juice N (46 mg/100 ml in 1977), which would be critical if such fruits were used as culls in cider manufacture. Lea and Beech (1978) have demonstrated in pot-grown cider apple trees that an increase of nutrient supply to the

TABLE V

Effect of herbicide/management treatments on fruit juice quality—Cox's Orange Pippin

Treatment (from 1974)	Fertilizer	Sp. gr. (degrees)		Acidity (mequiv./l)		Juice N (mg/100 ml)	
		A	B	A	B	A	B
1. Alley sward cut; tree row weed-free.	NPK	50	58	111	64	31	35
2. Alley sward controlled MH + 2,4-D; tree row weed-free.	NPK	54	56	107	63	26	31
3. As 2. Differential fertilizer.	NNPK	55	59	121	70	**14**	39
4. Alley and tree row overall white clover; free growing.	PPK	52	57	109	66	41	41
5. As 4, with strawberry clover.	NPK		59		67		29
6. Overall weed-free; simazine 2·2 kg a.i./ha/year.	NPK	56	52	110	63	33	37
7. Overall weed-free; periodic simazine 11·2 kg a.i./ha.	NPK	56	54	115	58	32	29
8. As 6, differential fertilizer.	PK		53		63		**26**

A = 1973. Treatment 3: overall sward controlled by MH (maleic hydrazide) + 2,4-D. NPK applied.
B = 1978.

TABLE VI

Effect of nitrogen fertilizer on yield in weed-free management

Cultivar	Mean annual yield, 1974-78 (tonnes/ha)		
	Treatment 6, with N		Treatment 8, without N
	NPK		PK
Cox's Orange Pippin	19·67 (31·4)	n.s.	17·22 (25·8)
Golden Delicious	34·61 (61·0)	n.s.	34·29 (61·5)

N and K 148 units (75 kg) and P 47 units (24 kg)/ha.
Yield in 1978 in brackets.
n.s. = not significant.

tree reduces bitterness and astringency in the final product, making the fruit unacceptable for cider manufacture, and even leading to the possibility of instability after bottling (Lea and Timberlake, 1977). Hence the Food and Beverages Division at Long Ashton is particularly interested in the impact of herbicide/management systems on apple juice characteristics not only because the cider industry accepts quantities of dessert fruit as culls, but also because commercial cider apple production is moving closer to the management methods employed by dessert growers.

The trials here reported show that the herbicides used in our current management systems are unlikely to have deleterious effects on juice quality, but that associated fertilizer inputs may need modification.

STORAGE QUALITY

With information from our trials as a basis, the following reviews the effect of herbicide/management systems on the retention of fruit quality during storage. Fruit samples from Long Ashton experiments have been sent since 1970 to East Malling Research Station for assessment and analysis. The Fruit Storage Division assessed N, P, K, Ca, Mg (mg/100 g fresh wt.), % dry matter and storage quality in air, and more recently, in controlled atmosphere storage. Both Cox and Golden Delicious are purposely over-stored to achieve maximum development of senescent disorders in an attempt to exaggerate treatment differences (Sharples *et al.*, 1975).

Only two clear treatment differences have emerged. By 1973, both Cox and Golden Delicious trees were suffering intense competition from the overall grass sward in Treatment 3. Table II shows that these trees were significantly smaller and less vigorous than trees with standard orchard

TABLE VII

Effect of fertilizer nitrogen in overall weed-free systems on leaf nitrogen (% dry wt.)

Cultivar	Treatment	1974	1975	1976	1977	1978
Cox's Orange Pippin	6. Overall weed-free; NPK fertilizer.	2·72 n.s.	2·77 n.s.	2·90 n.s.	2·84 n.s.	3·01 *
	8. Overall weed-free; PK only.	2·93	2·73	2·92	2·86	2·79
Golden Delicious	6. Overall weed-free; NPK fertilizer.	2·58 n.s.	2·25 n.s.	2·84 n.s.	2·24 n.s.	2·56 n.s.
	8. Overall weed-free; PK only.	2·63	2·39	2·79	2·33	2·57

*Significant at $P = 0{\cdot}05$. n.s. = not significant.

management. Leaf nitrogen was significantly less being only 1·76% in Golden Delicious compared with 2·35% (for standard orchard management) and 2·19% instead of 2·60% respectively in Cox. In both Cox and Golden Delicious, Treatment 3 produced the lowest yields. The fruit were small, highly coloured, low in nitrogen (Table V), high in calcium and potassium and were relatively free from bitter pit and storage disorders (Stott *et al.*, 1976), illustrating the accepted view of the interaction of these factors on storage disorders. Grass increases fruit phosphorus, lowers nitrogen, and if the fruit are small they have high calcium, good colour and store well, whereas in overall weed-free systems these trends may be reversed.

High fruit calcium has been stressed by many researchers as being the key to good fruit storage (see Atkinson *et al.*, 1980). The calcium content of fruit is related more to fruit size than to any other factor. Large apples are low in calcium and usually high in nitrogen (Sharples *et al.*, 1972; Perring, 1976). Hence treatments which lead to excessive availability of nitrogen and increased vigour can be detrimental to fruit storage quality because if they lead to a yield increase by increasing fruit size, storage will be bad, but if they lead to an increased number of average size fruit, storage will be reasonable. The weather and many other factors can easily tip the balance between these two alternatives, leading to seasonal variations in storage behaviour between treatments. Hence fruit size has a key role in fruit storage. Size is affected by yield, and it is through their effects on yield/fruit size that herbicide/management systems exert their main effect on fruit storage quality.

The other trend we found was that fruit from the overall herbicide treatments tended to be lower in phosphorus than fruit from herbicide-strip or overall grass systems, Golden Delicious being more responsive than Cox (Table VIII). Perring (1975) also noted a tendency for overall herbicides to lower fruit phosphorus levels and conversely for grass treatments, particularly with irrigation to increase them. The effect of grass in mobilizing phosphorus was noted by Rogers *et al.* (1948) and Wallace (1953) and when grass systems are converted to overall weed-free systems, a significant reduction in tree phosphorus has been found (Stott *et al.*, 1971). Though low phosphorus levels are not normally a problem in British orchards, overall herbicides could lead to a breakdown in stored fruit if phosphorus levels on the site are low, and particularly if high rates of nitrogen are also applied. Perring (1976) believes that Cox is at risk when fruit phosphorus falls below 9 mg/100 g. The minimum recommended value for successful storage of Cox is 11 mg/100 g (Waller, 1980). Table VIII shows that in some years, Cox fruit phosphorus in overall

TABLE VIII

Effect of herbicide/management treatments on fruit phosphorus (mg/100 g fresh fruit)

Treatment	1974		1975		1976		1977	
	Cox	Golden	Cox	Golden	Cox	Golden	Cox	Golden
1. Alley sward cut; tree row weed-free.	12·7	9·0	11·7	12·8	11·5	8·2	11·0	10·2
	*	n.s.	n.s.	***	n.s.	n.s.	n.s.	*
6. Overall weed-free simazine 2·2 kg/ha/year.	10·6	8·2	10·2	9·6	11·7	8·5	10·4	8·7

*,*** Difference significant at $P=0{\cdot}05$ and $P=0{\cdot}001$ respectively.
n.s. = not significant.

TABLE IX

Quality and yield of Cox's Orange Pippin, 1976

	Yield (tonnes/ha)	Marketable EEC grades					
		I,II		III		Waste	
		%	Tonnes/ha	%	Tonnes/ha	%	Tonnes/ha
1. Alley sward cut; tree row weed-free.	9·6	62	5·92	34	3·26	4	0·38
6. Overall weed-free simazine 2·2 kg/ha/year.	12·0	59	7·08	25	3·00	16	1·92

herbicide plots fell below this value, whereas that from the standard orchard management of grassed alleyways and herbicide tree rows did not.

In these trials Cox yields were heavy in 1975, light in 1976 and medium in 1977, but despite this variation, the fruit were low in calcium in all years (3-4 mg/100 g) and so all treatments were affected by bitter pit and internal breakdown. Treatments 6, 7 and 8, overall weed-free were worst in 1975, best in 1976 and 6 and 8 (but not 7) worst in 1977. In that year Treatment 7, maintained weed-free with periodic high rates of simazine (11·5 kg a.i./ha) was best.

These results illustrate the variability in the effects of herbicide/soil management treatments on storage quality that have been noted by many others (Perring, 1975; Sharples, 1980). Both Johnson and Johnson (1980) reporting ADAS trials and Perring (1980) stress the large orchard-to-orchard and year-to-year variation in storage quality, so it is unlikely that any system will be advocated or condemned on storage grounds alone. Further, cultivars differ greatly in their response. In our trials Golden Delicious, though containing very similar and low amounts of calcium to Cox showed few defects in any year and none that could be related to treatments.

Finally though our trials show that herbicide mediated weed-free management systems may lead to reductions in fruit quality, these treatments are significantly more profitable. All the fruit from the treatments listed in Table V is graded separately for sale and Table IX shows that in 1976, 16% of the fruit from one of the overall bare treatments was classed as waste (poor colour, incipient breakdown) compared with only 4% from the standard orchard management treatment. However, because the total yield from overall weed-free treatments is so much greater (Table III—Yields 1973-78) the market returns are greater, in fact £2500/ha greater in 1978 (Table III). Similar financial gains have been obtained in other years.

CONCLUSIONS

The use of herbicides has not produced noteworthy effects on fruit taint, size, colour, fruit juice or storage qualities. Their main effect has been to facilitate modern more profitable weed-free management systems. For these it is becoming increasingly clear that further savings can be made by reducing nitrogenous fertilizers to prevent fruit nitrogen rising to levels that can impair fruit colour and fruit storage quality. The aim should be to produce regular, heavy crops and then to obtain good storage by additional treatments (e.g. calcium sprays) and by modifying storage conditions. Leaf, fruit and soil nutrient levels can be monitored to establish trends so that remedial treatments can be applied if required.

ACKNOWLEDGEMENTS

We thank Dr L.F. Burroughs, Long Ashton, for the juice analyses and useful discussion, Mr T. Samuelson, East Malling, for the mineral analyses of the fruit and Miss Margaret Holgate, Long Ashton, for statistical analyses.

REFERENCES

Atkinson, D. and White, G.C. (1976). *Proc. 13th Br. Crop Prot. Conf. Weeds* **3**, 873-884.

Atkinson, D., Jackson, J.E., Sharples, R.O. and Waller, W.M. (eds) (1980). "Mineral Nutrition of Fruit Trees" ISHS Conference Canterbury, 1979. Butterworths, London.

Gormley, R., Robinson, D. and O'Kennedy, N.D. (1973). *J. Sci. Fd. Agric.* **24**, 227-239.

Jackson, J.E. (1978). *Acta Hortic.* **65**, 61-70.

Johnson, P.A. and Johnson, D.S. (1980). *In* "Mineral Nutrition of Fruit Trees" ISHS Conference, Canterbury, 1979 (D. Atkinson, J.E. Jackson, R.O. Sharples and W.M. Waller, eds), pp. 289-290. Butterworths, London.

Lea, A.G.H. and Beech, F.W. (1978). *J. Sci. Fd. Agric.* **29**, 493-496.

Lea, A.G.H. and Timberlake, C.F. (1977). *Rep. Long Ashton Res. Stn for 1976*, 132-133.

Luckwill, L.C. and Child, R.D. (1973). *Acta Hortic.* **34**, 213-220.

Perring, M.A. (1975). *J. hort. Sci.* **50**, 425-433.

Perring, M.A. (1976). *Rep. East Malling Res. Stn for 1975*, 169-170.

Perring, M.A. (1980). *In* "Mineral Nutrition of Fruit Trees" ISHS Conference, Canterbury, 1979 (D. Atkinson, J.E. Jackson, R.O. Sharples and W.M. Waller, eds), pp. 291-292. Butterworths, London.

Proebsting, E.L. (1970). *Proc. XVIIIth Int. Hort. Conf. Tel Aviv* **4**, 223-239.

Robinson, D.W. and O'Kennedy, N.D. (1978). *Scientia Hort.* **9**, 127-136.

Rogers, W.S., Raptopoulos, Th. and Greenham, D.W. (1948). *J. hort. Sci.* **24**, 228-269.

Sharples, R.O. (1980). *In* Mineral Nutrition of Fruit Trees" ISHS Conference, Canterbury, 1979 (D. Atkinson, J.E. Jackson, R.O. Sharples and W.M.Waller, eds), pp. 17-28. Butterworths, London.

Sharples, R.O., Hammond, J.I. and Langridge, I.W. (1972). *Rep. East Malling Res. Stn for 1971*, 72.

Sharples, R.O., Johnson, D.S. Hamer, J.I. Langridge, I.W. (1975). *Rep. East Malling Res. Stn for 1974*, 76.

Stott, K.G. (1967). *Rep. Long Ashton Res. Stn for 1966*, 66.

Stott, K.G. (1976). *Ann. appl. Biol.* **83**, 327-330.

Stott, K.G., Jarrett, R.M. and Corke, A.T.K. (1971). *Rep. Long Ashton Res. Stn for 1970*, 167-168.

Stott, K.G., Jarrett, R.M., Corke, A.T.K. and Sharples R.O. (1974). *Rep. Long Ashton Res. Stn for 1973*, 147.

Stott, K.G. Jefferies, C.J. and Belcher Ann (1976). *Rep. Long Ashton Res. Stn for 1975*, 30-31.

Stott, K.G., Ringner, A. and Stinchcombe, G.R. (1977). *Rep. Long Ashton Res. Stn for 1976*, 26-27.
Wallace, T.W. (1953). "Science and Fruit", pp. 140-161. University of Bristol.
Waller, W.M. (1980). *In* "Mineral Nutrition of Fruit Trees" ISHS Conference, Canterbury, 1979 (D. Atkinson, J.E. Jackson, R.O. Sharples and W.M. Waller, eds), pp. 383-394. Butterworths, London.

II.9 Pest Damage and Insecticide Performance in Vegetable Crops for Processing

G.A. WHEATLEY AND A.R. THOMPSON

National Vegetable Research Station, Wellesbourne, Warwick. CV35 9EF, England

INTRODUCTION

Crops grown for processing need to be well protected from pests to safeguard growers' investments and to ensure that supplies of high quality produce are sufficient to meet the demands of processors.

The most serious pests of peas, carrots, broad beans, beetroot, Brussels sprouts, dried bulb onions and green beans, the principal vegetables grown for processing in Britain, were well controlled by the insecticide treatments available 20 years ago. The success of the treatments can be attributed in retrospect mainly to the use of relatively persistent compounds, particularly organochlorines such as DDT, dieldrin, aldrin and γ-HCH, but it was only temporary. Accumulating residues in soil encouraged the development of resistant strains of some pests, led to undesirably high residues in root crops such as carrots and, by various routes, moved from the original sites of usage and became ubiquitously distributed in the environment. The use of these very persistent compounds was therefore curtailed.

Few, if any, of the replacement organophosphorus and carbamate insecticides introduced since 1960, when the problems of persistence were first becoming evident, seem likely to protect crop foliage for more than about two weeks, as judged from the study of the persistence and performance of systemic aphicides on Brussels sprouts by Suett and Padbury (1976) and in soil largely disappear or degrade within the season of use (Suett, 1975).

The elementary tactics of usage appropriate to the organochlorine compounds are not adequate, therefore, for the range of insecticides now available, nor is there any likelihood of more persistent compounds being either developed or approved for use in the future. To overcome the general lack of persistence of available insecticides, new strategies and tactics need to be devised to exploit their potential as fully as possible. It is necessary, therefore, to ensure a general understanding of the nature of pest damage for each pest/crop/market problem (Wheatley and Coaker, 1970), what insecticides can reasonably be expected to do or not to do, what constitute realistic targets for pest control systems and what factors can also be utilized to complement the performance of the insecticides.

CLASSIFICATION OF PEST DAMAGE

In relation to vegetable production, Wheatley and Coaker (1970) and Wheatley (1971a, 1974) recognized three main types of pest damage, each with specific implications for crops destined for processing. The three types are not seen as rigid categories but rather as forming a two-dimensional continuum relating the percentage of crop units (the carrot, sprout button, bean or pea seed or pod, potato tuber, etc) affected to the severity of damage on each unit. The broad categories were distinguished as follows:

TYPE I DAMAGE AFFECTING CROP ESTABLISHMENT

Pest damage to seeds, seedlings or young transplants reduces the numbers of plants which become established. With carrots, for example, incomplete, gappy stands result in irregular compensatory growth which can reduce the proportion of the crop acceptable for processing, though the total yield may be little affected. Type I damage may not affect many plants but usually any seed, seedling or young transplant even slightly damaged either dies or fails to yield marketable produce.

Traditionally, some allowance for this type of damage has been achieved by sowing an excess of seed to ensure that sufficient survive to produce a reasonable crop. Seed is no longer cheap, however, and increasingly is being sown to achieve a planned plant density. A very high level of protection of the seed is therefore essential whether in seed-beds or in the field.

TYPE II DAMAGE IMPAIRING CROP GROWTH

Some direct damage by pests to the unmarketed parts of plants can often be tolerated even when most, if not all, plants are infested. The level of infestation causing a detectable decrease in yield depends, however, on how well the crop is growing and consequently can differ widely from site-to-site and year-to-year. The main effect of the damage is to restrain plant growth. Only exceptionally do plants die after they have become established, though they may visibly wilt or show other signs of stress. In moist soil, for example, Brussels sprout plants can tolerate on their roots quite large infestations of cabbage root fly larvae (*Delia brassicae* Hoffmannsegg) whereas the plants would be severely affected in a drought.

The transmission of viruses by aphids, however, causes an indirect form of damage which can seriously reduce crop growth and yield, especially if an infection occurs early in the season. Carrot seedlings therefore need to be well protected from willow-carrot aphids (*Cavariella aegopodii* (Scop.)) entering the crop early in June, even though well-established plants can support appreciable numbers of the aphid without the yield of carrots being subsequently much affected.

Although some, but not too much, damage can be tolerated on the unmarketed parts of crop plants, infestations which depress yield also often impair the quality of the produce simultaneously so that the nett yield available for processing may become uneconomically small.

TYPE III DAMAGE AFFECTING THE MARKETED PRODUCE

By far the most serious type of damage is that occurring on the part of the crop destined for processing. It may take the form of severe or even only minor blemishes, or the pest organism or its debris such as frass or cast aphid skins may occur on or in the produce. Secondary pathogenic organisms can also enter wounds caused by pests and lead to rots especially if the produce is stored to spread the supply to processing plants. Many pests are implicated in the different forms of the Type III damage, for example, the pea moth (*Laspeyresia nigricana* (F.)), the carrot fly (*Psila rosae* F.), the cabbage root fly (*D. brassicae* Hoffmannsegg), the cabbage aphid (*Brevicoryne brassicae* (L.)) and cutworms (Noctuidae). The most frequent complaints of the inadequacies of present-day control measures generally relate to pests causing this type of damage.

Type III damage may occur on only a small proportion of the total number of crop units and is often so slight that its effect on the gross yield is not detectable. The quest for blemish-free produce has often been criticized in recent years with some justification because attempts to eliminate very minor blemishes purely for cosmetic purposes can lead to much unnecessary use of crop protection chemicals. However, the presence of a pest organism or its debris in a branded product on sale contravenes the law relating to the supply of pure and wholesome food and creates a very difficult quality control problem. Even a small proportion of slightly damaged crop units can require such an expensive grading-out operation preparatory to processing that a crop must be rejected. An extreme example is the occasional presence of cabbage root fly larvae in developing Brussels sprout buttons. They are indetectable until the buttons are cut open and existing treatments cannot effectively deal with an infestation if it affects more than about 1-2% of the buttons (Wheatley, 1974).

INSECTICIDE PERFORMANCE

Gair (1971) drew attention to the difficulty of achieving satisfactory levels of pest control on vegetable crops using the array of insecticides then available. Despite continuing evaluation of new compounds, the difficulties are not being resolved mainly because the more recently developed products are intentionally very much less persistent than their predecessors. They cannot therefore provide the lengthy protection from pest damage that is important if high levels of control are to be economically achieved. The problem is illustrated by the duration of carrot fly control on carrots possible with soil applications of dieldrin and phorate. Wright (unpublished data) incorporated 2·2 kg a.i. dieldrin ha^{-1} into a peaty-loam soil on 25 March, 1956. No further insecticide was applied and carrots were grown

TABLE I

Prolonged control of carrot fly on maincrop carrots by a single application of dieldrin, 2·2 kg a.i. ha^{-1}, incorporated into a peat soil on 25 March, 1956 (Wright, unpublished data)

Year	1956	1957	1958	1959	1960	1961	1962	1963	1964
Undamaged carrots (%)									
Treated	95	88	94	77	88	90	94	96	98
Untreated	0	23	47	33	20	30	22	11	73
Reduction in numbers of larvae (estimated %[a])	100	91	91	77	92	91	96[b]	98	95

[a]See Wheatley (1969). [b]In January 1963, 60% of the applied dose was still present in the soil.

TABLE II

Decline in the control of carrot fly during autumn after soil treatment with phorate, 2·2 kg a.i. ha^{-1}, applied by the bow-wave method when sowing into peat soil on 1 May, 1969

Date sampled (1969)	5 Aug.	7 Oct.	3 Dec.
Undamaged carrots (%)			
Treated	98	81	46
Untreated	81	51	28
Reduction in numbers of larvae (estimated %[a])	90	69	40

[a]See Wheatley (1969).

on the plots for nine successive seasons. With the exception of 1959, a very dry season, the single treatment gave good protection of the carrots from carrot fly damage throughout the experiment (Table I). Furthermore, on a mineral soil in a complementary experiment complete protection was achieved in all nine seasons.

In contrast to dieldrin, the performance of phorate deteriorates markedly as the second generation of carrot fly larvae develops, exemplified by the results of an experiment in 1969 on the same peaty-loam soil (Table II).

INSECTICIDAL EFFICIENCY AND EFFECTIVENESS

In an attempt to provide a better understanding of the performance of insecticides and to resolve the confusion arising between research workers, advisors and growers about the meaning of the phrase "control of a pest", Wheatley (1973, 1974) distinguished between the *efficiency* of the treatment as it affects a pest population and its eventual *effectiveness* in improving the yield of marketable produce.

The combination of insecticide, formulation, dose, application method and timing constitute a treatment with a particular potential biological activity (efficiency) against a pest that may be modified by indigenous factors such as climate and soil-type. Within broad limits, a treatment tends to decrease the numbers of individuals in a pest population by a characteristic percentage that is, for practical purposes, largely independent of the numbers of the pest present and hence, in terms of population dynamics, is a density-independent (catastrophic) regulating factor. It is referred to as a % reduction in the pest population (Thompson and Wheatley, 1977) equivalent to "% mortality" in laboratory toxicological studies and is interpreted as a quantitative estimate of the treatment efficiency.

Many other factors, however, determine the eventual yield or marketability of a crop. Assuming that no treatment is likely to be 100% efficient, a corollary of the density-independent action of a treatment is that larger numbers of a pest will survive treatment if the attacking pest population is large than if it is small. The numbers surviving interact with the plant density to determine the proportion of the total number of plants that will be attacked and the distribution of damage between plants. Wheatley and Coaker (1970) and Wheatley (1974) showed how this affected the interpretation of the performance of treatments of differing efficiencies in protecting a crop such as carrots against carrot fly. The model oversimplifies the problem but nevertheless shows clearly that, according to the level of pest infestation, treatments with particular efficiencies will vary in effectiveness in providing a marketable crop. Thus a treatment which is consistently efficient may be effective against a light-to-moderate attack of a pest but ineffective against a severe infestation.

EFFICIENCY AND EFFECTIVENESS OF TWO INSECTICIDES AGAINST CARROT FLY ON CARROTS

The necessity of considering efficiency and effectiveness as distinct stages in crop protection systems can be illustrated by referring to results obtained for two organophosphorus insecticides, phorate and terbufos, evaluated for control of carrot fly on carrots. In 1969, treatment of peaty-loam soil with phorate protected carrots satisfactorily from damage by the first generation of carrot fly up to August (Table II). The efficiency of the phorate treatment, and all others with organophosphorus or carbamate insecticides in the same experiment, declined as the second generation attack developed during the autumn (Wheatley, 1969), presumably as decreasing insecticide residues (Suett, 1975) became less active at the lower soil temperatures (Harris, 1971; 1972). Greatly improved protection cannot, however, be readily provided by increased doses of insecticides. Experiments

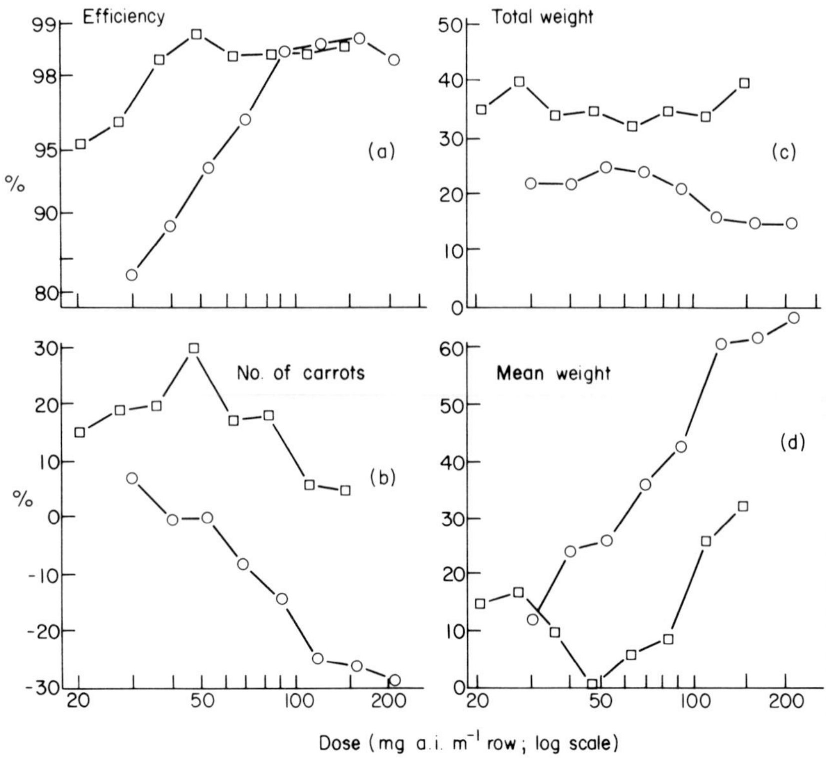

Fig. 1. Performance of bow-wave applications of phorate (○) and terbufos (□) to a sandy-loam soil for carrot fly control on carrots in 1972; (a), efficiency estimated as the % reduction in numbers of larvae (probit scale), and effectiveness relative to untreated carrots as; (b), % change in the number of carrots surviving; (c), % increase in total weight of carrots, and (d), % increase in the mean weight per carrot.

at Wellesbourne using a log-dose evaluation technique (Wheatley and Niendorf, 1969; Wheatley, 1971b; 1972; Thompson and Wheatley, 1977) have shown that carrot fly larvae are not completely controlled by treatments with phorate, or the more active terbufos, using doses even several times larger than those likely to be feasible in practice (Fig. 1a).

The numbers of carrots surviving the phorate and terbufos treatments also differed (Fig. 1b), implying a dose-related phytotoxic effect, particularly by phorate. At the recommended dose of about 60 mg a.i. m^{-1} of row, it seemed to have killed 30-40% of the seeds or seedlings, so that fewer survived than on the untreated plots. The insecticides had thus modified the "planned" plant density. It is well known that the total yield of carrots may change little over a wide range of plant density (Bleasdale, 1966). This was apparent with phorate and terbufos (Fig. 1c) but each insecticide also resulted in a different relationship between the dose applied and the mean weight of the carrots (Fig. 1d). Thus, apart from protecting the carrots from carrot fly, the insecticides influenced the proportions of the crop suitable for particular market outlets, including processing.

SPECIFICATIONS FOR PEST CONTROL

Using experience of the extent of damage to crops arising from moderate to severe pest infestations, the concept of insecticide efficiency and the effectiveness of treatments enables targets to be suggested for pest control measures. The examples shown in Table III are necessarily speculative but are intended primarily to illustrate the principle and to focus attention on some of the present shortcomings of treatments.

Some pests such as the cabbage root fly or the carrot fly may cause more than one type of damage and the performance of treatments used to control them may differ for each type. The expected severity of damage is based on experience with evaluation trials, surveys or verbal reports and is intended to portray the severity of attack that a control measure can reasonably be expected to have to counter. The observed effectiveness indicates the level of protection likely to be achieved by the best of present-day treatments against an attack of that severity, whereas the required effectiveness suggests damage tolerances for acceptance of produce for processing, the performance targets for the control measure. The likely efficiencies of the treatments were estimated from the stated severities of attack and the values for effectiveness, the discrepancies between the observed and the required efficiencies representing shortcomings of present insecticide treatments.

Although high levels of protection are needed against Type I damage, the duration of protection is quite short and, for most crops, can usually

TABLE III

A comparison of the performances of present insecticide treatments with the performances required to achieve suggested tolerance specifications (required effectiveness) for vegetable produce for processing

Problem					Insecticide	Performance			
Crop	Pest	Damage			treatment	Effectiveness		Efficiency	
		Type	Criterion	Severity		Observed	Required	Observed	Required
Broad bean	[a]bean seed fly	I	% plants killed	71	seed	8	5	93	96
Brussels sprout	cabbage root fly	I	% plants killed	45	soil	5	3	91	95
Carrot	carrot fly	I	% plants killed	50	soil	40	5	95	95
Brussels sprout	cabbage root fly	II	% loss of yield	30	soil	5	5	85	85
Carrot	willow-carrot aphid	II	% loss of yield	50	foliage	5	5	90	95
				50	soil	10	5	80	95
Pea	[b]pea moth	III	% pods damaged	40	foliage	12	1	75	98
Carrot, early	carrot fly	III	% roots mined	30	soil	5	2	86	94
Carrot, maincrop	carrot fly	III	% roots mined	50	soil	15	2	77	97
Carrot, late	carrot fly	III	% roots mined	70	soil	45	2	51	98
Brussels sprout	[c]cabbage root fly	III	% buttons infested	3	foliage	0·9	0·4	70	87

[a] Gould and Mayor (1975); [b] Legowski and Gould (1960); [c] Coaker and Ensor (1967).

be achieved by existing seed or soil treatments. The general phytotoxicity of organophosphorus and carbamate insecticides to seed or seedlings, as discussed for the carrot fly example (Fig. 1b), can result in zero effectiveness as assessed on the basis of seedling survival though they would have been well-protected against carrot fly. Type II damage is very variable and the effectiveness of pest control is difficult to assess at low-to-moderate levels of pest damage. However, lesser standards of pest control are needed than for Type I and III damage and most control measures do not fall far short of the requirements.

In contrast, the effectiveness for Type III damage problems must match stringent damage tolerances and virtually all present-day control measures are inadequate to counter moderate to severe pest damage to the marketed part of a plant. The discrepancies between the observed and required efficiencies are serious and usually cannot be corrected merely by adjusting the applied dose, or detail of an application method. They occur in the "diminishing returns" part of the response curve and so could require a ten-fold or greater improvement in performance of control measures to overcome them. Increasing attention must therefore be given to other means of suppressing pest populations to complement insecticides.

INTEGRATION OF CONTROL MEASURES

There are many other possible methods of suppressing pest populations. In isolation none is likely to be adequately effective but in combination they may help to suppress pest populations sufficiently to enable insecticide treatments to achieve more nearly the required effectiveness. The possibilities

TABLE IV

Advantage of integrating complementary population regulating factors; hypothetical examples for two pests of brassicas

Complementary regulating factors	Estimated efficiencies (%)	Numbers of pest	
		Cabbage root fly	Cabbage aphid
None	0	1000	1000
Farm hygiene	30	700	700
Deep winter ploughing	40	420	–
Crop rotation/movement	30	294	?
Partially resistant cultivar	66	98	233
Trapping	50	49	—
Insecticide	90	5	23
(Insecticide alone)	90	100	100

of utilizing traditional and new methods in integrated systems now need to be explored to determine feasible combinations.

Possible combinations of methods for controlling two pests of brassicas, the cabbage root fly and the cabbage aphid, are illustrated as a hypothetical example in Table IV. A percentage efficiency has been estimated for each regulating factor based on the limited information available at present, assuming that each is density independent. The cumulative effect should be a marked improvement on what can be achieved solely by an insecticide. Of particular interest are the contributions by even a partially resistant cultivar, supporting perhaps only one-third of the pest population found on more susceptible cultivars, and by trapping if present research can improve trap performance.

As improved levels of pest control on a field-by-field basis are not likely to be very effective there will probably be considerable organizational, as well as some unusual political, problems to overcome before an integrated control approach can be inaugurated on a sufficiently large scale in the principal vegetable-growing areas in Britain.

ACKNOWLEDGEMENTS

The authors gratefully acknowledge the collaboration of Professor D.W. Wright in permitting some of his unpublished data to be used in this paper.

REFERENCES

Bleasdale, J.K.A. (1966). *Ann. appl. Biol.* **57**, 173-182.
Coaker, T.H. and Ensor, H.L. (1967). *Proc. 4th Br. Insectic. Fungic. Conf.* **1**, 226-228.
Gair, R. (1971). *Proc. 6th Br. Insectic. Fungic. Conf.* **3**, 765-770.
Gould, H.J. and Mayor, J.G. (1975). *Pl. Path.* **24**, 245-246.
Harris, C.R. (1971). *J. econ. Ent.* **64**, 1044-1049.
Harris, C.R. (1972). *A. Rev. Ent.* **17**, 177-198.
Legowski, T.J. and Gould, H.J. (1960). *Pl. Path.* **9**, 119-126.
Suett, D.L. (1975). *ADAS q. Rev.* **19**, 125-138.
Suett, D.L. and Padbury, C.E. (1976). *Rep. natn. Veg. Res. Stn for 1975*, 77-78.
Thompson, A.R. and Wheatley, G.A. (1977). *Pestic. Sci.* **8**, 418-427.
Wheatley, G.A. (1969). *Proc. 5th Br. Insectic. Fungic. Conf.* **1**, 248-254.
Wheatley, G.A. (1971a). *World Rev. Pest Contr.* **10**, 81-93.
Wheatley, G.A. (1971b). *Proc. 6th Br. Insectic. Fungic. Conf.* **2**, 386-395.
Wheatley, G.A. (1972). *Rep. natn. Veg. Res. Stn for 1971*, 66.
Wheatley, G.A. (1973). *Proc. 7th Br. Insectic. Fungic. Conf.* **3**, 991-1004.
Wheatley, G.A. (1974). *Pestic. Sci.* **5**, 101-112.

Wheatley, G.A. and Coaker, T.H. (1970). *In* "Technological Economics of Crop Protection and Pest Control" *SCI Monogr.* No. 36, 42-55. Society of Chemical Industry, London.
Wheatley, G.A. and Niendorf, F.K. (1969). *Rep. natn. Veg. Res. Stn for 1968*, 75-76.

Section III
Quality at Harvest—Physiological Aspects

III.1 The Regulation of Gene Expression during the Ripening of Tomato Fruits

D. GRIERSON, G. TUCKER and N. ROBERTSON

University of Nottingham, Sutton Bonington, Loughborough, LE12 5RD, England

INTRODUCTION

A number of important changes in chemical composition occur during the ripening of tomato fruits (Hobson and Davies, 1971). These include: (i) loss of chlorophyll, when fruits may become pale green or white; (ii) synthesis of pigments such as β-carotene and lycopene, responsible for fruits turning yellow, orange and eventually red; (iii) decrease in starch content and accumulation of reducing sugars; (iv) changes in metabolism of organic acids such as citric acid and malic acid and accumulation of amino acids such as glutamic acid; (v) synthesis of various compounds associated with flavour development; and (vi) fruit softening caused by the hydrolysis of pectins and other components of the cell walls. These changes in composition result from a series of separate physiological or biochemical processes which take place in the cells of the fruit. The evidence that ripening can be viewed as a series of separate processes comes from studies which show that various aspects of ripening can be prevented or modified, for example, by mutation or by applying or removing growth active or inhibitory chemicals or gases, without affecting other aspects of ripening.

In climacteric fruits, such as tomatoes, ripening is stimulated by the respiratory climacteric which is itself "triggered" by ethylene. However, the basic changes associated with ripening such as pigment synthesis, sugar production, fruit softening and formation of flavour compounds occur slowly and without a respiratory rise in many fruits (Leopold and Kriedemann, 1975). Such changes in composition whether in climacteric or non-climacteric fruits, are brought about by alterations in the activity of specific enzymes required for pigment synthesis, sugar production and fruit softening.

Figure 1 outlines the way in which genetic information contained in the base sequence of the DNA of the cell is expressed. The essential features are: (i) the production of a mRNA (messenger-ribonucleic acid) copy of one strand of the DNA by the enzyme RNA polymerase II, (ii) the "processing" of mRNA and (at least for some mRNAs) attachment of a "cap" at the 5′ end of the molecule (Haugland and Cline, 1978) and addition of a (poly(A)sequence (polyadenylic acid) at the 3′ end (Grierson, 1976), (iii) binding of mRNA to ribosomes, (iv) polymerization of amino acids in the ribosome-mRNA-tRNA (transfer-RNA) complex to produce proteins specified by each particular mRNA, (v) the "trimming", folding and activation of the proteins to produce active enzymes. The net activity of enzymes can be measured by assay of cell extracts. However, a large number of factors can affect the amount or activity of a particular enzyme (potential control points are indicated in Fig. 1) and the precise way in which enzyme changes occur cannot be determined without more detailed studies.

There is considerable evidence that changes in enzyme activity are associated with fruit ripening (Sacher, 1973). Hulme *et al.* (1963) suggested that in apples an increase in malic enzyme and pyruvate carboxylase is associated with the climacteric. Hobson (1964) showed that during tomato ripening there is a rapid rise in polygalacturonase activity and Iki *et al.* (1978) have shown a similar rise in invertase activity. In addition cycloheximide, an inhibitor of protein synthesis on 80S ribosomes, retards ripening in pears (Frenkel *et al.*, 1968), citrus fruits (Riov *et al.*, 1969) and bananas (Brady *et al.*, 1970). Furthermore the enhanced incorporation of radioactive amino acids into proteins during the climacteric in avocado pear (Richmond and Biale, 1966), bananas (Sacher, 1966), apples (Hulme *et al.*, 1971) and tomatoes (De Swardt *et al.*, 1973) strongly suggested a mechanism involving protein synthesis. A similar stimulation of incorporation of radioactive precursors into RNA occurs just prior to ripening in apples (Looney and Patterson, 1967; Hulme *et al.*, 1971), pears (Richmond and Biale, 1967; Ku and Romani, 1970) and tomatoes (De Swardt *et al.*, 1973) suggesting that the synthesis of new RNA (ribonucleic acid) is also associated with ripening. Finally, Mettler and Romani (1976) have suggested that changes in tRNA may control the translation of messenger RNAs required for ripening.

In this paper we present evidence that suggests ripening involves (i) the regulation of mRNA synthesis, and (ii) the synthesis of enzymes (steps (1), (2) and (3) in Figs 1 and 2).

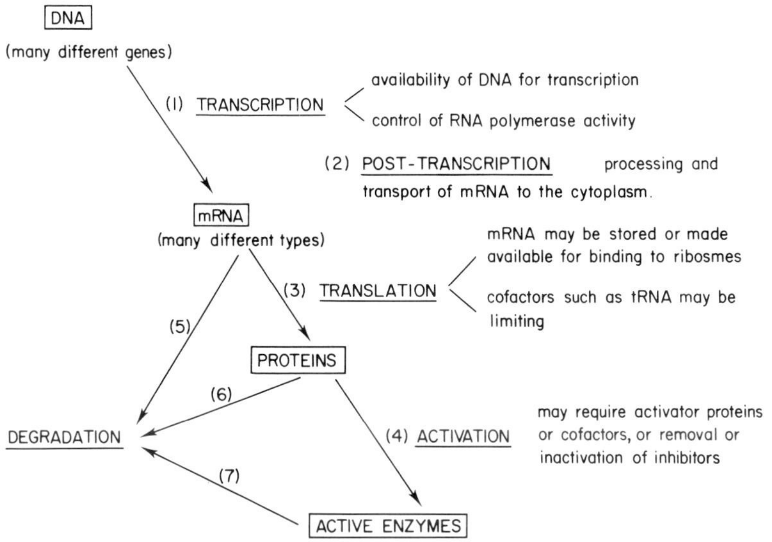

Fig. 1. Possible control points in gene expression.

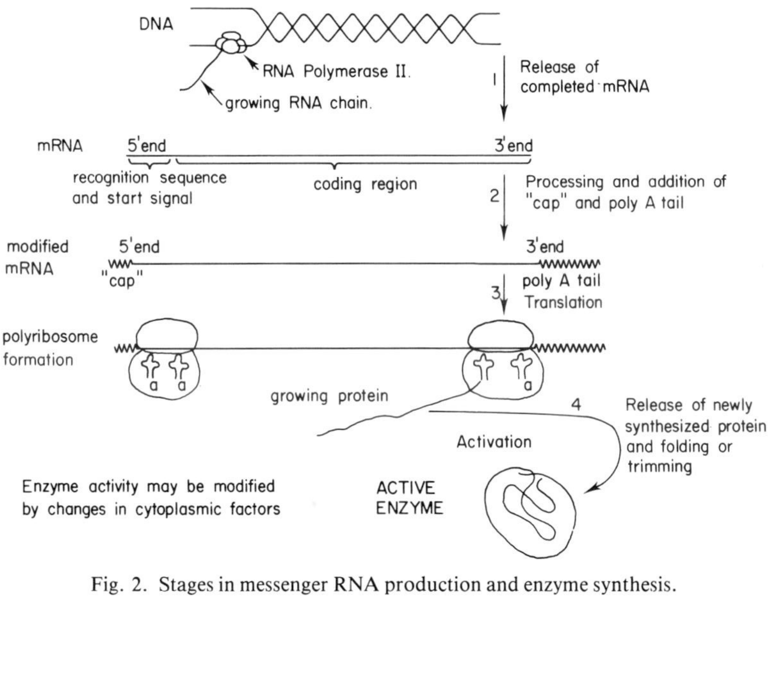

Fig. 2. Stages in messenger RNA production and enzyme synthesis.

METHODS

MESSENGER RNA STUDIES

These were carried out as described by Rattanapanone *et al.* (1978). The general procedure is outlined on p. 183. Tomato plants (*Lycopersicon esculentum* Mill.) of the GCR 145, GCR 118 and Minipopella (slow ripening) varieties were grown in soil in a heated glasshouse at temperatures between 15 and 25°C. The ages of the fruits at the onset of ripening varied according to variety (Rattanapanone *et al.*, 1977). Unripe refers to fruits 6-8 weeks old which have not yet started to synthesize lycopene. Ripening refers to fruit 1-2 weeks older in which lycopene synthesis is well established.

Extraction and purification of nucleic acids

Nucleic acids were extracted from outer pericarp tissue as previously described (Rattanapanone *et al.*, 1977). The nucleic acids, dissolved in 50 mM Tris/HCl pH 7·4, 2 mM Mg acetate, were incubated with DNase (Sigma) at 100 μg/ml for 30 min at 0°C. The solution as made up to 100 mM KCl and 10 mM Mg acetate and further purification was carried out by incubation at 10°C for 30 min with 200 μg/ml Proteinase K (Boehringer). After incubation the volume was increased 2-3 fold by the addition of 50 mM Tris/HCl pH 7·4 containing 10 mM Mg acetate and the solution deproteinized 3 times by shaking with an equal volume of $CHCl_3$/PhOH (1:1) (Grierson and Covey, 1975) followed by centrifugation. The supernatant was made 2% with NaOAc pH 5 and the RNA precipitated for 18 h at −20°C by the addition of 2·5 volumes EtOH.

Poly(A)-containing RNA was purified by oligo (dT)-cellulose chromatography (Grierson and Covey, 1975).

Nucleic acid concentration was estimated by measuring the A_{260nm}. A 40 μg/ml solution was assumed to have an $A_{260} = 1$ using a 1 cm light path.

In vitro *protein synthesis*

Synthesis was carried out using an 80S ribosome system prepared from wheat germ (obtained from Bar-Rav Mill, Tel Aviv, Israel). Preparation and use of the wheat germ S.30 system followed the procedure of Marcu and Dudock (1974). The radioactive amino acid was methionine-[^{35}S].

Acrylamide gel electrophoresis

In vitro synthesized proteins were fractionated by electrophoresis on 15% slab gels (Laemmli, 1970) containing 0·2% SDS and 8 M urea. Detection of methionine-[^{35}S] proteins was by fluorography (Bonner and Laskey, 1974).

Scheme for the purification of messenger-RNA and assay by in vitro *protein synthesis*

Homogenize tomato tissue in detergent-phenol mixture and extract and purify total nucleic acid by phenol deproteinization and ethanol precipitation.

↓

Redissolve nucleic acid, centrifuge through 2 M LiCl solution. Pectins, DNA and tRNA remain in solution and mRNA and rRNA precipitate.

↓

Redissolve mRNA and rRNA in salt solution. Pass through a column of cellulose which has oligo-deoxythymidylic acid covalently attached. Messenger-RNA rich in polyadenylic acid binds to the column while rRNA passes through. After washing the column, the poly(A)-containing RNA is eluted at low ionic strength.

↓

Messenger-RNA is precipitated with ethanol, washed and dissolved in water ready for *in vitro* protein synthesis. About 1-2% of the total RNA is in this fraction.

↓

Homogenize wheat germ in buffer. Remove cell wall material, nuclei and other organelles by centrifugation. Remove low molecular weight compounds by gel filtration. This yields a cell-free extract containing ribosomes, tRNA and enzymes necessary for protein synthesis.

→

Mix mRNA with wheat germ extract. Add amino acids, including [^{35}S]-methionine, plus other substrates and cofactors.
The types of radioactive proteins synthesized are determined by the source and properties of the mRNA.

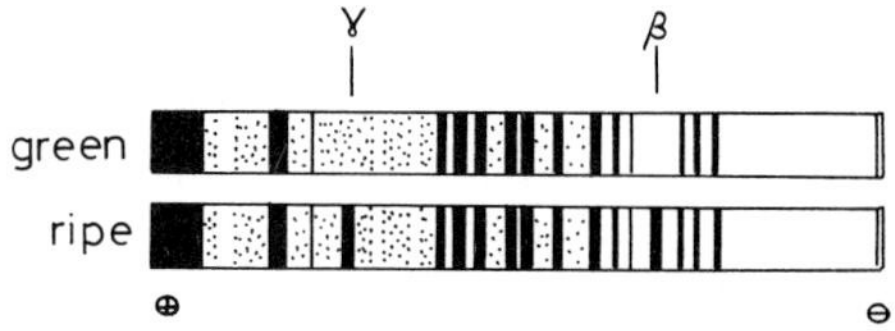

Fig. 3. Comparison of *in vitro* translation products of mRNAs from green and ripening fruit of the GCR 145 variety of tomato. Samples were denatured in 2% SDS, fractionated in 15% polyacrylamide slab gels and the radioactive proteins detected by fluorography. The protein bands marked β and γ have molecular weights of about 46000 and 20000 respectively (from Rattanapanone *et al.*, 1978).

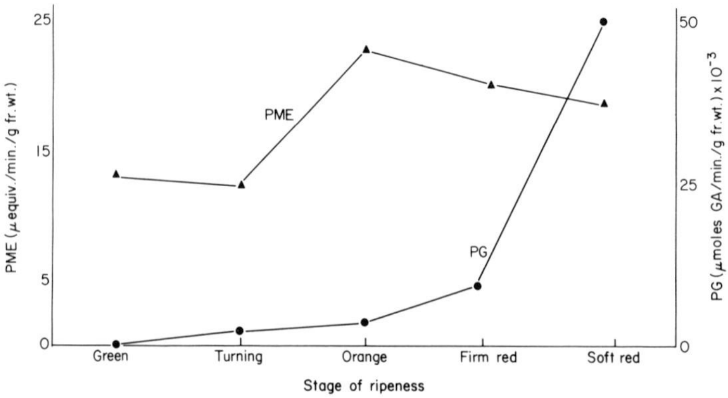

Fig. 4. Changes in activity of pectin methylesterase (PME) and polygalacturonase (PG) during ripening. PME = ▲——▲; PG = ●——●. (Tomato cultivar Potentate.)

ENZYME STUDIES

Tomatoes (cv. Ailsa Craig) were used, since many ripening mutations have been introduced into this variety for experimental purposes (Darby, 1978). Pectin methylesterase (PME) and polygalacturonase (PG) activities were extracted by the method of Pressey and Avants (1973). An outline of the purification procedures is given on p. 185. Enzyme assays for measurement of total extractable activity were carried out after step 3. Pectin methylesterase was assayed in a final volume of 10 ml 0·3 mM NaCl, pH 8·0, containing 0·5% pectin (Sigma Chemical Co.) at 25°C. The pH was maintained at 8 by continuous titration with 1 M NaOH using a pH meter and a micrometer syringe. Results were calculated as μequivalent NaOH/min. Polygalacturonase was assayed in a final volume of 1 ml 0·3 M NaCl, pH 3·8, containing 0·5% polygalacturonic acid (Cambrian Chemicals) at 25°C. The production of reducing groups was measured by the method of Nelson (1944). Invertase was assayed in a final volume of 1 ml 200 mM sucrose pH 4·6 at 25°C. Reducing groups were measured by the method of Nelson (1944).

Scheme for the purification of pectin methylesterase and polygalacturonase

Tomato pericarp is cut into thin slices and suspended in distilled water. Homogenize in a Polytron for 30-60 s. Spin for 10 min at 2400 **g**.

STEP 1

Supernatant discarded ← → Pellet resuspended in 1 M NaCl and adjusted with NaOH to pH 6·0. Stir 3 h at 4°C. Spin 10 min at 2400 **g**.

STEP 2

Pellet discarded ← → Supernatant filtered and solid ammonium sulphate added to give 75% saturation. Spin 20 min at 10000 **g**.

STEP 3

Supernatant discarded ← → Pellet resuspended in 0·15 M NaCl pH 6·0. Dialysed for 12 h against 0·15 M NaCl pH 6·0. Spin out slight precipitate.

STEP 4

Run supernatant on DEAE-A50 column in 0·15 M NaCl pH 6·0. This separates two isoenzymes (I and II) of each of PME and PG. These isoenzymes are collected by precipitating with 75% saturated ammonium sulphate

STEP 5

PGI, PGII, PMEI and PMEII are run separately on a G100 column. Enzymes are again collected by precipitating with 75% ammonium sulphate.

STEP 6

PGI, PGII further purified by adsorption on to DEAE-A50 column at pH 8·0. PGI elutes with 0·5 M NaCl, PGII elutes with 0·1 M NaCl

STEP 7

PMEII run on SDS gels gives one polypeptide band.

STEP 8

PGII run on SDS gels gives one polypeptide band.

RESULTS AND DISCUSSION

EVIDENCE FOR CHANGES IN mRNA RELATED TO TOMATO FRUIT RIPENING

We have shown previously that there is a marked increase in the rate of labelling of RNA just prior to lycopene synthesis and the onset of ripening in tomatoes. The main types of RNA synthesized during this period are rRNA and tRNAand smaller quantities of mRNA. There is no evidence for preferential synthesis of mRNA during ripening (Rattanapanone *et al.*, 1977), although this does not mean that no significant change in mRNA occurs. This is firstly because mRNA always represents a small proportion of total cell RNA and secondly because it is the *type* of mRNA which is important for the control of protein synthesis. However, because mRNA forms such a small part of the total cell RNA, and because there are many different types of mRNA, it is rather difficult to study. We have been able to take advantage of the presence of sequences of polyadenylic acid (poly(A)) attached to plant mRNA (Grierson, 1976; Covey and Grierson, 1976; Higgins *et al.*, 1976) in order to purify tomato mRNA. The quality of the mRNA can then be studied by allowing it to direct the synthesis of proteins *in vitro*. This is one of the very few procedures which make it possible to study changes in the types of mRNA present in cells. The procedure is outlined on p. 183. Messenger-RNA is used to direct the synthesis of proteins in a wheat-germ extract. These proteins are made radioactive by the addition of [^{32}S]-methionine to the reaction mixture. The radioactively labelled proteins are then fractionated by electrophoresis in polyacrylamide slab gels in the presence of sodium dodecyl sulphate (SDS). The proteins in the gel are detected by fluorography: the slab gel is impregnated with diphenyloxazole and then dried and placed in contact with an X-ray film. Radioactive proteins emit β-particles which cause the diphenyloxazole molecules to emit light, which is detected by the X-ray film.

A comparison of the protein product synthesized in response to mRNA from green and ripe tomato fruits is show in Fig. 3. The radioactive proteins are referred to by molecular weight, which is calculated with reference to marker proteins fractionated in the same gel. The presence or absence of a particular protein band is due to the presence or abscence of the corresponding mRNA coding for that protein. There are clear differences in the proteins produced by translation of mRNA from green and ripe fruits, which clearly demonstrates for the first time that changes in mRNA quality occur during ripening. Further, more detailed analysis, may reveal other less dramatic mRNA changes. At the moment we are not certain whic enzymes these new mRNAs, synthesized during ripening, code for. Evidenc related to some possible candidates is discussed below.

EVIDENCE FOR THE SYNTHESIS OF SPECIFIC ENZYMES DURING RIPENING

We have been studying three enzymes which increase in activity during ripening. These are pectin methylesterase, which removes methyl groups from the pectins of the plant cell wall; polygalacturonase, which hydrolyses the resulting demethylated pectin to smaller fragments and invertase, which converts sucrose to glucose and fructose. Changes in the extractable activities of these enzymes during ripening are shown in Figs 4 and 5. Pectin methylesterase activity is already present in unripe fruit, therefore one may ask what the significance is of the further increase in activity during ripening. The explanation appears to be that there are two isoenzymes of pectin methylesterase (PMEI and PMEII) (first detected by Pressey and Avants (1972) and corresponding to isoenzymes I and IV according to their nomenclature). Chromatography on DEAE-cellulose separates PMEI from PMEII (Fig. 6). The properties of PMEI and PMEII are summarized in Table I. During ripening the increase in activity of PME is associated with a dramatic change in the ratio of PMEI and PMEII (1:4 in green fruit to 1:10 in orange fruit). The predominance of isoenzyne II in ripening fruit is quite characteristic and may indicate the requirement for this isoenzyme for the initiating of softening.

In Fig. 5 the increase in extractable polygalacturonase (PG) is compared with the increase in invertase activity. Polygalacturonase, unlike PME (Fig. 4), is virtually absent from unripe fruit. It was first shown to increase during ripening by Hobson (1964). The increase in extractable PG activity

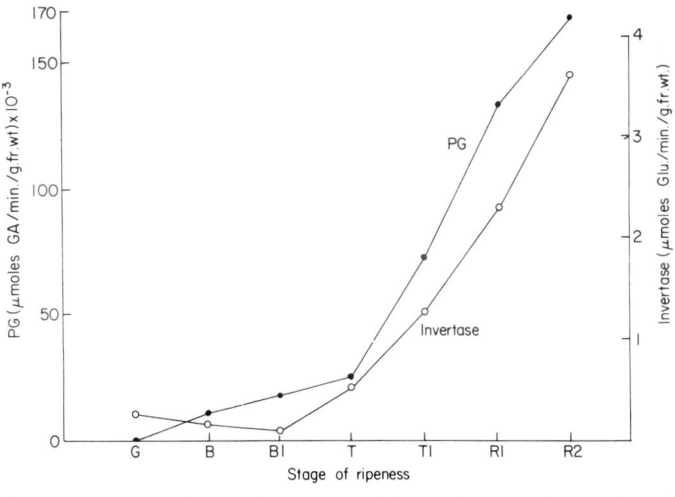

Fig. 5. Changes in activity of polygalacturonase (PG) and invertase during ripening. PG = ●——●; Invertase ○——○. Stages of ripening are: G = green, B = breaker, T = turning, T_1 = turning/orange, R_1 = ripe, R_2 = overripe. (Tomato cultivar, Ailsa Craig.)

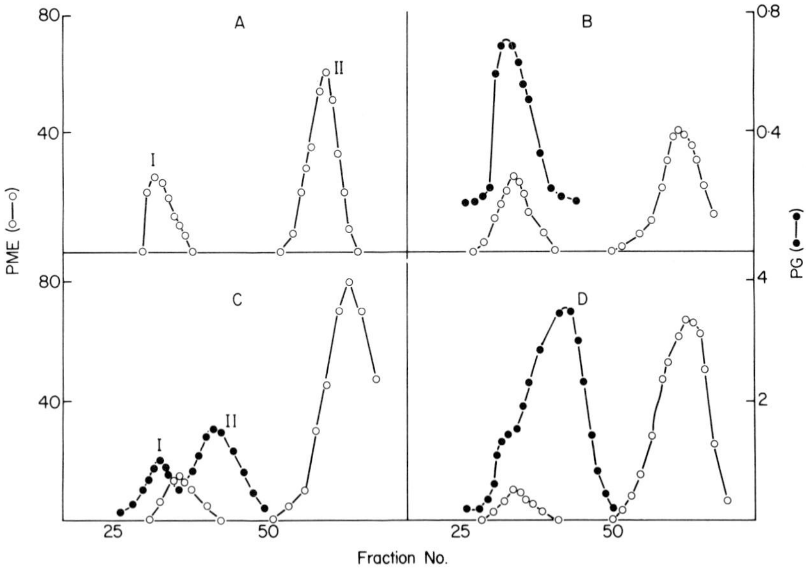

Fig. 6. Ion exchange chromatography of isoenzymes of pectin methylesterase (PME) and polygalacturonase (PG) at different stages of ripening. PME = (○——○), PG = (●——●). Isoenzymes I and II are labelled in Figs 6A and 6C. Figs 6A-6D correspond to stages G, B, T_1 and R_2 respectively shown in Fig. 5.

TABLE I

Some properties of pectin methylesterase isoenzymes I and II

	I	II
Approximate molecular weight	50 000	23 000
Temperature for 50% inactivation	56°C	62°C
pH optimum	7·6	7·6

is due to the appearance of first one and then a second isoenzyme (PGI and PGII). These two isoenzymes can be distinguished by DEAE—Sephadex chromatography (Pressey and Avants, 1973), as shown in Fig. 6. In *rin* (= "ripening inhibited") mutant tomatoes, which remain yellow and do not soften, no PG activity can be extracted whereas in *Nr* (= "never ripe") mutant tomatoes, which do not soften very much, only PGI is normally extracted.

The molecular weights, pH optima and heat stabilities of the two PG isoenzymes are shown in Table II. Studies on the change in activity of the

TABLE II

Some properties of polygalacturonase isoenzymes I and II

	I	II
Approximate molecular weight	100 000	40 000
Temperature for 50% inactivation	78°C	57°C
pH optimum	3-4	3-4

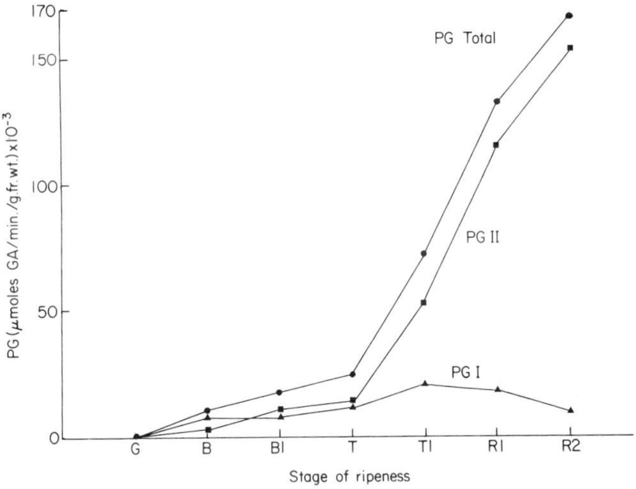

Fig. 7. Contribution of isoenzymes I and II to the increase in polygalacturonase activity during ripening. Stages of ripening are as shown in Fig. 5.

two PG isoenzymes during normal ripening shows that although PGI appears first it does not continue to accumulate. The main increase in PG is due to isoenzyme II (Fig. 7). We are currently trying to find out whether PGII is an entirely different enzyme or whether it arises by conversion of PGI.

Recently Iki *et al.* (1978), using an antibody to invertase, have obtained results suggesting that the amount of invertase enzyme protein increases at the same rate as enzyme activity during ripening, suggesting that the increase in invertase is due to synthesis of new protein molecules and not simply to the activation of pre-existing inactive enzyme. We have purified PMII and PGII (an outline of the purification procedure for PGII is given in the Methods section) and are preparing antibodies for use in similar studies. We have already shown that if ripening tomato fruits are injected with radioactive amino acids, and PGII is subsequently extracted from the fruit and purified, that the enzyme is radioactive. This suggests that the increase in PGII activity is due to enzyme *synthesis* and not activation.

The unanswered question is whether the synthesis of these enzymes depends upon the translation of pre-existing mRNA or whether new mRNA is required, as suggested by the results in Fig. 3. The similarity between the molecular weights of the β and γ bands in Fig. 3 (from translation of mRNA from ripening fruit) and the estimated molecular weights of PMEII and PGII has not escaped our notice. It is hoped that further work will enable us to relate ripening-associated increase in specific mRNAs to the synthesis of known enzymes.

CONCLUSIONS

These results show that the problem of the regulation of specific genes during ripening is open to direct biochemical attack. Alterations in enzyme activity seem to involve both mRNA synthesis and enzyme synthesis. Many further details of the process remain to be worked out and other mechanisms may also operate for particular enzymes. Specific mutants of ripening should prove useful in further studies.

The conversion of a mature green tomato to a ripe fruit involves a transition from one physiological or biochemical state to another. In this sense ripening is a change in "physiological age". The studies described in this paper were initiated in order to understand more fully what controls ripening. Of course, we already know how to tell when a fruit is ripe, but the results suggest that there are changes in mRNA and protein synthesis which *cause* ripening. One might expect that similar detailed studies of other processes involving a change in "physiological age", but with less obvious symptoms for visual detection, would provide biochemical "markers" for measuring "physiological age".

ACKNOWLEDGEMENTS

We thank Dr G.E. Hobson and Dr M.J.C. Rhodes for helpful discussions and Mr L.A. Darby for very generously providing us with seeds of cvs Ailsa Craig and Potentate tomatoes, together with isogenic lines incorporating the *rin* and *Nr* mutations. Part of this work was supported by the Agricultural Research Council.

REFERENCES

Bonner, W.M. and Laskey, R.A. (1974). *Eur. J. Biochem.* **46**, 83-88.
Brady, C.J., Palmer, J.K., O'Connell, P.B.H. and Smillie, R.M. (1970). *Phytochemistry* **9**, 1037-1047.
Covey, S.N. and Grierson, D. (1976). *Eur. J. Biochem.* **63**, 599-606.
Darby, L.A. (1978). *Hort. Res.* **18**, 73-84.
De Swardt, G.H., Swanepoel, J.H. and Dubenage, A.J. (1973). *Z. Pflanzenphysiol.* **70**, 358-363.
Frenkel, C., Klein, I. and Dilley, D.R. (1968). *Pl. Physiol.* **43**, 1146-1153.
Grierson, D. (1976). *In* "Molecular Aspects of Gene Expression in Plants" (J.A. Bryant, ed.) pp. 53-108. Academic Press, London and New York.
Grierson, D. and Covey, S.N. (1975). *Planta* **127**, 77-86.
Haugland, R.A. and Cline, M.G. (1978). *Pl. Physiol.* **62**, 838-840.
Higgins, T.J.V., Zwar, J.A. and Jacobsen, J.V. (1976). *Nature, Lond.* **260**, 166-168.
Hobson, G.E. (1964). *Biochem. J.* **92**, 324-332.
Hobson, G.E. and Davies, J.N. (1971). *In* "The Biochemistry of Fruits and their Products" (A.C. Hulme, ed.) Vol. 2, pp. 437-482. Academic Press, London and New York.
Hulme, A.C., Jones, A.D. and Wooltorton, L.S.C. (1963). *Proc. R. Soc.* (B), **158**, 514-535.
Hulme, A.C., Rhodes, M.J.C. and Wooltorton, L.S.C. (1971). *Phytochemistry* **10**, 749-756.
Iki, K., Sekiguchi, K., Kurata, K., Tada, T., Nakagawa, H., Ogura, N. and Takehana, H. (1978). *Phytochemistry* **17**, 311-312.
Ku, L.L. and Romani, R.J. (1970). *Pl. Physiol.* **45**, 401-407.
Laemmli, U.K. (1970). *Nature* (*New Biol.*) **227**, 680-685.
Leopold, A.C. and Kriedemann, P.E. (1975). "Plant Growth and Development." McGraw Hill, New York.
Looney, N.E. and Patterson, M.E. (1967). *Phytochemistry* **6**, 1517-1520.
Marcu, K. and Dudock, B. (1974). *Nucleic Acid Res.* **1**, 1385-1397.
Mettler, I.J. and Romani, R.J. (1976). *Phytochemistry* **15**, 25-28.
Nelson, N. (1944). *J. biol. Chem.* **153**, 375-380.
Pressey, R. and Avants, J.K. (1972). *Phytochemistry* **11**, 3139-3142.
Pressey, R. and Avants, J.K. (1973). *Biochim. biophys. Acta* **309**, 363-369.
Rattanapanone, N., Grierson, D. and Stein, M. (1977). *Phytochemistry* **16**, 629-633.
Rattanapanone, N., Speirs, J. and Grierson, D. (1978). *Phytochemistry* **17**, 1485-1486.
Richmond, A. and Biale, J.B. (1966). *Pl. Physiol.* **41**, 1247-1253.
Richmond, A. and Biale, J.B. (1967). *Biochim. biophys. Acta* **138**, 625-627.
Riov, J., Monselise, S.P. and Kahan, R.S. (1969). *Pl. Physiol.* **44**, 631-635.
Sacher, J.A. (1966). *Pl. Physiol.* **41**, 701-708.
Sacher, J.A. (1973). *A. Rev. Pl. Physiol.* **24**, 197-224.

III.2 Differentiation of Storage Cells in Seeds of Legumes

JOHN A. BRYANT

Department of Plant Science, University College, P.O. Box 78, Cardiff, CF1 1XL, Wales

INTRODUCTION

The term "quality" used in description of a crop embodies at least two different concepts. Firstly, the term refers to the acceptability of the crop to the consumer or processor. Since a crop may be used in a number of different ways, acceptability itself may mean different things to different users. Users of fresh vegetables for freezing or direct consumption have different criteria for acceptability from those applied by users of vegetables for bulk food or bulk extraction of vegetable products, such as oil and protein. Secondly, the term "quality" refers to nutritional quality, the value of the crop as food. Clearly, acceptability to the crop user and nutritional quality are not necessarily mutually exclusive, and indeed may often be complementary. However, in some instances, the maximization of the acceptability of a crop for use as a fresh vegetable, for example, may well not lead to a maximization of the crop's nutritional value. This may occur because maximum consumer acceptability and maximum nutritional yield occur at different stages in the plant's development and differentiation. These problems are well illustrated by crops in the family Fabaceae (legumes), which are used in a variety of ways, including direct consumption, freezing, processing for bulk extraction of oils, protein and starch, and increasingly importantly, as major components in human nutrition (Heiser, 1973; Davies, 1977; Pate, 1977; Rickerd, 1977; Evans and Gridley, 1979).

FRUIT AND SEED DEVELOPMENT IN GRAIN LEGUMES

Typically, the fruits of leguminous plants are boat-shaped pods ("legumes"), each of which contains between one and several individual seeds.

Growth and expansion of the pods starts very soon after fertilization of the ovules, and the expansion phase of pod growth is more or less complete before any significant growth of the seeds occurs. This is illustrated for the field pea, *Pisum arvense*, in Fig. 1. The pod then inflates by differential growth of inner and outer layers of cells, to form a hollow envelope in which the seed may expand and develop. In the garden pea (*P. sativum*) and field pea (*P. arvense*) and probably in most other grain legumes, the end of the expansion phase of the pod is marked by high levels of starch, sucrose, protein and soluble nitrogen within the tissues of the pod (Pate and Flinn, 1977). This is clearly relevant to consumer acceptability in crops where pods as well as seeds are consumed, e.g. runner bean, *Phaseolus coccineus*. At about the time at which pod expansion is complete, layers of cells beneath the epidermis become sclerified (thickened with lignin), imparting a greater mechanical strength to the pod (Fahn and Zohary, 1955; Flinn, 1974), and giving it a fibrous texture which may not be acceptable to the consumer. Selection of strains in which this particular phase of differentiation is very much delayed, or reduced in extent, has clearly been important in improving crop quality for direct consumption or freezing.

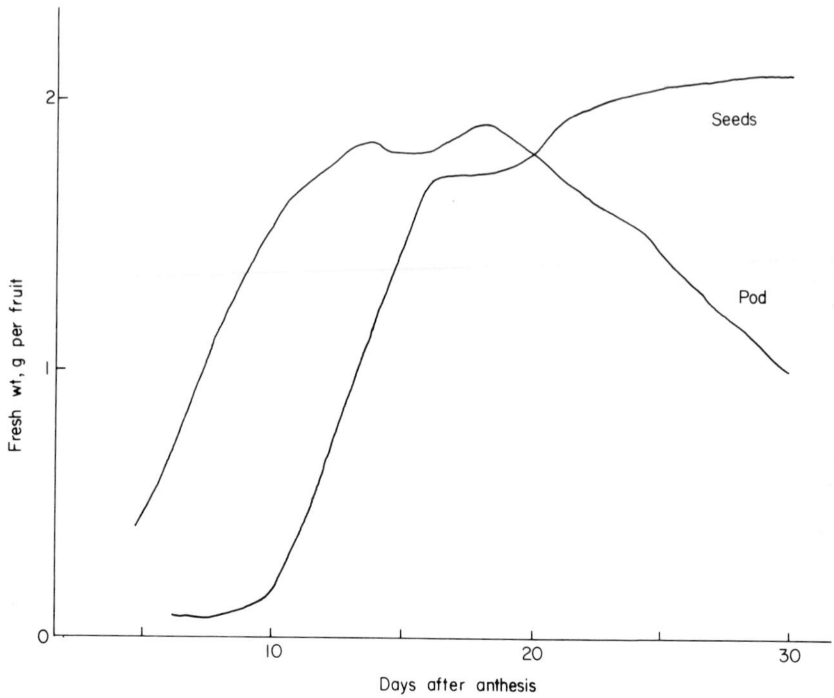

Fig. 1. Growth of pods and seeds in field pea, *Pisum arvense* (redrawn from Pate and Flinn, 1977).

As pod growth slows down, the growth rate of the seeds therein increases rapidly. This growth occurs almost entirely by the expansion of cells which have already arisen by a series of rapid cell divisions (Reeve, 1948; McKee *et al.*, 1955; Bain and Mercer, 1966; Walbot *et al.*, 1972; Smith, 1973; Dure, 1975). The embryo has by this stage become anchored in the endosperm sac (Walbot *et al.*, 1972; Pate and Flinn, 1977) and the cotyledons, which are the main sites of storage of protein, lipid and carbohydrate, expand rapidly. The expansion phase is accompanied by vacuolation of the storage parenchyma cells (Millerd, 1975). As the cotyledons expand, they fill the space occupied by the liquid endosperm (Kapoor, 1966), absorbing sucrose, amino acids, ammonium salts and organic acids from the endosperm (Pate and Flinn, 1977). The cells of the outer epidermis of the cotyledons differentiate as transfer cells, facilitating absorption of nutrients from the nucellus (Pate and Gunning, 1972). As the cotyledon cells expand, so the nutrient reserves are laid down within them (Fig. 2). The major reserves in terms of dry weight are protein and starch, with lipids and phytins also making a significant contribution (Pate, 1977; Sutcliffe and Bryant, 1977; Lott and Buttrose, 1978). The proportions of each component

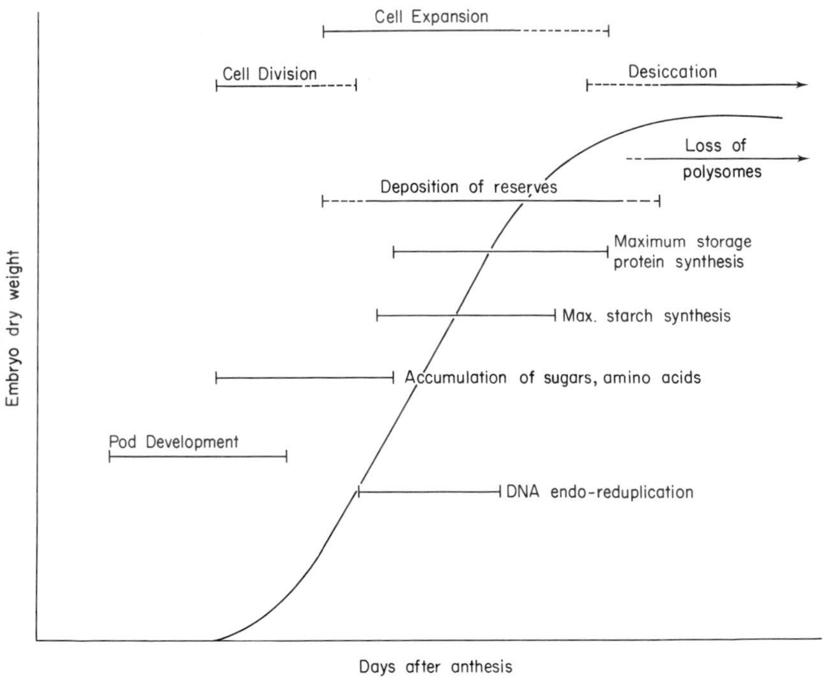

Fig. 2. Deposition of storage protein during cell expansion in garden pea, *Pisum sativum* (redrawn from data in Millerd and Spencer, 1974).

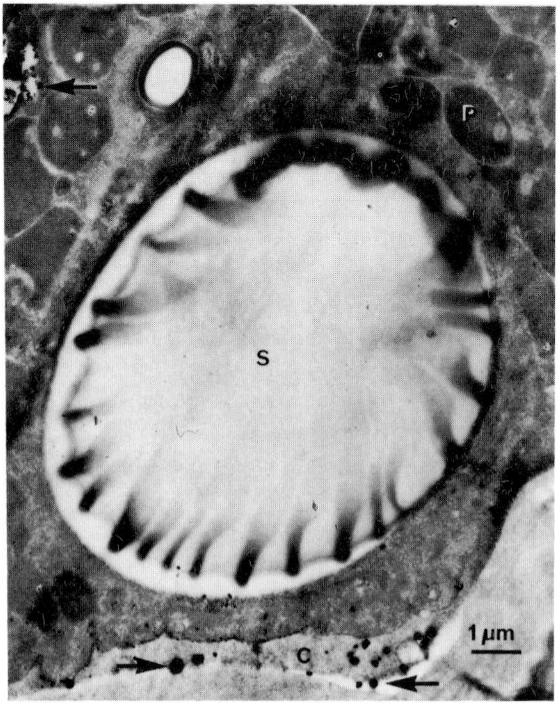

Fig. 3. Part of storage parenchyma cell from a cotyledon of *P. sativum*. c = cytoplasm; s = starch grain; p = protein body. The cells have been stained for acid phosphatase activity (arrows) which is apparent in protein bodies, cytoplasm, and at the cytoplasm-wall interface (from Bowen and Bryant, 1978).

laid down vary considerably between species. For example, average values (as per cent of dry weight) for pea, *P. sativum*, are protein, 20%, starch 35-45% and lipid, 1-2%, whilst average values in soya bean, *Glycine max*, are protein 35%, starch, 15% and lipid, 18% (Heiser, 1973; Davies, 1977; Pate, 1977; Sutcliffe and Bryant, 1977; Evans and Gridley, 1979). The protein and phytin are deposited in protein bodies which are 1-2 μm in diameter; as the protein bodies accumulate, so they replace the original vacuoles of the cell (Millerd, 1975; Lott and Buttrose, 1978). Lipid is laid down in lipid bodies which are approx. 0·1 μm in diameter, and starch forms very large starch grains, 10-20 μm in diameter, which completely fill the amyloplasts in which they are deposited (Pate and Flinn, 1977). In the fully expanded state, the storage cells are very large (approx. 200 μm diameter) and are so packed with storage products, that the cytoplasm is confined to a thin layer at the cell periphery (Fig. 3). In nature, the final stage of development is dehydration. For crops which are consumed as fresh vegetables, or used for freezing, harvesting occurs before dehydration commences, but for very many purposes, dried seeds are required, and the dehydration process is allowed to go to completion before harvest.

The storage parenchyma cells are thus very unusual, and represent a pattern of differentiation which is not seen elsewhere in the legume plant (although small quantities of nutrient reserves are laid down in other parts of the embryo). The next section of this paper deals with selected features of the biochemistry of this differentiation process, concentrating particularly on those aspects which affect quality.

BIOCHEMICAL ASPECTS OF DIFFERENTIATION IN STORAGE PARENCHYMA

SUGAR AND STARCH

In most legumes, starch synthesis and deposition are initiated during the early phase of cotyledon expansion but reach maximum rates either late in the expansion phase or after the cells have ceased expanding (Bain and Mercer, 1966; Dure, 1975; Pate and Flinn, 1977). The starch accumulates at the expense of soluble carbohydrates (sugars), firstly from the liquid endosperm, and then from maternal tissues (Dure, 1975; Pate and Flinn, 1977). The amount of soluble carbohydrate present in the cotyledons at any one time is a reflection of the balance between the supply of sugars to the cotyledons, and its utilization in starch formation; typically, soluble carbohydrate content reaches a peak during the cell expansion phase (Turner *et al.*, 1957; Dure, 1975; Pate and Flinn, 1977). Selection of a crop for maximum sweetness for the fresh or frozen vegetable market thus militates against yield in terms of total dry weight (Fig. 2).

PROTEIN

Studies of a wide range of legume species indicate that there are two major types of proteins, albumins and globulins, in the cotyledons (Millerd, 1975). The albumins (water-soluble proteins), are in the main the "working" proteins of the cell, such as enzymes, whilst the globulins (soluble in salt solutions) are the storage proteins, and are not found anywhere in the plant other than in the cotyledons, and to a much lesser extent in the embryo axis (Millerd, 1975; Ashton, 1976). This is thus a particularly interesting example of the well-known principle that differentiation of specific cell types involves the expression of specific genes coding for specific proteins (Bryant, 1976).

There are two major types of globulin, easily separable by size. The larger type has a sedimentation coefficient of 11-12 *s* and the smaller type 7-8 *s* (Millerd, 1975; Derbyshire *et al.*, 1976). Some species contain a third, minor component, with sedimentation coefficient of 2-3 *s* (Millerd, 1975). The names given to the different types of globulin vary according to the

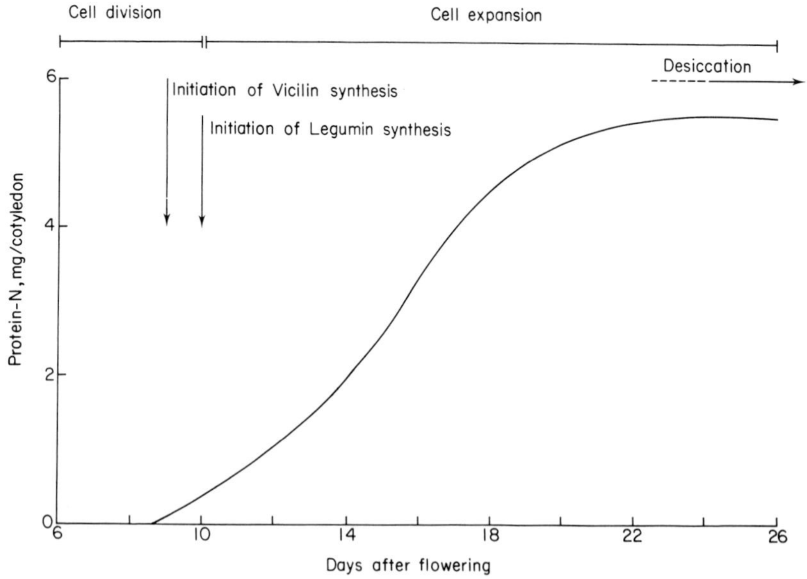

Fig. 4. A generalized scheme for seed development and differentiation in legumes (redrawn from Dure, 1975).

TABLE I

Content of essential amino acids, plus cysteine, of seed proteins of *P. sativum*, and of 'standard' animal protein. Data are moles per cent of total amino acid content

Amino acid	Albumins	Globulins		Animal protein
		Vicilin	Legumin	
Isoleucine	4·8	5·1	4·0	4·6
Leucine	5·1	9·2	8·1	8·3
Lysine	7·7	7·9	4·9	6·2
Methionine	1·0	0·22	0·65	1·9
Phenyl alanine	6·3	6·2	4·9	5·4
Threonine	4·6	3·4	2·9	4·9
Tryptophan	2·4	—	1·1	2·1
Valine	5·4	4·6	4·6	5·8
Cysteine	5·4	0·35	0·71	1·1

species in question. In *Vicia* and *Pisum*, for example, the 11-12 *s* component is called legumin and the 7-8 *s* component is called vicilin. Within each species, the 11-12 *s* component is a single type of protein, consisting of three subunits, whilst the 7-8 *s* component actually consists of several very similar proteins, each of which has four subunits (Millerd, 1975; Derbyshire *et al.*, 1976; Beachey *et al.*, 1978). The timing of the deposition of these major proteins is illustrated in Figs 2 and 4.

Examination of the amino acid compositions of the globulins illustrates that they are poor in content of the sulphur amino acids, cysteine and methionine (Table I). This can lead to serious deficiencies in instances where legumes provide the bulk of dietary protein, since methionine is an essential amino acid which cannot be manufactured in the human body. Ironically, therefore, selecting for high yield in legumes means selecting for poorer quality protein, since high-yielding strains are rich in storage proteins (globulins), which are much poorer in sulphur amino acid content than the albumins (Table I). There have been no reports of attempts to modify the sulphur amino acid content by increasing the sulphur status of the plant, and indeed, the studies of globulin synthesis in cell-free systems suggests that such an approach would not be worthwhile (see p. 201). However, it is clear that in several species of *Lupinus*, sulphur-deficiency can cause the globulin fraction to be even poorer than normal in sulphur amino acid content (Gillespie *et al.*, 1978). This is because sulphur deficiency leads to an overproduction of one globulin, conglutinin β, which is very poor in sulphur amino acid content, at the expense of conglutinins α and γ, which contain larger (although not ideal) proportions of sulphur amino acids.

At present then, increased sulphur content in protein, and hence increased nutritional quality of protein, can only be achieved by selecting for albumins and against globulins. Since the albumin content of a seed usually approaches a maximum long before globulin deposition is complete (Millerd, 1975), such a selection can be achieved by harvesting relatively early during the phase of deposition of reserves. This is of course normal practice when crops are destined for the table or freezer, and thus legumes used in this way contain protein of greater nutritional quality than legumes harvested at maturity. However, it must be stressed that early harvesting also results in reduced protein yields, and is therefore not applicable in situations where bulk protein is required.

LECTINS AND PROTEASE INHIBITORS

Storage parenchyma cells of legumes may contain a number of potentially harmful proteins, including lectins and protease inhibitors (Liener, 1977). Lectins or phytohaemagglutinins are proteins which recognize the antigenic determinants of blood cells, causing agglutination; they are also mitogenic agents, causing unscheduled cell division in cultured animal cells. Although both these effects could be potentially very serious for consumers, neither seems to occur following ingestion of legume protein, suggesting that processing, cooking and digestion inactivate lectins.

Protease inhibitors are proteins which inhibit the proteases of the digestive tract. The effect of these inhibitors on the utilization of legume proteins by man has not been determined. It is possible that their presence causes legume proteins to be inherently less digestible than proteins from other plant sources (Evans and Gridley, 1979) although Liener (1977) suggests that normal processing or culinary techniques should denature the inhibitors.

HYDROLYTIC ENZYMES

Developing and mature seeds of legumes contain significant levels of a wide variety of hydrolytic enzymes (Wilson, 1975; Ashton, 1976; Sutcliffe and Bryant, 1977). Some of these, such as protease (Ashton, 1976) and acid phosphatase (Bowen and Bryant, 1978) are actually located in the protein bodies. The presence of these enzymes causes problems both with legumes harvested in the immature state for direct consumption or freezing, and with legumes harvested in the dry state. When legumes are harvested in the immature state, there is a wound response to damage sustained during harvest, and wilting occurs because of the severance of the pod from its water supply. Both the response of a plant to wounding and the response to wilting involve increases in the activity of hydrolytic enzymes (Bryant, 1969; Leopold and Kriedemann, 1975; Wilson, 1975). Further, both responses involve increases in respiration rate, and therefore increased loss of respiratory substrates such as carbohydrates (ap Rees, 1966; Bryant, 1969; Leopold and Kriedemann, 1975), leading in turn to a decrease in the acceptability of the crop to the consumer.

For dried seeds, the problem is a different one. Many of the hydrolytic enzymes show vast increases in activity, either through activation, or *de novo* synthesis, during seed *germination* (Sutcliffe and Bryant, 1977). These processes are initiated by hydration, and so any increase in the water status of the seed between harvest and use, may lead to losses of reserves by the action of hydrolytic enzymes.

PROSPECTS FOR THE FUTURE

In discussing the prospects for improvements in the quality of legumes at harvest, attention is focused on three areas.

(a) UNIFORMITY OF CROP

The use of legumes as fresh vegetables or as vegetables for freezing, coupled with the need for mechanical harvesting means that ideally a crop stand should exhibit synchronous development, with all pods at comparable stages of differentiation. Typically, legumes do not show this, but instead exhibit a pattern of sequential flowering and pod-filling. The progress being made in selecting for synchronous development and differentiation is dealt with elsewhere in this Symposium (see Marx, G.A. pp. 67-81).

(b) PROTEIN QUALITY

The ability to isolate the messenger-RNA molecules which direct the synthesis of globulins, and to translate those mRNA molecules *in vitro* has led to a greater understanding of the molecular control mechanisms underlying protein deposition (Sun *et al.*, 1975; Beachey *et al.*, 1978; Evans *et al.*, 1979; Spencer and Higgins, 1979). It is clear from these studies that in an individual cultivar of a given species, it is difficult, if not impossible to increase the sulphur amino acid content of individual polypeptide chains by biochemical manipulation. The constraints on sulphur amino acid content in an individual component of the globulin fraction are not biochemical but genetic. However, individual components of the globulin fraction differ in their content of sulphur amino acids (Table I), and in the long run, examination of the factors controlling the ratio of the various globulin components should be worthwhile (Millerd, 1975).

(c) USE OF "NEW" CROPS

The winged bean, *Psophocarpus tetragonolobus*, a tropical member of the Fabaceae, exhibits a number of significant differences from the pattern of seed development and differentiation described above (Gillespie and Blagrove, 1978). Firstly, the seed pods (and indeed the plant's leaves) are rich in protein. Secondly, the storage proteins in the seed differ from those of other legumes in that there are two major components, with sedimentation coefficients of 6 *s* and 2 *s*, and an 8 *s* minor component. Thirdly, the sulphur amino acid content is significantly higher than in other legume proteins. Further, the protein and lipid contents of the seed rival those seen in soya bean (*Glycine max*). These features have led to the suggestion that *P. tetragonolobus* could be widely used as a crop with high nutritional quality in the tropics.However, as pointed out by Gillespie and Blagrove (1978), a good deal of agronomic research will be... "needed to transform the winged bean into a commercial crop."

REFERENCES

ap Rees, T. (1966). *Aust. J. biol. Sci.* **19**, 981-990.

Ashton, F.M. (1976). *A. Rev. Pl. Physiol.* **27**, 95-117.

Bain, J.M. and Mercer, F.V. (1966). *Aust. J. biol. Sci.* **19**, 49-67.

Beachey, R.N., Thompson, J.F. and Madison, J.T. (1978). *Pl. Physiol.* **61**, 139-144.

Bowen, I.D. and Bryant, J.A. (1978). *Protoplasma* **97**, 241-250.

Bryant, J.A. (1969). Nucleic acid metabolism in carrot root tissue. Ph.D. thesis, University of Cambridge.

Bryant, J.A. (1976). *In* "Molecular Aspects of Gene Expression in Plants" (J.A. Bryant, ed.), pp. 217-248. Academic Press, London and New York.
Davies, D.R. (1977). *Sci. Prog., Oxf.* **64**, 201-214.
Derbyshire, E., Wright, D.J. and Boulter, D. (1976). *Phytochemistry* **15**, 3-24.
Dure, L.S. (1975). *A. Rev. Pl. Physiol.* **26**, 259-278.
Evans, A.M. and Gridley, H.E. (1979). *Curr. Ad. Plant Sci.* **11**(1), Commentaries Pl. Sci. No. 32, 1-17.
Evans, I.M., Croy, R.R.D., Hutchinson, P., Boulter, D., Payne, P.J. and Gordon, M.E. (1979). *Planta* **144**, 455-462.
Fahn, A. and Zohary, M. (1955). *Phytomorphology* **5**, 99-111.
Flinn, A.M. (1974). *Physiologia Pl.* **31**, 275-278.
Gillespie, J.M. and Blagrove, R.J. (1978). *Aust. J. Pl. Physiol.* **5**, 357-369.
Gillespie, J.M., Blagrove, R.J. and Randall, P.J. (1978). *Aust. J. Pl. Physiol.* **5**, 641-650.
Heiser, C.B. (1973). "Seed to Civilisation". Freeman, San Francisco.
Kapoor, B.M. (1966). *Genetica* **37**, 557-568.
Leopold, A.C. and Kriedemann, P.E. (1975). "Plant Growth and Development". McGraw-Hill, New York.
Liener, I.E. (1978). *In* "Plant Proteins" (G. Norton, ed.), pp. 117-140. Butterworth, London.
Lott, J.N.A. and Buttrose, M.S. (1978). *Aust. J. Pl. Physiol.* **5**, 89-111.
McKee, H.S., Robertson, R.N. and Lee, J.B. (1955). *Aust. J. biol. Sci.* **8**, 137-163.
Millerd, A. (1975). *A. Rev. Pl. Physiol.* **26**, 53-72.
Millerd, A. and Spencer, D. (1974). *Aust. J. Pl. Physiol.* **1**, 331-341.
Pate, J.S. (1977). *In* "The Physiology of the Garden Pea" (J.F. Sutcliffe and J.S. Pate, eds), pp. 469-484. Academic Press, London and New York.
Pate, J.S. and Flinn, A.M. (1977). *In* "The Physiology of the Garden Pea". (J.E. Sutcliffe and J.F. Pate, eds), pp. 431-468. Academic Press, London and New York.
Pate, J.S. and Gunning, B.E.S. (1972). *A. Rev. Pl. Physiol.* **23**, 173-196.
Reeve, R.M. (1948). *Am. J. Bot.* **35**, 591-602.
Rickerd, J.P. (1977). *Sci. Dimens., Ottawa* **9**(4), 4-7.
Smith, D.L. (1973). *Ann. Bot.* **37**, 795-804.
Spencer, D. and Higgins, T.J.V. (1979). *Curr. Ad. Plant Sci.* **11**(3), Commentaries Pl. Sci. No.34, 1-15.
Sun, S.M., Buchbinder, B.U. and Hall, T.C. (1975). *Pl. Physiol.* **56**, 780-785.
Sutcliffe, J.F. and Bryant, J.A. (1977). *In* "The Physiology of the Garden Pea" (J.E. Sutcliffe and J.F. Pate, eds), pp. 45-82. Academic Press, London and New York.
Turner, J.F., Turner, D.H. and Lee, J.B. (1957). *Aust. J. biol. Sci.* **10**, 407-413.
Walbot, V., Clutter, M. and Sussex, I.M. (1972). *Phytomorphology* **22**, 59-68.
Wilson, C.M. (1975). *A. Rev. Pl. Physiol.* **26**, 187-208.

III.3 Scientific Control of Ripening in Tomato Fruit

PETER W. GOODENOUGH and E. JOHN WRIGHT

University of Bristol, Department of Agriculture and Horticulture, Long Ashton Research Station, Long Ashton, Bristol BS18 9AF, England

INTRODUCTION

In 1971 Drs G.E. Hobson and J.N. Davies wrote that storage investigations should be allied to analytical and organoleptic tests to ensure that the quality of the fruit does not diminish too rapidly. However, since that time the scientific control of ripening has not progressed very far. This paper discusses the problems of tomatoes while leaving out most of the storage work on pome fruits (see Knee, M., and Sharples, R.O. this Symposium, pp. 341-352).

The problem facing the scientist who investigates any of the control mechanisms in plants is very complex, but thanks to the development of molecular biology we can now begin to unravel the mechanisms which control changes in plant parts. Although some way behind discoveries in animal and bacterial systems, the study of the molecular aspects of regulation in plants represents the most important major area of research at the moment. The main developments are centred upon the possible control of gene expression, in our context the control of gene expression which changes an unripe fruit to a ripe fruit. Although the term "ripe" has no scientific status, it has come to mean a collection of biochemical changes which generally leads to a fruit which is at its most pleasant when eaten. Evolution has selected fruit which develop in this way and human selection rather than natural selection has refined the process. However, we now want to understand the fine biochemical control so that we can induce even more variation in the ripening pattern. But in order to pinpoint the areas of control we must know the enzyme changes that occur and where to look for control of these changes. The possible points of control of gene expression are shown in an earlier paper to this Symposium (Grierson *et al.*, pp. 179-191).

The use of inhibitors such as cycloheximide and puromycin (inhibitors of protein synthesis) and actinomycin D and 6-methyl purine (inhibitors of RNA synthesis) is vital in this research but their use is greatly criticized for their many side effects in the experimental system. However, if control of fruit ripening is to be achieved the control of the well known physiological processes must be known. The involvement of ethylene seems to be at the point of transcription and can be inhibited by cycloheximide. When the physiology of fruits which have a climacteric rise in respiration (Kidd and West, 1925) has been investigated in detail the events which give rise to ripening are usually synchronous. Thus, softening of the wall, disappearance of starch and malic acid, appearance of higher concentrations of monosaccharides and citric acid, loss of chlorophyll and appearance of other pigments are all occurring at the same time under the influence of ethylene. In the pome fruits, *de novo* synthesis of several enzymes involved in these changes has been largely proved but not in other fruits (Frenkel *et al.*, 1968).

The ripening of fruits relies on a trigger; in climacteric fruit it is ethylene which modifies protein synthesis in the transcriptional mode and produces a series of enzyme changes. These seem to be synchronous but at least one report has indicated how a separation, in time, is brought about in the ripening events (Dostal and Leopold, 1967).

MEANS OF CONTROLLING RIPENING

The control of these biochemical changes must hinge on stopping the trigger mechanism acting. Although recent evidence confirms the transcriptional control of ripening (Grierson *et al.*, this Symposium pp. 179-191; and Rattanapanone *et al.*, 1978) there is no detailed evidence on the mechanism that stimulates ethylene formation. One must divide the fruits into those of tropical and of temperate origin. The pome fruits of temperate origin can, in some cases, have ripening retarded by lowering the temperature which decreases the velocity constant of the reactions. However, fruits of tropical origin suffer a progressive change in their physiology at low temperatures (Lyons, 1973; Raison and Chapman, 1976). The activation energy of the mitochondria changes such that the Arrhenius plot develops a break at a certain temperature which eventually leads to severe chilling injury.

Thus, reducing the temperature below a minimum of 12°C (for tomatoes for instance) does not produce any significant gain. However, genetic variation is such that a breeding programme which has as its aim the slowing down of ripening after harvest is a possibility. Taking the tomato as an example, recent work has resulted in the incorporation of the *nor* or non-ripening gene into a commercial fruiting type. The mutants are still able to change colour, although not to such a great extent as our normal fruit, and respiratory behaviour and ethylene production is much reduced (Fig. 1). The cell wall degrading enzymes are also much reduced in the *nor*-mutant and the fruit remains firm. Details of sugar and acid changes are not given but the originators of this work (Tigchelaar and McGlasson, 1978) state that applications of growth regulators which normally induce fruit ripening are ineffective in inducing mutant fruit to ripen and say further that primary genetic control does not occur through known plant hormones. This statement presumably refers to our lack of knowledge about the triggering of ethylene and subsequent events in ripening. We can see that the F_1 hybrid between mutant and normal fruit is able to extend the shelf-life of tomato fruit considerably (Tigchelaar and McGlasson, 1978).

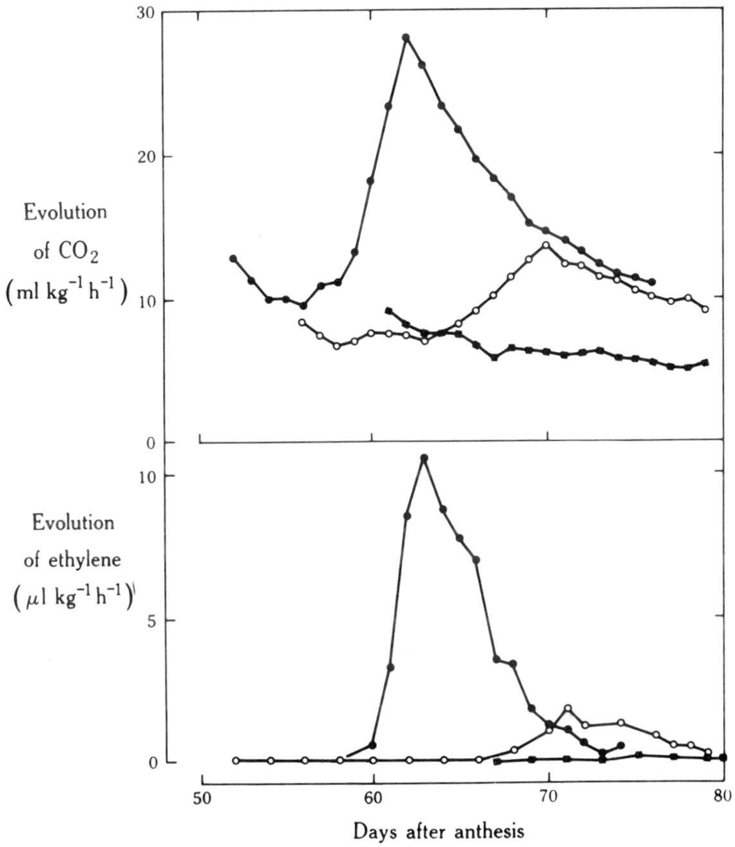

Fig. 1. Evaluation of ethylene and carbon dioxide from normal fruit during ripening, compared with a non-ripening mutant and the F_1 fruit from a cross between them (from Tichelaar and McGlasson, 1978).
Key: ●, normal; ○, F_1 hybrid; ■, non-ripening mutant.

If we turn to ways of controlling ripening during storage, then the biochemical reactions such as increase in respiration and evolution of ethylene must be slowed down. In the tomato the increase in respiration and increase in carbon dioxide evolution seems to come mainly from the greater activity of the citric acid cycle as opposed to decarboxylation of malate to pyruvate as found in the pome fruits. However, there has been little work on the relative amounts of carbon dioxide evolved by the two pathways during the climacteric and an active malic enzyme EC 1.1.1.40 has been shown to be present during ripening of the tomato (Hobson, 1974). In order to slow down the rise in respiration, either decreasing the concentration of reactants decreases the velocity at constant temperature or accumulation of the products of the reversible reactions progressively decreases the net rate of change, until a state of dynamic equilibrium is reached in which there is zero change. This obviously means that if oxygen concentration is lowered and carbon dioxide concentration is increased, then respiration is slowed down and the whole process is retarded. Before any practical work was done on the concept of retarding respiration, the law of mass action was the theoretical basis for respiration reduction. It was assumed that respiration was the oxidation of a sugar through at least partly reversible steps to give carbon dioxide and water (Burton, 1974). When converted to practical systems the theory worked well in the case of pome fruits where normally there could be a long storage phase in the absence of any special storage environment but has not been widely applied in the case of fruits like the tomato.

CONTROL OF RESPIRATION RATE

The cytochrome oxidase enzymes have the highest affinity for oxygen of plant enzymes and published figures of K_m values (substrate concentrations giving half maximal activity of the enzyme) towards oxygen, range from 3×10^{-6}M to 7×10^{-8}M in plant potato tubers. Other oxygen requiring enzymes have a much lower affinity and polyphenoloxidase has K_m values in the region of $1{\cdot}5$-6×10^{-4}M for oxygen. These low affinity oxidase enzymes will be affected much more quickly by depleted oxygen than the cytochrome oxidase proteins and oxidation of phenolics will slow down before production of adenosine triphosphate is affected and electron flow retarded. The influence of other oxidations (particularly phenolics) on the climacteric is not known but in cultivated mushrooms the uptake of oxygen during maturation is affected by an increase in polyphenoloxidase activity (Goodenough, 1978). The concentration of both carbon dioxide and oxygen will not necessarily be uniform throughout the tissue and several investigations of the distribution in pome fruits have shown a gradient from the periphery to the centre (Fidler and North, 1967; Brandle, 1968).

This would probably not be so in the case of the tomato but no investigations have, as yet, been undertaken. If we consider an external atmosphere of 3% oxygen and 5% carbon dioxide it is thought the internal atmosphere ranges from 1-2% oxygen and 6-7% carbon dioxide. In areas of limited diffusion, which probably do not exist in the tomato, then less than 1% oxygen and about 8-9% carbon dioxide may occur. The first figure for oxygen concentration gives adequate oxygen status for cytochrome *c* oxidase but the latter figure does not. There would be no possibility of low affinity oxygenase enzymes functioning. In pome fruits the respiratory climacteric is cancelled out under reduced oxygen conditions (Knee, 1971) and evolution of carbon dioxide and ethylene only occur at a very low concentration; this system employed a flow-through of gases rather than a static situation. Subsequent to long-term storage under reduced oxygen conditions a normal rise in respiration does not occur. When the low temperature of pome fruit storage is taken into account along with the low oxygen concentration, then the K_m may change at low temperatures to a value above the oxygen concentration. However, this remains to be tested in tomatoes stored at 12°C.

TOMATO FRUIT RIPENING IN CONTROLLED ATMOSPHERES

Originally the recommendations of Kidd and West (1933) were 5% oxygen and 5% carbon dioxide at 12°C. Subsequently tomatoes were stored at 3% oxygen and zero carbon dioxide (Parsons *et al.*, 1970) or 2·5% oxygen and 5% carbon dioxide (Eaves and Lockhart, 1961). However, few studies of compositional change have been included in these studies. There has been no attempt to follow the development of enzymes intimately involved in ripening changes. In this area the activities of polygalacturonase and invertase have not been monitored although Salunkhe and Wu (1973) using a number of oxygen and nitrogen concentrations examined sugar, acid and starch concentrations as well as colour change. As the oxygen content was lowered the time taken for ripening to occur in store increased. When colour change began, synchronous changes in starch degradation, monosaccharide appearance also began. The information on monosaccharide change is in direct contrast to Goodenough and Thomas (1980) who found that, at the end of eight weeks' storage in 2·5% oxygen and 4% carbon dioxide, monosaccharide and organic acid concentrations had fallen to lower values than found in newly harvested mature-green or red fruit. Of six cultivars tested at least one was exceptional and did not react by losing monosaccharides and organic acids during storage.

METHODS

The following experiment was designed to follow the behaviour of mature-green tomato fruit during storage in 5% carbon dioxide and 5% oxygen at 12°C. This concentration of gases and reduced temperature should allow a respiratory climacteric to occur in a very minor way. The time scale of changes is up to eight weeks rather than several months as in the case of pome fruits. Figure 2 shows the relationship of the main parameters measured. Accepting that theoretically the cytochrome oxidase enzymes could function normally, the concentrations of substrates, starch, glucose, fructose, citric acid and malic acid were measured. The enzymes phosphorylating glucose in the 6-position, isomerizing the phosphate from position 1 to position 6, and the enzyme converting glucose-6-P to fructose-6-P were measured electrophoretically. These enzymes enabled the storage compound starch and also glucose to be converted to phosphorylated intermediates for use in glycolysis. The contribution of the pentose-phosphate pathway is to provide reducing power as NADPH and to produce D-ribose-5-phosphate for synthesis of nucleic acid, and the first enzymes of this pathway (glucose-6-phosphate and 6-phosphogluconate dehydrogenase) were measured in order to show whether the 27% of glucose supposed to go by the pathway (Wang *et al.*, 1962) was constant during storage. The activity of two Krebs cycle enzymes, malate dehydrogenase and isocitrate dehydrogenase, were measured to show if the citric acid cycle operated during storage. The malate decarboxylation system using malic enzyme was similarly tested.

Finally NADPH and NADH (cytochrome *c* reductase) dehydrogenase were tested to review what concentrations of NAD and NADP are being regenerated during storage and the efficiency of the electron transport chain.

The fruit was sampled for these enzymes and substrates during normal ripening on the vine and after four, six and eight weeks of storage (mature green fruit only stored).

EXPERIMENTAL

Fruit for experimental work was grown at the National Vegetable Research Station, Wellesbourne, and the cultivation has been described previously (Thomas *et al.*, 1977). Some of the cultivars used were bred in Israel as long storage life fruit and were F.72, F.103, F.1649, and finally Naama. Sleaford Abundance was the normal control fruit. The cultivar Sonatine was also used for storage after normal growth in a commercial glasshouse.

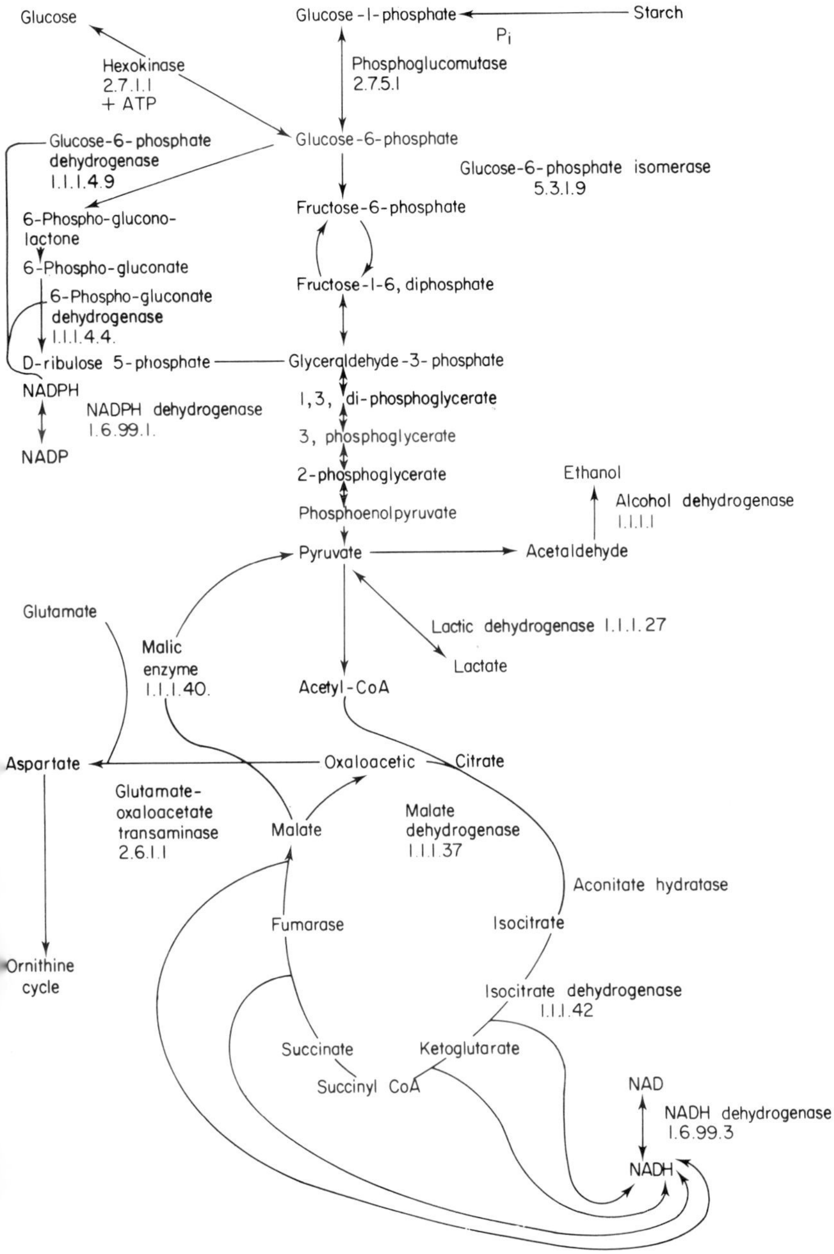

Fig. 2. General metabolic sequence chart showing enzymes that were examined during ripening and storage and the substrates that the enzymes act upon. Numbers refer to enzyme nomenclature list.

RESULTS

The fruits could be divided into two sets on the basis of normal ripening. Sonatine and Sleaford Abundance ripen off the vine normally while the four cultivars from Israel ripen much more slowly off the vine, the change in colour being particularly slow. Results for cultivars other than Sonatine are given in a later publication.

SOLUBLE AND INSOLUBLE SOLIDS

The change in aqueous ethanolic soluble and non-soluble solids is shown in Fig. 3. In some cultivars a higher solids content, before storage has been reported. However, there was a greater decrease in solids during storage of outdoor-grown cultivars than Sonatine but this stopped between 6-8 weeks of storage.

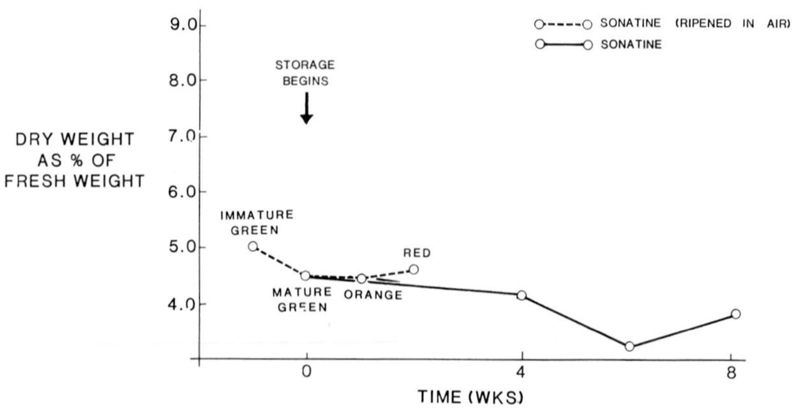

Fig. 3. Sum of soluble and insoluble solids (expressed as total dry weight) of Sonatine during storage at 12°C. The dry weight of Sonatine during ripening is also shown. All fruits were stored at the mature-green stage. Standard errors are encompassed within the points.

The reason for the greater fall in total dry matter in outdoor-grown cultivars compared to Sonatine was the greater decrease in the solvent insoluble material. Starch decreased similarly in all cultivars.

STORAGE POLYSACCHARIDES AND MONOSACCHARIDES

The normal changes during the climacteric, increase in glucose and fructose with a decrease in starch, occur in Sonatine ripened off the plant (Fig. 4a and b; dashed line). However, during the first four weeks of storage of mature-green fruit the same changes took place even though the colour change was retarded. After four weeks of storage a decrease in glucose and fructose concentration started. The initial increase in monosaccharides could be accounted for by the hydrolysis of starch.

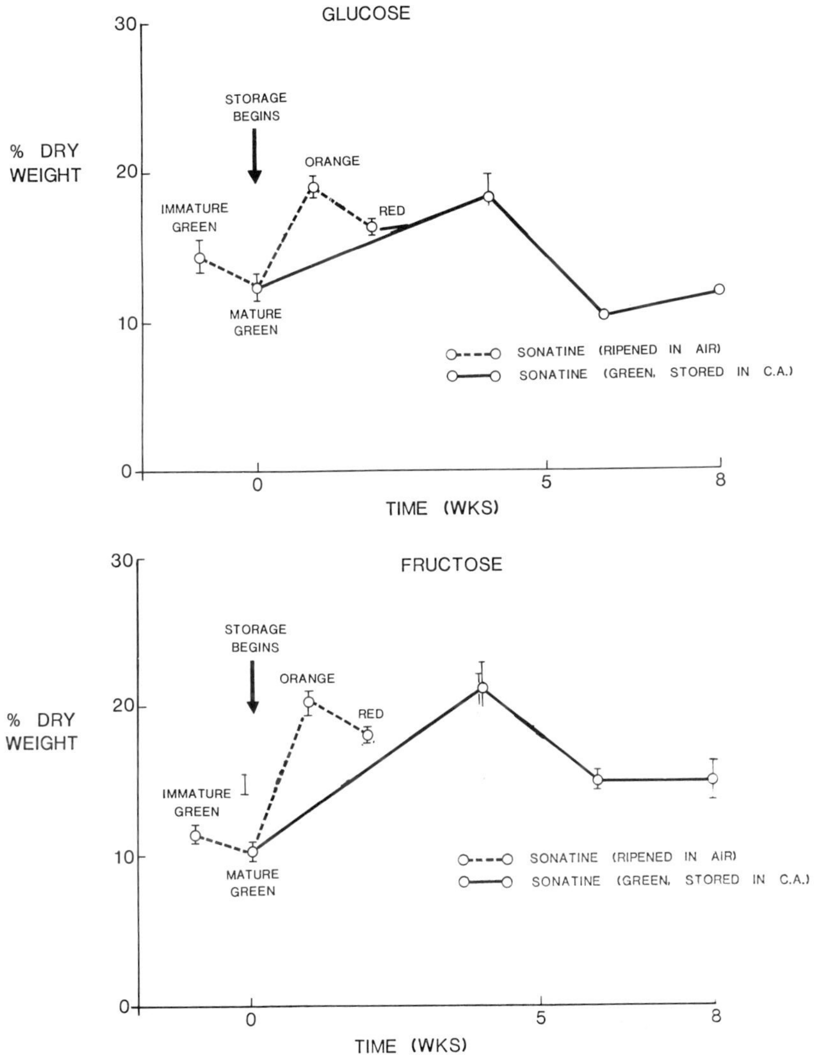

Fig. 4.(a) Glucose concentration (% dry weight) during normal ripening (dashed line) and eight weeks of storage. Sonatine also shown during ripening at 23°C. All storage at 12°C. Bars represent standard errors of the mean. (b) Fructose concentration, conditions as in Fig. 4a.

It was also found that in some cultivars there was a rapid and continuing rise in glucose concentration during the first six weeks of storage which was followed by a fall to initial values. The increase in Sonatine was highly statistically significant. However, fructose did not rise so quickly and only became significantly higher in tissue concentration after six weeks' storage (Fig. 4a and b).

ORGANIC ACIDS

The citric and malic acids were the most prominent acids in the ripening fruit, citrate increasing and malate decreasing (Fig. 5a and b; dashed line).

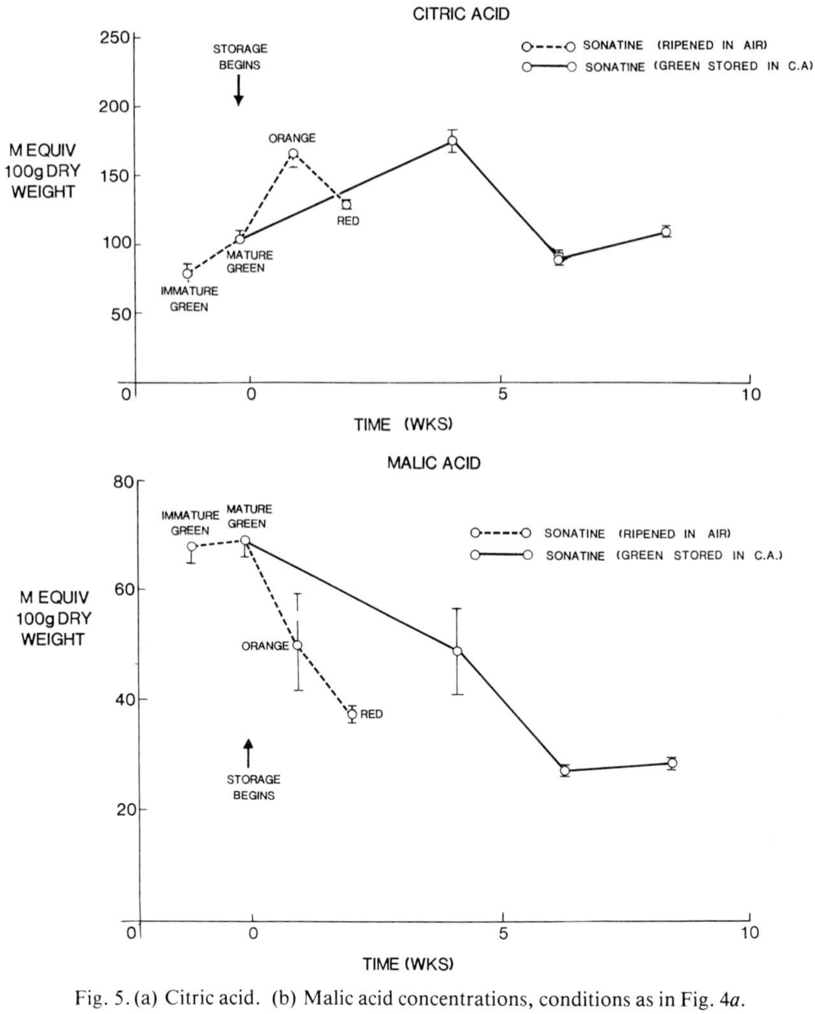

Fig. 5. (a) Citric acid. (b) Malic acid concentrations, conditions as in Fig. 4*a*.

This is possibly due to the activity of malate dehydrogenase (malate decrease) being greater than the activity of enzymes utilizing citrate. As the primary control point of the citric acid cycle is the condensation of acetyl CoA with oxaloacetate to yield citrate (citrate synthase), increased formation of oxaloacetate from malate will lead to increased citrate concentrations if the isocitrate dehydrogenase reaction (under allosteric control) does not increase in step. However, the production of carbon dioxide may be increased by decarboxylation of malate to pyruvate. There is an active malic enzyme EC 1.1.1.40 in tomato fruit and loss of malate may occur by this pathway.

After four weeks' storage the citrate had increased as in normal development and the malate had decreased (Fig. 5a and b; solid line). There was a fall subsequently in citric as well as malic acid. Thus there was normal control of organic acid catabolism during the first four weeks of storage as if ripening took place at a retarded rate. However, there was no colour change in the fruit and it can be proposed that a reduced respiration rate and evolution of ethylene was present. It may be that this is not the whole story as, in pome fruits, the decarboxylation of malate by malic enzyme may be more readily reduced by increased carbon dioxide content of the atmosphere than normal citric acid cycle respiration. Thus, in tomato fruit without such a well proven decarboxylating enzyme, respiration may rise during storage to a rate which is higher than in pome fruits under similar conditions.

COLOUR CHANGE

At four weeks of storage at 12°C the colour of all the fruit was still light green; however, the outdoor-grown cultivars were more bleached than Sonatine. When kept at 23°C in ambient atmospheres the fruit turned red as shown in Fig. 6.

ENZYMES OF SUGAR PHOSPHORYLATION IN CV. SONATINE

The enzymes were separated on polyacrylamide gels, stained histochemically for enzyme activity and peak areas estimated using a gel scanning device on a double-beam recording spectrophotometer. It was found that hexokinase, EC 2.7.1.1 (glucose→glucose-6-phosphate) decreased during storage until at eight weeks only trace activities could be detected. During normal ripening little change was detected (Fig. 7a). During his investigation of isoenzymes during ripening, Hobson (1974) did not include hexokinase. We have shown this enzyme has two main isoenzymes with two subsidiary enzyme forms. This enzyme was active throughout ripening and not only when starch was being metabolized.

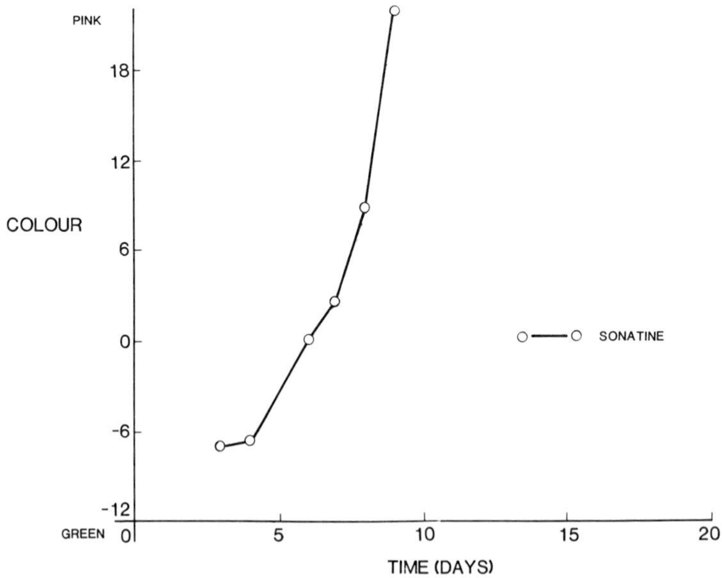

Fig. 6. Change in colour of fruit after 20 days storage in controlled atmospheres. Ripened at 23°C in air. Colour units +49·9 strong red; +22·4 light red; −15·0 light green.

The enzyme phosphoglucomutase EC 2.7.5.1 (glucose-1-phosphate→glucose-6-phosphate) seemed to increase in activity during storage and glucose-6-phosphate isomerase EC 5.3.1.9 (glucose-6-phosphate→fructose-6-phosphate) decreased slightly after eight weeks' storage. Thus phosphorylation of glucose seems to be retarded after eight weeks; this can be correlated with a reduced utilization of glucose and also a reduced production of ATP (see below, NADH dehydrogenase) (Fig. 7b). The four main isoenzymes previously reported for glucose-6-phosphate isomerase were detected but the six weak enzyme forms already documented (Hobson and Davies, 1976) were only seen in mature-green fruit.

ENZYMES OF THE PENTOSE-PHOSPHATE PATHWAY IN CV. SONATINE

Both 6-phosphogluconate dehydrogenase EC 1.1.1.44 (6-phosphogluconate→D-ribulose-5-phosphate) and glucose-6-phosphate dehydrogenase EC 1.1.1.49 (glucose-6-phosphate→6-phosphoglucose-lactone) decrease during storage, largely between 6-8 weeks. Again this would lead to reduced utilization of glucose, lowered production of reducing power and ribulose units for nucleic acid production (Fig. 8a and b). As in earlier work there were main isoenzymes but four were very clear as opposed to the three shown in Hobson (1974). There were usually two subsidiary isoenzymes.

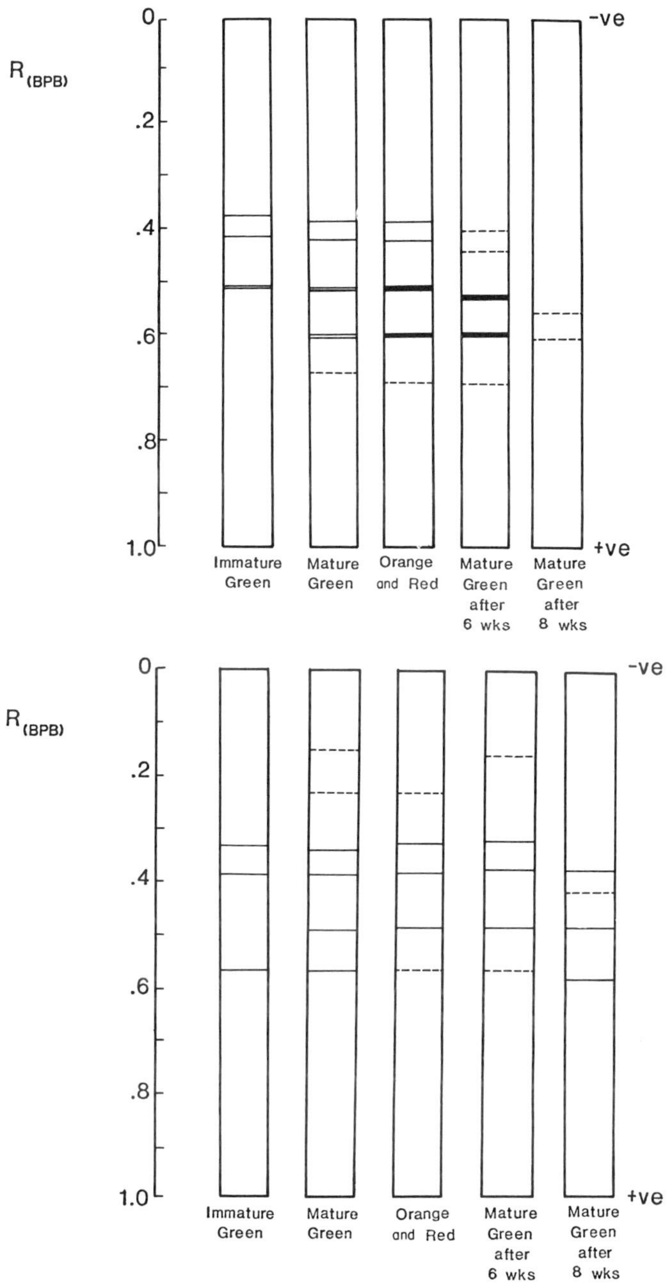

Fig. 7.(a) Electrophoretograms showing bands of hexokinase EC 2.7.1.1 activity at various stages of ripeness and also during storage of mature green tomatoes for 6 or 8 weeks. (b) As Fig. 7a, except glucose phosphate isomerase EC 5.3.1.9.
$R_{(BPB)}$ = Relative mobility using bromophenol blue as standard.

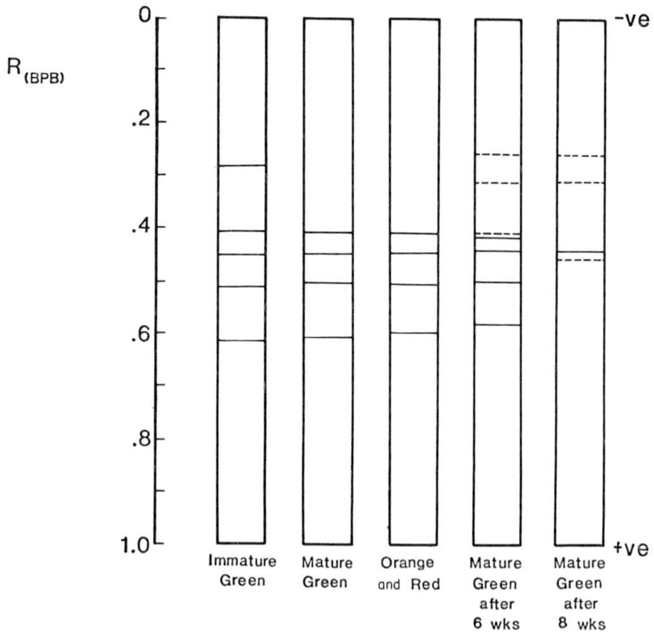

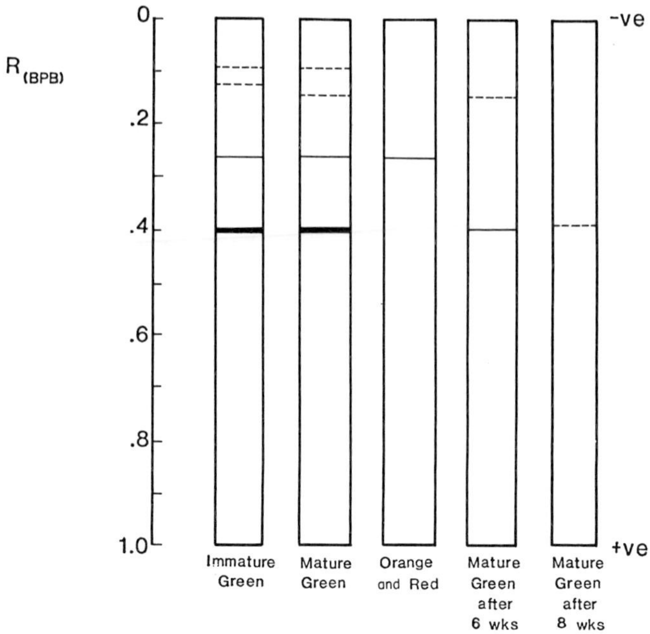

Fig. 8(a) As Fig. 7a, except 6-phosphogluconate dehydrogenase EC 1.1.1.44. (b) As Fig. 7a, except glucose-6-phosphate dehydrogenase EC 1.1.1.49.
Key as Fig. 7.

ENZYMES OF THE KREBS CYCLE AND MALATE DECARBOXYLATION IN CV. SONATINE

Malate dehydrogenase EC 1.1.1.37 (malate→oxaloacetate), isocitrate dehydrogenase EC 1.1.1.42 (isocitrate→ketoglutarate) and malic enzyme EC 1.1.1.40 (decarboxylation of malate to pyruvate) did not seem to alter activity during storage and were highly active after eight weeks' storage (Fig. 9).

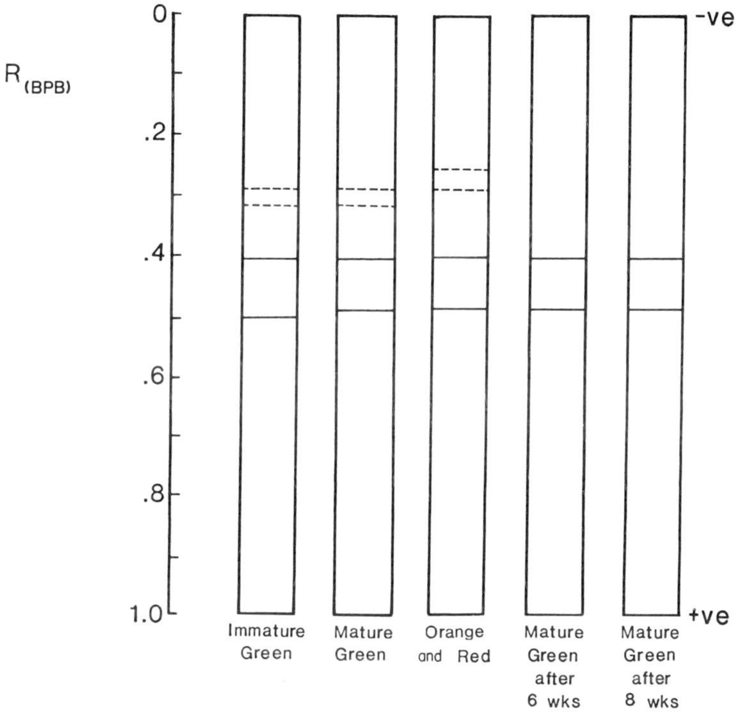

Fig. 9. As Fig. 7a, except malate dehydrogenase EC 1.1.1.37. Key as Fig. 7.

ENZYMES OF THE ELECTRON TRANSPORT CHAIN IN CV. SONATINE

NADH dehydrogenase and NADPH dehydrogenase EC 1.6.99.3 and EC 1.6.99.1 ($NADH_2$ or NADPH + acceptor (FMN)→NAD^+ or $NADP^+$ + reduced acceptor ($FMNH_2$)) are most active at the mature-green stage and fall in activity to a very low concentration at eight weeks. As NADH dehydrogenase is the first enzyme in the electron transport chain then an apparent fall in activity, due either to reduced synthesis of the enzyme or inhibition of activity (unlikely after polyacrylamide separation), will greatly reduce the generation of ATP and deplete the overall rate of respiration and ability of the fruit to ripen normally after storage (Fig. 10a and b).

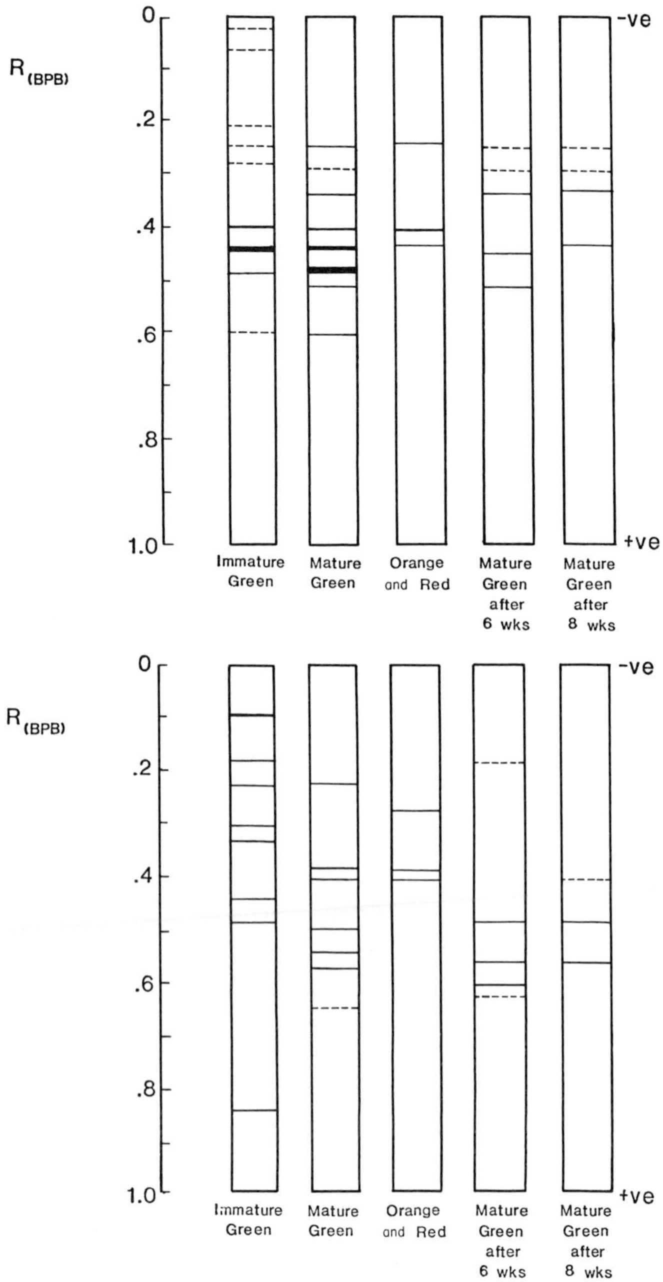

Fig. 10.(a) As Fig. 7a, except NADH dehydrogenase EC 1.6.99.3. (b) As Fig. 7a, except NADPH dehydrogenase EC 1.6.99.1.
Key as Fig. 7.

ENZYMES IN THE ISRAELI CULTIVARS

All the enzymes mentioned previously for Sonatine were also monitored during storage of the Israeli cultivars. Although the rate of change in organic acids, monosaccharides and starch was similar in both Sonatine and the other cultivars, this was not so when enzyme activities were compared. In the case of the NADH dehydrogenase activities, there was very little decrease in number and concentration of the individual isoenzymes. The number and position of bands was different from that found in Sonatine and these are to be described in a later publication. However, the ability of these cultivars to maintain either protein synthesis or retard the decrease in enzyme half-life prevents ripening and subsequent deterioration setting in as fast as it does in Sonatine. Long-life during storage can then be shown to depend on the way in which vital enzyme activities are preserved. This could be by either an efficient protein translation or by transcription or by enhanced half-life of proteins (turnover of enzymes).

DISCUSSION

In the experiments described the cultivars can be grouped in several different ways. Firstly there is the distinction of the growth conditions, one type of result is based on plants grown in a controlled environment (glasshouse) and subjected to higher temperatures than the second group of cultivars. The second group were grown in the field in a moderate English summer (maximum 23±2°C). Another distinction which can be drawn is the genetic selection of these cultivars. Sonatine was selected as a high yielding cultivar whereas the cultivars from Israel were selected to be long-storage life fruit with slow colour change but not incorporating a non-ripening gene. There are other ways of grouping the cultivars based on soluble solids content or sugar:acid ratios but these usually fall into the latter classification described above. The amount of dry weight contributed by soluble solids in the outdoor-grown cultivars was about the same as Sonatine but insoluble solids were greater in the outdoor-grown fruit. This appears to be a feature of tomato cultivars grown in the field for processing (Goodenough and Thomas, 1980) and transcends the environmental factors, such as either being grown in a protected system or in the field. The values for concentration of citric acid in unstored Sleaford Abundance are twice those found in F.103, but monosaccharides are similar initially whereas Sonatine has only 66-70% of the amount of monosaccharides in F.103 and Sleaford Abundance. It was possible to show that the concentrations of monosaccharides in mature-green Sleaford Abundance and F.103 were 1·65% and 2·25% of the fresh weight respectively. After four weeks' storage this had risen to 1·94% and 2·94% of the fresh weight respectively. This can be compared with figures of around 3·0% of fresh weight as monosaccharides for ripe glasshouse fruit (Davies and Winsor, 1969) and cv.Cal Ace grown and ripened in the field in the United States (Stevens *et al.*, 1979). However, these values had fallen to 1·46% monosaccharides for F.103 after eight weeks' storage and 1·05% in Sleaford Abundance. At this stage of storage the monosaccharide content of both cultivars is well below that of ripe fruit harvested in the field or glasshouse. As the values quoted after eight weeks' storage did not increase when the fruit were ripened, it can be seen that the Israeli cultivars obviously have the advantage over normal commercial cultivars as far as the two main taste components are concerned.

From the data in Figs 4a,b and 5a,b it can be seen that an increase in glucose, fructose and citric acid concentrations with a fall in malic acid concentrations occurred after four weeks of storage while at that time the fruit had not changed colour. This is a separation of the normal processes occurring during ripening of the fruit and may indicate that a lower oxygen content would be needed to prevent the redistribution of sugars and acids before the fruit are removed from store.

It has been quite rigorously shown that ripening involves synthesis of polygalacturonase, invertase and possibly the enzymes synthesizing lycopene (see Grierson *et al.*, this Symposium p. 187). Many other enzymes seem to fall in activity at this time. The enzymes controlling starch degradation seem to be present throughout ripening (Clements, 1970) as do those enzymes involved in regulation of citric and malic acid concentration that have been investigated. The work described here indicates that enzymes concerned in glycolysis and the citric acid cycle are present during long-term storage and there is no apparent change in numbers of isoenzymes. However, a change in content of organic acids implies a change in the reaction rate of several enzymes. Recent work on phosphofructokinase (Rhodes, personal communication) has tried to elucidate the regulation of this enzyme in relation to the proposal of Chalmers and Rowan (1971) that the changes in ripening are largely regulated at this point of glycolysis. There are very few reports on the regulation of other enzymes in the glycolytic and Krebs cycle in tomato but in other fruit malic enzyme is regulated by concentrations of organic acids.

During tomato fruit ripening, control of organic acid concentrations must be vested in regulation of the activity of some of the enzymes of the citric acid cycle and this type of regulation is clearly differentiated from enzyme synthesis. Synthesis produces new cell-wall degrading enzymes and possibly enzymes forming lycopene. This theory envisages a two-stage regulation of tomato fruit ripening. Stage I seems to be involved with regulation of enzymes of carbohydrate catabolism. Stage II, slightly behind in normal ripening, involves synthesis of mRNA and then enzyme protein. The appearance of polygalacturonase is Stage II controlled (see Grierson *et al.*, p. 186 this Symposium) but it is still in doubt whether colour development is enzyme synthesis or regulation.

This difference between the two stages is clearly shown in prolonged storage where a considerable time elapses between changes in regulation and the second stage. This is in contrast to the work of Salunkhe and Wu (1973) and similar to the two-stage theory of Tigchelaar *et al.* (1978). However, the latter workers, on the basis of results with non-ripening mutants, state that synthesis of enzyme protein is the first step with subsequent ripening events depending on the hydrolases formed *de novo.* We feel that the experiments on control of ripening by controlled atmospheric storage show that the regulation of enzymes occurs before *de novo* enzyme synthesis, but further experiments on the formation of polygalacturonase and invertase during storage are needed. Similarly, the role of zero carbon dioxide and low oxygen concentration could be assessed to improve control of the first stage of ripening.

Further research into methods of improving control of ripening should help both the industry and ourselves to understand the fundamental problems of fruit ripening.

REFERENCES

Brandle, R. (1968). *Ber. schweiz. bot. Ges.* **78**, 330-364.

Burton, W.G. (1974). *Ann. appl. Biol.* **78**, 149-168.

Chalmers, D.J. and Rowan, K.S. (1971). *Pl. Physiol.* **48**, 235-240.

Clements, R.L. (1970). *In* "The Biochemistry of Fruits and Their Products" (A.C. Hulme, ed.), Vol. I, pp. 159-177. Academic Press, London and New York.

Davies, J.N. and Winsor, G.W. (1969). *J. hort. Sci.* **44**, 331-342.

Dostal, H.C. and Leopold, A.C. (1967). *Science NY,* **158**, 1579-1580.

Eaves, C.A. and Lockhart, C.L. (1961). *J. hort. Sci.* **36**, 85-92.

Fidler, J.C. and North, C.J. (1967). *J. hort. Sci.* **42**, 189-206.

Frenkel, C., Klein, I. and Dilley, D.R. (1968). *Pl. Physiol.* **43**, 1146-1153.

Goodenough, P.W. (1978). *Phytochemistry* **17**, 633-636.

Goodenough, P.W. and Thomas, T.H. (1980). *Ann. appl. Biol.* **94**, 445-455.

Hobson, G.E. (1974). *Phytochemistry* **13**, 1383-1390.

Hobson, G.E. and Davies, J.N. (1971). *In* "The Biochemistry of Fruits and Their Products" (A.C. Hulme, ed.) Vol. II, pp. 437-482. Academic Press, London and New York.

Hobson, G.E. and Davies, J.N. (1976). *J. Sci. Fd Agric.* **27**, 15-22.

Kidd, F. and West, C. (1925). *Rep. Fd Invest. Bd 1924*, 26-34.

Kidd, F. and West, C. (1933). *Rep. Fd Invest. Bd 1932*, 209-211.

Knee, M. (1971). *J. Sci. Fd Agric.* **22**, 368-371.

Lyons, M.S. (1973). *A. Rev. Pl. Physiol.* **24**, 445-466.

Parsons, C.S., Anderson, R.E. and Penney, R.W. (1970). *J. Am. Soc. hort. Sci.* **95**, 791-794.

Raison, J.K. and Chapman, E.A. (1976). *Aust. J. Pl. Physiol.* **3**, 291-299.

Rattanapanone, N., Spiers, J. and Grierson, D. (1978). *Phytochemistry* **17**, 1485-1486.

Salunkhe, D.K. and Wu, M.T. (1973). *J. Am. Soc. hort. Sci.* **98**, 12-14.

Stevens, A.M., Kader, A.A. and Albright, M. (1979). *J. Am. Soc. hort. Sci.* **104**, 40-42.

Thomas, T.H., Gray, D. and Drew, R.L.K. (1977). *Ann. appl. Biol.* **87**, 117-121.

Tigchelaar, E.C. and McGlasson, W.B. (1978). *CSIRO Fd Res. Q.* **38**, 22-24.

Tigchelaar, E.C., McGlasson, W.B. and Buescher, R.W. (1978). *HortScience* **13**, 508-513.

Wang, C.H., Doyle, W.P. and Ramsey, I.C. (1962). *Pl. Physiol.* **37**, 1-7.

Section III
Quality at Harvest—Harvesting and Handling

III.4 The Selection of Apples and Pears for Long-term Storage

C.D. WALKER

MAFF, ADAS, East Malling Research Station,
East Malling, near Maidstone, ME19 6BE, England

INTRODUCTION

In the UK we grow dessert apples and pears towards the northern limit for their successful commercial culture and in consequence the consistent production of good quality fruit is never easy. Thus in 1978 our orchards produced too many small apples, the result of a generous fruit-set combined with a cool summer. Possibly our farming methods were at fault. It is also possible that we have been lulled into assuming that Cox's Orange Pippin will always command a market premium, and that we failed to pay enough attention to fruit thinning and other cultural items. But above all the season demonstrated the prime importance of climate, site and soil, and how essential it is for us to limit our apple and pear production to the really suitable locations. Whenever we neglect these basic rules the penalty is small crops, poor quality fruit and a completely non-competitive situation in relation to the rest of the EEC. The control of fruit quality starts at the planning stage before we even plant a tree.

Naturally we must also assist nature by designing a good orchard where the combination of tree density, rootstock and cultivar is so balanced that the grower obtains the best answer for his particular circumstances. We then add our skills in terms of husbandry, pest and disease control and the detailed management that go to make up a good orchard where we can look forward to harvesting high quality fruit. This is the traditional area for ADAS operations and we pay particular attention to ensuring that our fruit advisers can help growers with these basic decisions. We are dealing with a perennial crop and therefore cannot afford the luxury of mistakes.

In considering harvest quality we must assume that all these vital hurdles have been successfully overcome and we are left with the harvest operation and how to ensure that what looks a good quality apple on the tree can be turned into a useful profit. I would like, therefore, to concentrate on one or two areas where ADAS, in conjunction with East Malling Research Station, is trying to help the fruit grower to control the harvest operation more effectively and assist him to make the best possible use of his fruit, particularly in relation to long-term storage.

THE PICKING DATE

This is an important but surprisingly controversial subject since there is no one infallible test that will tell the fruit grower when his fruit has reached the ideal stage to be picked. It is generally impossible to harvest all the crop at the optimum moment because of management problems, so compromise is always necessary, particularly with our main cultivars where large orchard areas are involved. The harvest for any one cultivar must often start early and finish late, even though we know that such harvest action may lead to storage problems. The grower then attempts to control matters by marketing early or late-picked fruit as soon as possible and he reserves the cream of the crop in terms of both fruit quality and picking date for long-term storage. The classic example of this dilemma was seen in 1978 with Bramley's Seedling where it is very easy to leave things too late. Such fruit always stores badly, it is more subject to storage breakdown and inevitably goes into early senescence. Similarly, late-picked Cox rapidly become yellow and over-mature, whilst fruit that is picked too early may be susceptible to shrivelling and the disorders, scald and core flush.

The determination of the correct picking date is therefore vital. It cannot yet be called a science, but today, in addition to the normal signs of ripeness such as colour and the ease of picking from the tree, there are one or two tests that we now employ to help the grower make the right decision.

One test is to try to determine when the fruit is about to reach its climacteric. Briefly this is the time when the apple has reached is maximum development and it is beginning the final ripening process. At this stage there is a sudden rise in the respiration rate, the apple becomes fully ripe and if picking is delayed, over-mature. For long-term storage most cultivars, apart from the very late keepers, are best picked just prior to this period. The problem is to know when the climacteric is due since picking gangs must be assembled, bulk bins laid out and the general harvest operation set in motion. The measurement of respiration rates during the two or three weeks before harvest will at least give us some indication of when we are approaching this important period. This is a laboratory test and is now being used at East Malling to help solve this particular problem.

Two other possible scientific aids exist. One is to measure the internal production of ethylene which increases as the fruit nears its climacteric, and the other is to try to monitor the changes in starch content within the fruit since it is during the climacteric period that the starch starts to change to sugar. The determination of internal ethylene concentrations is again a laboratory technique but the approximate measurement of starch content by the Starch Iodine Test is a useful tool for both grower and the ADAS adviser.

THE STARCH IODINE TEST FOR FRUIT MATURITY

This test is particularly useful with Conference pears. The idea is to start measuring the starch content about two to three weeks before the estimated picking date by cutting pears in half and then immediately treating the cut surface with a solution of 1% iodine in 4% potassium iodide. The starchy areas show up with a blue stain. During the latter stages of growth the starch content reaches a maximum and then about 10 days before the climacteric it begins to disappear; when the area of blue staining has decreased by about one-third, the fruit is in an ideal state for long-term storage and should be picked. This test gives an excellent guide for harvesting Conference pears but is less reliable with other cultivars. In addition each cultivar shows its own particular starch pattern. We also use but do not completely rely on the test with Cox, but with late cultivars such as Golden Delicious or Crispin, it is not so effective since they are picked well before their climacteric.

In 1978 using information obtained from all these tests together with informed opinion, East Malling and ADAS sent out weekly notices to all fruit advisers and the main commercial interests giving guidance on the correct picking dates for apples and pears destined for long-term storage. Naturally the individual grower must use such information with discretion since the situation of the orchard, particularly its height above sea level will have an effect, and one date cannot be accurate for the whole country.

THE ADAS FRUIT ANALYSIS SERVICE

We grow fruit to the best of our ability and we safeguard quality by careful handling and picking at the correct date. In spite of these precautions growers know that heavy losses can occur in long-stored fruit from disorders such as bitter pit and internal breakdown, and that these losses will occur in fruit that from visual appearance looked perfect when it was placed in store.

As a result of many years of research we now know that such breakdown disorders are liable to occur when the chemical composition of the fruit is out of balance; particularly in relation to the main nutrients N, P, K, Ca and Mg. Soil and leaf analysis help to determine whether the whole tree's nutrition is satisfactory but neither of these methods will give us worthwhile information about the fruit at harvest.

The ADAS fruit analysis service is now available to assist the fruit grower who must decide which samples of fruit are best suited for long-term storage, which samples could be improved by pre-storage drenches and which samples are unlikely to keep and should be marketed as soon as possible. The service which was first introduced in 1971 has proved its worth; it is now available to all apple growers and in 1978 about 1400 samples were submitted for analysis.

To date most of the research in England has concentrated on our main cultivars Cox and Bramley's Seedling, but the technique can also be used with Spartan and Egremont Russet and in time it may become possible to make recommendations for other cultivars.

It is well known that fruit nutrient status can change quite rapidly around the harvest date. Thus heavy rainfall may produce a burst of vegetative growth and an increase in fruit size, effects that will lead to a depression in fruit calcium levels. The analysis results will, therefore, only be valid for fruit analysed within a week or two of the harvest date and close co-operation between the fruit grower and the ADAS Laboratory is essential if the best use is to be made of the available facilities.

We prefer the fruit grower to book laboratory time early in the summer. He can then be given instructions covering the sampling procedure and can make prior arrangements to bring the fruit to the laboratory on the agreed date. Essential features of the system are:

(i) Sample as close to harvest as possible, allowing one week for the results to be processed.

(ii) Up to 14 days before harvest the changes in fruit nutrient status are not usually significant. The ADAS monitoring system where certain orchards are kept under close observation will advise when this is not the case and of any necessary adjustments in interpretation of the results that should be made.

(iii) If samples are taken more than a fortnight away from harvest it is advisable to re-sample a proportion at harvest. The original samples will give the relative order of risk and the second sampling will confirm the degree of that risk.

(iv) To avoid disappointments and delay, it is best for the grower and laboratory to agree on a mutually convenient date for the samples to be sent in for analysis.

THE INTERPRETATION OF THE RESULTS

Harvest fruit analysis can be used to predict the storage potential and to rank the keeping quality of apples from various orchards. It cannot guarantee that disorders will not occur but it should indicate the degree of risk of the fruit succumbing to physiological disorders, particularly bitter pit and breakdown.

The fruit grower is, therefore, provided with an analysis of his fruit, together with some guide notes on how the figures should be interpreted. He must then use his own judgement on which fruit is worth storing or he can discuss further with ADAS.

The system now operates smoothly and in addition to directly helping the individual grower it is gradually building up a large bank of information which we hope can be used to sharpen up the interpretations and extend the system to other long-storing cultivars. The present recommendations are set out in the Appendix.

Finally it must be re-emphasized that while such aids help to keep stored fruit in good condition, the real factors affecting quality lie back in the orchard; the climate, site and soil, and above all the ability of the fruit grower.

APPENDIX—THE INTERPRETATION OF FRUIT ANALYSIS

The standards in Table I represent the optimum concentration for the long-term storage of apples picked at the correct time, cooled rapidly and stored at the recommended temperature. They refer specifically to Cox and Bramley (treated with a suitable anti-scald chemical) and should only be used for other cultivars with caution.

NUTRIENTS

Nitrogen (*N*). As the N content increases the fruit becomes more susceptible to rotting, while loss of firmness, poor skin finish, poor red colour can also result. Above 80 mg N/100 g the risk of these disorders is high in Cox. In Bramley, high N leads to more green colour but also a risk that firmness may be adversely affected.

Phosphorus (*P*). As P falls below 11 mg/100 g in Cox, there is an increased risk of fruit losing its firmness and developing breakdown particularly at low Ca levels. In Bramley, the P threshold for breakdown appears to be lower.

TABLE I

Optimum concentrations (mg/100 g fresh weight) of nutrients for long-term storage of apples

	N	P	K	Mg	Ca (min.)
Cox's Orange Pippin	50-70	11·0 (min.)	130-160	5·0	4·5[a] 5·0[b]
Bramley's Seedling	60 (max.)	9·0 (min.)	105-115	5·0	4·5[a] 5·0[b]

[a]For controlled atmosphere (CA) storage (Cox in 2% O_2 until late March: Bramley in 8% CO_2 until May).
[b]For storage in air at recommended temperatures.

Potassium (*K*). The risks of bitter pit, Gloeosporium rotting and core flush increase as the level of K increases. Bitter pit is greater the lower the Ca level in relation to the K content. Generally fruit flavour and acidity increase with increasing levels of K. Thus, if Ca levels are adequate (>5·5 mg/100 g), high levels of K may be advantageous in terms of fruit quality.

Magnesium (*Mg*). High levels of Mg increase the susceptibility of fruit to bitter pit, especially where Ca levels are marginal.

Calcium (*Ca*). Calcium levels of >5 mg/100 g are necessary to maintain high quality throughout long-term storage. The storage potential will be modified however in relation to the other elements, especially K and P. However, *disorders* associated with *low calcium* (bitter pit in particular) are more likely to occur in apples of a given Ca status which are stored in air rather than in CA. Thus fruit which meets the standards for N, P, K and Mg, and has Ca levels in the range of 4-5 mg/100 g should not be rejected for mid-term CA storage as they are unlikely to develop commercially significant levels of bitter pit or breakdown: this much lower risk is reflected by the dual CA standards given in Table I.

The Ca level should also be considered in relation to fruit K as a guide to the risk of bitter pit and susceptibility to Gloeosporium. If K is high and Ca marginal, the risk of bitter pit is increased. As a rough guide, a K/Ca ratio in excess of 30:1 in air-stored Cox or CA-stored Bramley, or 25:1 in air-stored Bramley is likely to lead to commercially important losses due to bitter pit. Where Ca levels are marginal (3·5-4·0 mg/100 g) and fruit P is also low (<9 mg/100 g), both Cox and Bramley are more susceptible to breakdown. In such cases, Cox should be marketed early and Bramley stored at a higher temperature (average 40°F; 4·5°C) and sold slightly earlier.

The incidence of senescent breakdown is greater in late-picked fruit for any given levels of P and Ca.

Susceptibility to low temperature breakdown in Cox stored at 38°F (air) and 39°F (2% O_2), although less common than in Bramley at 39°F, is also partly due to low Ca and P concentrations.

The standards given for Ca and K levels in Cox also apply to Egremont Russet, and preliminary data for Spartan indicates that fruit Ca levels should exceed 4·0 mg/100 g for storage beyond December.

III.5 The Design of New Harvesting Machines—Criteria in the Engineer's Solution

W. BOA

National Institute of Agricultural Engineering,
Wrest Park, Silsoe, Bedford, MK45 4HS, England

INTRODUCTION

In the development of new machines and methods for harvesting fruit or vegetables for the fresh market or for processing, the engineer seeks to improve on the rates of working and operating costs of existing methods. It is also desirable that he should be able to demonstrate in monetary terms that his development is likely to be advantageous to growers who may buy it.

At the outset it appears that costing the operation of harvesting machines will present no insuperable problems. Rates of working can easily be measured and provide the starting point for calculating operating costs. Capital and labour costs are also fairly simple to determine and there are established bases on how they should be charged. Losses are also measurable and their value can be assessed in accordance with production costs. Difficulties only arise when an attempt is made to include loss of product quality in the calculations.

Quality assessment frequently depends on subjective judgements against undefined standards. It is therefore unrealistic to attempt to translate changes in it into monetary terms. It is usually therefore treated by engineers as a separate entity and is best considered under three headings: (i) crop aspects; (ii) mechanical damage; and (iii) product contaminants.

This paper seeks to analyse under these headings the criteria against which the engineer judges how changes in machine design can effect improvements in product quality. It also gives examples of instances where engineers working on close co-operation with plant scientists, growers and processors have achieved results which none could have achieved alone.

CROP FACTORS

It is commonly believed by engineers that the development of satisfactory harvesting machines would proceed more rapidly if suitable crop cultivars were bred. To some extent this belief may be justified by citing the example of the NIAE Universal Vegetable Harvester (Boa, 1966) which was very simple, produced low crop damage and relied on the existence of F_1 hybrid and other cultivars suited to once-over harvesting.

Even maturing may be taken as a pre-requisite for successful mechanized harvesting of most vegetable and fruit crops, and screening new cultivars with suitable growth habits to ensure that they produce satisfactory once-over yields at acceptable harvest dates is likely to continue to be a very important contribution towards successful mechanized harvesting. Vining peas are a well-known example where annual trials are conducted to assess the once-over harvest yield. Similar trials have also been undertaken commercially to assess the capability of tomato cultivars to produce 90% mature fruit within the number of degree-days available between seeding and optimum harvest date (Stilwell and Portas, 1977).

Breeding cultivars specially for machine harvesting is obviously a possibility and strawberry breeders in the UK who have been keeping themselves aware of the engineer's requirements since about 1966 are now offering seedlings for machine harvesting trials (Kemp, 1978). According to Strobel *et al.* (1969) most American plant breeding programmes in tomatoes and peppers have been involved in varying degrees with the perfection of cultivars suitable for mechanical harvesting. The first tomato cultivar which was satisfactory for mechanical harvesting was introduced in 1962 after 19 years of development, of which about five years was spent in the development of a concept of the characteristics desirable in such cultivars.

Breeding specialized cultivars for machine harvesting, however, has not always been successful. Trials with potato cultivars which had been selected because the tubers remained firmly attached to the haulm showed early signs of success (Boa, 1967) but were later abandoned in favour of methods likely to be successful more quickly (Carruthers, 1973). In this connection it is noteworthy that engineering development and plant breeding operate on very different time scales which makes co-operation between the two disciplines difficult when a problem is to be solved. The plant breeder is likely to take at least ten years to develop and test a new cultivar to the stage where it can be released for commercial use. He then expects it to remain successful for a further ten years. The engineer rarely has more than five years to develop a new machine and is likely to be surprised if it is not superseded after five years in commercial use. It is possible that the example given of tomato harvester development was successful because mechanical harvesting would have been completely impossible without new cultivars.

Long-term co-operation between plant breeders and engineers is, however, to be encouraged and, provided the plant breeder restricts himself to general topics such as once-over harvest yields and resistance to mechanical damage, is likely to be very useful. If, however, the plant breeder seeks to solve more specialized problems he may often find that the engineer has solved the problem and produced another before the new cultivar is ready for release.

Although co-operation between the plant breeder and the engineer aimed at improving the quality of mechanically harvested fruit and vegetables is likely to be successful only in a few instances, there are several examples where co-operation with plant physiologists and horticulturists has been very successful and others where success in the future seems very likely.

One example where successful co-operation has solved a problem in a vegetable crop is methods of reducing clod populations in potato crops (Shotton, 1970) to give lower damage at harvest. For the future, research on fluid drilling of pre-germinated seeds (Salter, 1978), on more precise methods of dry seeding (Bufton, 1977) and on transplanting (Boa *et al.*, 1980) may all be keys to allow a wider range of vegetable crops to be once-over harvested.

Experiments on in the row spacing of blackcurrants, which were carried out at Luddington Experimental Horticulture Station (Ingram, 1973), were of great assistance in the development of harvesters without excessive losses of fruit. Similarly, spacing experiments with strawberries (Woodward, 1972) suggested how once-over harvests can be timed without incurring the penalty of including too much immature and overmature fruit in the sample. For the future it is possible that the canopy method of growing raspberries (Dunn *et al.*, 1976) now being investigated at East Malling may enable mechanical harvesters to work with minimal fruit and cane damage.

A frequently quoted example of a horticulturist helping to improve the quality of machine-harvested top fruit concerns cherries. It was demonstrated that reducing the number of branches in the trees reduced fruit damage during harvesting by shaking operations without reducing the yield (Larsen, 1969). For the future, meadow orchards and hedgerow orchards pioneered by Long Ashton give hope that means can be found of harvesting apples without excessive damage (Child and Le Flufy, 1979) (Fig. 1).

MECHANICAL DAMAGE

Mechanical damage is probably the most important quality facet that affects the engineer who attempts to develop fruit and vegetable harvesting machinery. Its importance cannot be over-stressed and extends to almost every crop and all crop uses. With crops that are to be processed, damage can increase preparation costs; with those that are to be stored, lesions and breakages allow excessive evaporation and provide openings for organisms causing rot and taints. Damage to crops that are to be sold fresh detracts from their appeal and reduces prices.

Fig. 1. The NCAE/LARS over-row apple harvester combing fruits from a narrow hedgerow of apple trees.

Fig. 2. The SIAE dual web potato harvester.

One indication of the importance of damage is given in a Potato Marketing Board Survey (1974) which showed that 22% of the crop was so seriously damaged by harvesting and transport to the store that it fell below the Board's ware standard. With carrots the position is no better. An ADAS survey (Berks, personal communication) showed that flesh and break damage can cause rejection of 60% of the yield and that about half of the damage occurs at the harvester. It has also been shown (Derbyshire, 1971) that mechanized harvesting causes so much damage to ware carrots that they cannot be stored.

The causes and possible cures of mechanical damage have been most extensively researched in the potato crop and researchers in the UK, the

USA and the USSR have all produced their own classifications for external damage which are not interchangeable. An added complication is that the term "Bruising" is used in the UK to mean internal damage and in the USA to mean a type of external damage. Internal damage is referred to in the USA as "black spot".

The classifications of damage used in the UK are: (i) scuffed; (ii) peeler damage (<1·5 mm deep); (iii) severe damage (>1·5 mm deep); and (iv) bruised. Each of these classifications can to some extent be attributed to specific design faults of present day harvesters. The development engineer armed with test data listing the amounts of these various kinds of damage caused by his machine therefore has some hopes of improving its performance.

It is, unfortunately, possible to combine the data into "indices". The Potato Marketing Board did so in their 1973 survey producing a Damage Index which was calculated as follows: percentage scuffed ×1, plus percentage peeler damage ×3, plus percentage severe damage ×7. They also produced a Bruising Index derived from the total number of bruises expressed as a percentage of the number of potatoes in the sample. No evidence was produced in the report to justify the use of the three multiplying factors and there is no logical reason for believing that a harvester which inflicts one bruise on each of 20% of the tubers is better than another which inflicts slightly more than two bruises on each of 10% of them. Furthermore, since the engineer needs all the information that he can get if he is to improve his machine, it is very doubtful whether such treatment of the data serves any useful purpose.

Over the past 20 years, attention has been paid to methods of reducing potato damage by improved cultivations. Since the harvester is required to sift about 30 tonnes/ha of potatoes from about 1500 tonnes of soil it is clear that reducing the clod population is likely to allow a less severe sifting to be used in the harvester. Because clods are likely to be created during post-planting operations and chemical weed control had then become an acceptable alternative, Bleasdale (1962) suggested experiments in which potatoes were grown in flat beds with no post-planting operation except blight spraying. Wayman (1969) investigated methods of mechanizing the planting and harvesting operations of such a system. He found that its use resulted in fewer clods being lifted than with conventional ridges, but the yield did not compensate for the increased seed rate or for the difficulties in harvesting the crop. More recently, experiments at Terrington EHF (Shotton, 1970) have shown lower tuber damage, probably as a result of reduced clod populations when the crop is grown in 910 instead of 760 mm rows. Pascal and Sheppard (1977) have also shown that secondary cultivation based on a rotary harrow produced fewer clods than either a rotary cultivator or a reciprocating harrow.

Stones have an effect on the level of mechanical damage because they impact on the tubers as they pass through the harvester. Whitney (1978) showed that stone levels up to 200 tonnes/ha could be reduced to 7-10 tonnes/ha in the ridge by stone removal or windrowing. Harvester speed was increased by 43% after windrowing and by 22% after stone removal. In both cases severe mechanical damage was reduced by over 30%.

Ever since the first potato harvesters were introduced in the early 1950s, engineers have been seeking methods of improving them. Some of the modifications have been aimed primarily at increasing output rates, but these have usually also sought not to increase damage. Others have sought only to reduce mechanical damage.

The effect of share design on damage is two-fold. The amount of soil that it lifts determines to some extent the severity of the mechanisms needed to separate the potatoes and its method of feeding the soil on to the next stage of the harvester, usually a conveyor web, determines to some extent the damage caused by that component.

Bailey (1957) measured the characteristic growth habits of ridge-grown potatoes in the early 1940s. Hawkins (1957) then used a ground-driven concave disc share which had a profile of cut similar to Bailey's ideal shape. It had the added advantage that it inverted the ridge on to a spider which was the main soil separating component of his harvester. Almost all of the damage caused by the harvester was of the scuffing type. Unfortunately, the share could not be adapted to suit 910 mm rows. Hawkins and Ofield (1964) then devised a power-driven double disc share which could be used in wider ridges and yet lifted only the soil profile specified by Bailey. Accurate control of share depth is also desirable to reduce the volume of soil taken into the harvester, but preliminary tests of a commercially available unit (Carlow and McRae, 1977) have showed that there is need for more consideration to be given to reference points if accurate control is to be achieved.

The elevator webs of harvesters where the soil is separated from the potatoes have received a considerable amount of attention, but not all of the research findings have yet been incorporated in commercial harvesters. Brecka (1970) reported that two-stage webs with the first web running at a higher speed gave improved separation without increased damage but advocated that alternatives be sought for the universally used eliptical agitators. McGechan (1977) showed that horizontal acceleration is more effective and less likely to cause damage than vertical acceleration and thus suggested that radical changes in harvesters were desirable. Butson *et al.* (1977) confirmed earlier findings that damage increases as the ratio of web speed to ground speed becomes larger, suggesting the need for easy web speed adjustment with conventional present-day harvesters.

Possibly the development most likely to reduce damage from potato harvesters employing elevator webs is the dual web devised by J.C. Hawkins of NIAE and developed by Carruthers (1977). The dual web has an outer web with a rod gap of 51 mm and an inner one of standard pitch. The outer web is not agitated and carries the largest most damage prone tubers to the picking table. The inner web not having to carry material larger than 51 mm is a more effective sieve and requires less agitation than normal. Thus both large and small potatoes are handled more gently than in conventional machines (Fig. 2).

The propensity of potatoes to become damaged when they are dropped on a hard surface has long been understood (Green, 1956) but satisfactory methods of reducing drop heights between conveyors are more recent (Johnson *et al.*, 1974). Similarly the trailer-filling elevators of harvesters have been fitted with adjustments to reduce the height of drop into the trailer for many years but severe damage attributable to the harvester delivery elevator was reported in 9% of the case studies of the Potato Marketing Board Survey (1974). A simple automatic control that ensures that the drop height does not exceed 200 mm has been described by McRae (1974).

Although carrots are just as prone to damage as potatoes and are almost exclusively harvested by machine, very little research has been undertaken into methods of reducing damage. Sharp and Farrent (1975) measured and classified the crop damage that occurred during mechanical harvesting operations on 11 East Anglian farms over two years and used the data to identify features of machine design having the greatest bearing on root damage. Then by a series of experiments under more closely controlled conditions they examined their findings in more detail to see if recommendations could be made for modifications to the design or operation of current harvesters (Sharp and Farrent, 1976).

In the controlled experiments the machine factors investigated were share depth, elevator bar gap, degree of agitation and elevator web speed. Mechanical damage was subdivided into three classes; surface damage, flesh damage and broken roots but, because little more than 10% of the surface damage was attributable to the machine factors it was disregarded in the results. As with earlier potato damage research, the data was presented as damage indices with surface and flesh damage being scored 0-3. Thus, for a sample of 100 roots the maximum flesh damage score was 300. The proportion of roots free from each class of damage was therefore recorded and quoted with the damage indices.

As was to be expected the survey showed that stone population had a considerable effect on damage. The controlled experiments showed that the most easily corrected cause of broken roots was the practice of setting the share too shallow. This also, by reducing soil intake, increased flesh damage. With the share set at the correct depth, increased soil separation was required and the experiments showed that this could be achieved by increasing the speed of the first elevator web and using a wider bar gap. Increasing the number and eccentricity of the agitators, particularly those in the first elevator web also increased soil separation. Rather unexpectedly it was found that increased agitation either by increasing the number or the severity of the agitators reduced flesh damage except when the agitators nearest to the share were increased in severity. The only penalty likely to result from the adjustments recommended in the report was increased loss of small carrots through the wider bar gap but this was to some extent nullified by increasing the speed of that web and increasing the share depth.

Although a very considerable volume of research and development has been undertaken in the USA into mass methods of harvesting orchard fruit, very little has been done in the UK. Preliminary trials on commercial holdings by NIAE staff in 1968 showed that both trunk and branch shaking are effective methods of harvesting fruit (Sharp, 1969).

On one holding where a grower was using a branch shaker of his own design to harvest Bramley's Seedling apples, damage was considered to be no worse than that of typical hand-picking. Damage to plums for processing when a trunk shaker was used was higher than that of a hand-picked crop of similar ripeness. In the case of the branch shaker nearly all the damage was caused by fruit striking branches as it fell through the tree. With the trunk shaker, half the damage was attributable to the design of the catcher. It is possible that most of this could have been eliminated by the use of a more elaborate catcher like the inverted umbrella developed by CNEEMA in France (Holt and Boa, 1970).

Continued investigations into trunk shaking (Sharp, 1972) showed that over half the fruit was detached from the tree with some horizontal displacement and one-third of it landed more than half a metre horizontally from its position in the tree. Least damage occurred with trees where the trunk was offset to one side of the fruiting wood and it was concluded that tree shaping experiments should precede further machine trials.

In Holland, Van der Werken (1973) devised a continuous moving shaking and catching machine for dessert apples grown as slender spindle bushes. The unit straddled the row of trees and incorporated three separate fruit catching levels and a continuously moving tree shaker. Although it gave encouraging results in trials, the work had to be abandoned because there was no commercial interest in manufacture. The work now being undertaken at Long Ashton (Child and Le Flufy, 1979), although using similarly shaped trees, does not rely on shaking and should provide a simpler and more commercially acceptable design.

PRODUCT CONTAMINANTS

In grain harvesting it is normal to judge the efficiency of a harvester by, among other factors, the cleanliness of the sample that it produces. In the vining of peas for processing such judgements are acceptable because leaf fragments can quickly produce taints and off-flavours (Gane, 1966). Similarly with potato harvesting it is obviously undesirable to transport large quantities of soil to the store and, where automatic stone separation is being used, it is important that dirt tare should be kept to less than 35% (McRae, 1973).

With some other vegetables and fruit it is clearly undesirable to judge the efficiency of the harvester by reference to the cleanliness of the sample it produces and allowances must be made for the operation that is to follow harvesting. Thus, for example, during the development of the NIAE Universal Vegetable Harvester (Boa, 1966) it was argued that the product should not be trimmed on the harvester because outer leaves and other crop debris would act as a cushion during transport. The counter argument that additional transport is required is not always borne out in practice: with Brussels sprouts, for example, the inclusion of some crop debris in the sample is likely to allow larger unit loads to be transported to the packing shed without damage (Boa, 1973).

In the mechanized harvesting of soft fruit crops, some crop debris in the sample can be advantageous for a variety of reasons. With blackcurrants the presence of some strig in the sample is advantageous in that it acts as a cushion during transportation and also makes for easier juice extraction. At the present stage of the development of mechanized strawberry harvesting it appears likely that removing the calyces will be a packing shed operation. For satisfactory operation of the decapping machines it will then be desirable for the harvester to leave some pedicel attached to the berries (Holtman *et al.*, 1977), because fruit which cannot be decapped is likely only to be suited for purée or juice stock (Ledbebuhr *et al.*, 1977).

Even with orchard fruit there appear to be good reasons for not attempting to produce a clean sample in the harvester. La Belle (1969) showed that with mechanical harvesting of cherries it is more expensive, because of the damage caused during the operation, to try to remove unripe fruit and stems manually at the farm instead of mechanically at the processing factory.

SUMMARY AND CONCLUSIONS

Because it is difficult to attach true costs to the effects of reduced quality, engineers usually consider product quality separately from the other headings under which they seek to improve on the performance of existing harvesting machines and methods.

Progress towards better harvesting machines is likely to result from co-operation between engineers and plant scientists but much can be done by the engineer alone to reduce the amount of damage done by harvesting machines. It is, however, unrealistic to judge the performance of fruit and vegetable harvesting machines in terms of the cleanliness of the sample they produce. In many cases the harvester should be regarded as one part of a system and judged accordingly.

REFERENCES

Bailey, P.H. (1957). *J. agric. Engng Res.* **2** (2), 146-151.

Bleasdale, J.K.A. (1962). *Rep. nat. Veg. Res. Stn for 1961*, 37-38.

Boa, W. (1966). *J. and Proc. Inst. agric. Engrs*, **22**, 101-110.

Boa, W. (1967). Dep. Note 213/CD, *Natn. Inst. agric. Engng.* *

Boa, W. (1973). *ADAS q. Rev.* **8**, 157-163.

Boa, W., Cox, E.F., Birkenshaw, J.E., Foster, D.K. and Senior, D. (1980). *Expl. Hort.* **31**, 26-34.

Brecka, J. (1970). *Sb. zemed. Tech.* **16**, 105-117.

Bufton, L.P. (1977). *Acta Hortic.* **72**, 135-157.

Butson, M.J., Hamilton, A.J., Lock, G.S. and Sheppard, B.W. (1977). Dep. Note SIN/222, *Scot. Inst. agric. Engng*, pp. 19.*

Carlow, C.A. and McRae, D.C. (1977). Dep. Note SIN/224, *Scot. Inst. agric. Engng, Penicuik.* *

Carruthers, J. (1973). Dep. Note SSN/177, *Scot. Inst. agric. Engng*, pp. 9.*

Carruthers, J. (1977). *Pot. Res.* **20**, 53-61.

Child, R.D. and Le Flufy, M.J. (1979). *Grower,* **91**, 27-30.

Derbyshire, D.M. (1971). *Comml Grower*, 703.

Dunn, J.S., Stolp, M. and Lindsay, G.G. (1976). Paper 76-1543, Am. Soc. agric. Engrs, St. Joseph.

Gane, A.J. (1966). *J. and Proc. Inst. agric. Engrs*, **22**, 91-100, 110-111.

Green, H.C. (1956). *J. agric. Engng Res.* **1**, 56-62.

Hawkins, J.C. (1957). *J. agric. Engng Res.* **2**, 14-24.

Hawkins, J.C. and Ofield, R.J. (1964). *J. agric. Engng Res.* **9**, 298-300.

Holt, J.B. and Boa, W. (1970). Dep. Note MP/064/1320, *Natn. Inst. agric. Engng.* *

Holtman, J.B., Hansen, C.M., Ledbebuhr, R.L., Clary, C.D., Pierson, T. and Bradford, L.J. (1977). Paper No. 77-1030, Am. Soc. agric. Engrs, St. Joseph.

Ingram, J. (1973). *Rep. Luddington exp. hort. Stn for 1972*, 59-61.

Johnson, L.F., Peterson, C.L., Gentry, J.P. and Bailey, E.M. (1974). Paper No. 74-1509, Am. Soc. agric. Engrs, St. Joseph.

Kemp, I.M. (1978). "Strawberries for Processing". Proceedings of Meeting, Isle of Ely College, Wisbech, pp. 9.

La Belle, R.L. (1969). Processing machine-harvested fruits and vegetables. *In* "Fruit and Vegetable Harvest Mechanization : Technological Implications". (B.F. Cargill and G.E. Rossmiller, eds), Vol. 1, 143-152. East Lansing, Rural Manpower Cent. (RMC Rep. No. 16).

Larsen, R.P. (1969). Cultural practices for cherry mechanization. *In* "Fruit and Vegetable Harvest Mechanization: Technological Implications". (B.F. Cargill and G.E. Rossmiller, eds), Vol. 1, 687-697. East Lansing, Rural Manpower Cent. (RMC Rep. No. 16).

Ledbebuhr, R.L., Hansen, C.M., Bradford, L.J. and Newell, J. (1977). Paper No. 77-1529, Am. Soc. agric. Engrs, St. Joseph.

McGechan, M.B. (1977). *J. agric. Engng Res.* **22**, 229-245.

McRae, D.C. (1973). *J. and Proc. Inst. agric. Engrs,* **28**, 43-47.

McRae, D.C. (1974). *Pot. Res.* **17**, 138-151.

Pascal, J.A. and Sheppard, B.W. (1977). Dep. Note SIN/244, *Scot. Inst. agric. Engng*, pp. 8.*

Potato Marketing Board (1974). National Damage Survey, 1973.

Salter, P.J. (1978). *Acta Hortic.* **83**, 245-249.

Sharp, J.R. (1969). Dep. Note R/5/69, *Natn. Inst. agric. Engng.* *

Sharp, J.R. (1972). Dep. Note M/182/1320, *Natn. Inst. agric. Engng.* *

Sharp, J.R. and Farrent, J.W. (1975). Dep. Note VSF/611/1320. *Natn. Inst. agric. Engng.* *

Sharp, J.R. and Farrent, J.W. (1976). Dep. Note VSF/622/1320. *Natn. Inst. agric. Engng.* *

Shotton, F.E. (1970). *Terrington Expl Husb. Farm, Annual Review*, 15-19.

Stilwell, M.R. and Portas, C.A.M. (1977). *Acta Hortic.* **72**, 201-210.

Strobel, J.W., Hanna, G.C., Smith, P.G., John, C.A., Reynard, G.B., Wann, E.V. and Leeper, P.W. (1969). Breeding developments for fruit and vegetables. *In* "Fruit and Vegetable Harvest Mechanization: Technological Implications" (B.F. Cargill and G.E. Rossmiller, eds), Vol. 1, 239-268. East Lansing Rural Manpower Cent. (RMC Rep. No. 16).

Van der Werken, J. (1973). *Jaarverslag. Inst. v. Tuinbouwtechniek, Wageningen*, 67-70.

Wayman, J.A. (1969). *Eur. Potato J.* **12**, 200-214.

Whitney, B.D. (1978). Tech. Note 189 E/C, *E. Scot. Coll. agric.*

Woodward, J. (1972). *Rep. Luddington Exp. hort. Stn for 1971*, 96-99.

*These Departmental Notes can be obtained from the Librarian, National Institute of Agricultural Engineering, Silsoe, Bedfordshire, UK.

Section IV
Post-harvest Changes—Physiological Effects

IV.1 Membrane Changes Underlying Post-harvest Deterioration in Quality

E.W. SIMON

Department of Botany, The Queen's University of Belfast, Belfast, Northern Ireland

INTRODUCTION

Once they are harvested, fruits and vegetables can only have a limited life. How long they will last depends on their stage of development; the life of leafy vegetables will be shortened if they are approaching senescence at harvest, and the same is true for fruit like pears and bananas if they are already ripe when picked. Assuming that the material is harvested at the proper time, what then determines its storage life? Apart from possible depredation by insects or micro-organisms, the major problem remains the inevitable drift towards senescence (Simon, 1977a). This paper is concerned with other types of deterioration involving membrane damage, in particular calcium deficiency and chilling injury. Deterioration due to these agencies may already be under way at harvest, the symptoms persisting into the post-harvest period, and possibly becoming worse.

CALCIUM DEFICIENCY

A number of disorders that may blemish fruits and vegetables, detracting from their quality, are not induced by extreme environmental conditions such as low temperature or high carbon dioxide concentration, nor is any causative micro-organism present. The symptoms were for many years ascribed to "physiological disorders" and it has only recently become evident that many of them are in fact due to calcium deficiency. In an important review article Shear (1975) showed that 35 such disorders were associated with a localized deficiency of calcium, and that conditions such as nutrition, growth rate and moisture supply which modified the severity of the disorders, had corresponding effects on the calcium status of the tissues affected.

Although calcium is abundant in most soils, and symptoms of a foliar deficiency of calcium are seldom encountered, there may be "an inadequate level of calcium in the fruits, storage roots, or tubers of many plants, or to the heart leaves of cabbage, lettuce, and other compact leafy vegetables" (Shear, 1975). Such localized deficiencies are a consequence of the mechanisms involved in the translocation of calcium in plants (Marschner, 1974; Redmond, 1975). Calcium is supplied to leaves through the transpiration stream in the xylem, but once located there it remains in place, little if any being redeployed subsequently to supply fruits or the heart leaves of Brassicas. Their only source of supply remains the xylem, but because they are relatively bulky with a low surface/volume ratio, transpiration is slow and hence the supply of calcium to them is far below the amount delivered to leaves.

Just how a low calcium level leads to the expression of the observed symptoms is not entirely clear, although two mechanisms can be outlined here: more details are given by Simon (1978).

ENHANCED MEMBRANE PERMEABILITY

One of the factors controlling the permeability of cell membranes is the calcium content of the tissue. A supply of calcium is necessary both for the uptake of solutes by plant roots and their retention within root cells (Epstein, 1972; Shay and Hale, 1973). As an example, some experiments on the leakage of carbohydrates out of the roots of cotton seedlings may be cited (Christiansen *et al.*, 1970). Chilling the seedlings to 5°C, or exposing them to an anaerobic atmosphere caused leakage to begin, unless 10^{-5} M calcium sulphate was present. The addition of 0·002 M EDTA (which sequesters divalent ions) induced very rapid leakage from the roots. The authors ascribe the action of calcium to a physical effect on the membrane rendering it less permeable, but the precise mechanism involved remains uncertain.

Evidently the membranes of cells low in calcium become leaky and must in consequence lose their turgor. As in tissues treated with chloroform or ether such a loss of membrane integrity allows cell fluids to invade the air spaces between the cells. The tissue becomes heavy, water-soaked and transparent or "glassy" in appearance. In due course the water may evaporate away leaving a more or less shrunken necrotic area, a pattern of events that corresponds to the aetiology of disorders like leaf necrosis, tipburn of lettuce and internal browning of Brussels sprouts.

BURSTING OF CELLS AND MEMBRANES

One of the distinguishing features of plant, as opposed to animal cells, is the possession of a wall, tough enough to withstand the hydrostatic pressure exerted by osmosis. The magnitude of this pressure depends on the water potential of the medium outside the cell, and of course, on the concentration of solutes within the cell. It is this pressure exerted by osmosis and countered by the relatively inflexible wall that accounts for cell turgor and hence the rigidity of thin leaves and other tissues that are without much lignin. When such a cell is placed in water, osmosis will cause a further entry of water, unless the pressure due to osmosis is already balanced by turgor pressure. Such at least is the received wisdom of elementary texts.

However, when slices of the pulp of soft, fleshy fruits like apple, tomato, plum and grape are placed in water, 90% of their electrolytes and soluble carbohydrates diffuse out into the water within 2 or 3 h. At the same time the tissue becomes permeable to Evan's blue, a dye that does not enter living cells. Apparently these cells burst when they are placed in water, the wall rupturing along with the underlying plasma membrane, so allowing loss of solutes and entry of the dye (Simon, 1977b). The tendency of the cells of apple to burst in water is something that develops as the fruits mature, and it may well be that our efforts to breed fruits with a soft, sugary flesh has resulted in the appearance of cells clothed in walls too weak to withstand the pressures exerted if they are exposed to water. In the intact fruit the cells will be protected from bursting in part by the pressure exerted by the surrounding cells, all of them ultimately encased within the skin of the fruit, and in part perhaps by the presence of media of low water potential in the apoplast.

If cell-bursting is involved in the aetiology of calcium deficiency disorders it follows that calcium must strengthen cell walls so that they can withstand the turgor pressure that may be generated inside, rather than expanding until they burst open. It is known that calcium ions inhibit cell elongation and in the past it has been supposed that calcium renders cell walls more rigid by some kind of cross-linking reaction (see Hall, M.A. p. 53, this Symposium). However, recently Cleland and Rayle (1977) have shown that subsequent exposure to calcium does not make living, auxin-treated coleoptile sections any less extensible, and they propose that calcium inhibits the H^+-induced wall loosening process. In effect low calcium levels permit the amount of wall loosening that normally accompanies the orderly growth of cells in tissues but may allow cells placed in water to balloon out, becoming so thin that they finally burst.

A first indication of the possible importance of turgor pressure in calcium deficiency disorders comes from the observation that apples, tomatoes and other fruit may crack and split if they are short of calcium; the severity of the cracking is reduced if the fruits are first treated with calcium. This cracking of fruits is generally attributed to the osmotic uptake of water, entering as liquid through the skin, or as vapour through the lenticels. The uptake of water and the swelling that causes the fruit to crack are clear indications that individual pulp cells must enlarge under the turgor pressure to which they are subjected. If the cells enlarge, it is but a small step to suppose that they may ultimately burst. The work of Bramlage and Shipway (1967) on water core provides a clear indication that events may proceed in this sequence in apple. Water core is a condition in which the fruits become water-soaked, the intercellular spaces filling with a dilute solution fed into the apple through the phloem. This fluid may disperse during storage, being absorbed by the cells, but often the disappearance of water core is accompanied by internal breakdown, the pulp becoming spongy in texture and brown in colour following the oxidation of polyphenols. The loss of turgor and change in colour are evidence that the orderly compartmentation of the tissue into separate cellular units has now broken down. Blossom-end rot of tomatoes probably develops in the same way, cell breakdown leaving an extensive water-soaked region which later collapses as the water evaporates off, leaving a massive sunken area.

ENHANCING CALCIUM LEVELS TO AVERT INJURY

The incidence of calcium deficiency disorders could be lessened if some way were found of altering the distribution of calcium within the plant so that a higher proportion reached the parts destined for market. It may now be possible to achieve this aim for Brussels sprouts and produce crops free from internal browning, a disorder in which more or less extensive regions of the young leaves within the sprout become necrotic and discoloured. The outermost leaves are perfect and hide the disorder from view; it can only be seen when the bud is cut open to reveal the young leaves inside. These leaves are so protected and enfolded by the older ones that their local environment is one of constantly high humidity in which transpiration must be greatly restricted if not halted altogether. They are therefore deprived of a continuing supply of calcium. In an interesting study on cabbage plants it has been shown that ^{45}calcium is transported from the rooting medium to the inner leaves if xylem sap is forced up to the leaves by a root pressure flow (Palzkill *et al.*, 1976; Palzkill and Tibbitts, 1977). The stomata of cabbage leaves do not shut at night, but a root pressure flow can be forced upon the plant by ensuring that the whole plant, including the outer leaves, is placed under conditions of humidity high enough to inhibit transpiration. It should therefore be possible to protect sprouts from internal browning by saturating the atmosphere around them at night.

It is more difficult to alter the natural pattern of distribution of calcium within an apple tree and most attempts to raise calcium levels in the fruits have relied on spraying the trees or dipping the apples in calcium chloride solution after harvest and prior to storage (Redmond, 1975).

CHILLING INJURY

Many plants are injured by exposure to temperatures well above freezing point. Such chilling injury to a growing plant may reduce its final yield, but in addition the crop itself may be injured if it is chilled in the field during transport or in cold store.

Lyons (1973) has reviewed the horticultural aspects of chilling injury. Certain of the symptoms are only skin-deep (spots, pits or discolouration), but others involve a more or less extensive breakdown of internal tissues which may only be visible when the fruit is cut open. Biale (1950) lists five types of symptoms seen in chilled citrus fruit, and there is no doubt that chilling temperatures induce equally complex syndromes in other fruit as well. In addition to the visible symptoms, chilled fruit tend to be more susceptible to decay and they may fail to ripen properly.

Most chilling-sensitive fruits and vegetables must be exposed to low temperatures for a matter of days or weeks before injury becomes apparent. The lower the temperature and the longer the exposure, the more severe the injury. An additional factor governing symptom expression is the temperature that follows the chilling experience, transfer to the warm hastening the appearance of symptoms. There is at present no reliable quantitative method for assessing the severity of chilling damage to fruit and vegetables, so that it has been difficult to assess the critical temperature below which chilling injury will appear. In general terms it can be said that the low temperature limit is around 0-4°C for apple, about 8°C for citrus, avocado and pineapple and 12°C for banana (Wilkinson, 1970).

The molecular mechanism underlying chilling injury is now becoming clear following the discovery of a difference in the response to temperature of mitochondria from chill-sensitive and chill-resistant species (Lyons, 1973; Raison and Chapman, 1976). The basic observation is a marked rise in the activation energy for mitochondrial oxidations at chilling temperatures. As a result, mitochondrial oxidations proceed at a relatively slow pace compared with glycolytic reactions and toxic or inhibitory products such as ethanol, acetaldehyde and keto acids begin to accumulate, ultimately it is supposed, causing death of the tissue. The change in activation energy for mitochondrial oxidations occurs at about the temperature at which a fruit develops symptoms of chilling injury, but there is still a need for critical studies to test this relationship for a range of chilling-sensitive tissues.

Electron spin resonance studies show that in chilling-sensitive species low temperatures have a direct impact on membrane lipids, causing them to undergo a phase transition from a liquid-crystalline to a gel-like condition. This phase change has two consequences. It causes a change in the conformation of the oxidase enzymes embedded in the membrane, so raising their activation energy in the cold; in addition it induces cracks or channels to appear in the membrane, rendering it more permeable, allowing solutes to leak out of the cells when they are placed in cold water. It may be that, as in cucumber leaves, leakage only becomes apparent at temperatures below that at which a change in oxidation rates becomes apparent (Minchin and Simon, 1973); if so there might be a range of temperatures over which the different symptoms of chilling injury would become manifest.

In summary, the hypothesis that guides most research on chilling injury revolves around the membrane phase transition which leads to an imbalance of metabolism with an accumulation of toxic products, a reduced synthesis of ATP and a decline in the efficiency of compartmentation as membranes become more permeable.

REFERENCES

Biale, J.B. (1950). *A. Rev. Pl. Physiol.* **1**, 183-206.

Bramlage, W.J. and Shipway, M.R. (1967). *Proc. Amer. Soc. hort. Sci.* **90**, 475-483.

Christiansen, M.N., Carns, H.R. and Slyter, D.J. (1970). *Pl. Physiol.* **46**, 53-56.

Cleland, R.E. and Rayle, D.L. (1977). *Pl. Physiol.* **60**, 709-712.

Epstein, E. (1972). "Mineral Nutrition of Plants: Principles and Perspectives." Wiley and Sons, New York.

Lyons, J.M. (1973). *A. Rev. Pl. Physiol.* **24**, 445-466.

Marschner, H. (1974). *Neth. J. agric. Sci.* **22**, 275-282.

Minchin, A. and Simon, E.W. (1973). *J. exp. Bot.* **24**, 1231-1235.

Palzkill, D.A. and Tibbitts, T.W. (1977). *Pl. Physiol.* **60**, 854-856.

Palzkill, D.A., Tibbitts, T.W. and Williams, P.H. (1976). *J. Amer. Soc. hort. Sci.* **101**, 645-648.

Raison, J.K. and Chapman, E.A. (1976). *Aust. J. Pl. Physiol.* **3**, 291-299.

Redmond, W.J. (1975). *Commun. Soil Sci. Plant Analy.* **6**, 261-272.

Shay, F.J. and Hale, M.G. (1973). *Pl. Physiol.* **51**, 1061-1063.

Shear, C.B. (1975). *HortScience* **10**, 361-365.

Simon, E.W. (1977a). *Ann. appl. Biol.* **85**, 417-421.

Simon, E.W. (1977b). *J. exp. Bot.* **28**, 1147-1152.

Simon, E.W. (1978). *New Phytol.* **80**, 1-15.

Wilkinson, B.G. (1970). *In* "The Biochemistry of Fruits and their Products" (A.C. Hulme, ed.), Vol. 1, pp. 537-554. Academic Press, London and New York.

IV.2 Hormonal Changes During Senescence, Ripening and Regrowth of Stored Vegetables

T.H. THOMAS

National Vegetable Research Station, Wellesbourne, Warwick, CV35 9EF, England

INTRODUCTION

It has been estimated that at least "40 per cent of the unconsumed produce harvested in the world is lost through post-harvest deterioration and disease" (Salunkhe, 1974). Much of the deterioration in store is pathological resulting from either primary infection in the field or secondary infection through tissues damaged during harvesting and handling. Considerable losses may also be physiological in origin due to ripening, senescence, abscission and eventual regrowth of tissues. There is an obvious need to delay these ageing processes until the harvested crop is consumed. Inhibition of the vital metabolic processes can be achieved in different crops for periods of days, weeks or even months by chemical treatments and/or storage at low temperatures, under controlled gas atmospheres or low gas pressures. However, storage problems and treatments are very different depending on the nature of the produce to be stored and this variation in performance has stimulated much interest in the post-harvest physiology and biochemistry of fruit and vegetables.

Ripening and senescence are very familiar phenomena and many of the chemical changes which occur to induce alterations in flavour, texture and odour have been well characterized for a range of crops. Conversely, the endogenous factors controlling and regulating such changes remain obscure and few investigations have been made of the hormonal control of growth in stored vegetable crops other than potatoes and tomatoes. There is little doubt that the whole ageing process, like other growth processes, is subject to hormonal control and to control by nutritional changes within and between individual organs. Progressive changes in metabolic relationships result in the exchange of metabolites and the establishment of new sources and sinks for nutrients. For example, the development of the reproductive structures within stored white cabbage will accelerate the ageing of the outer leaves by acting as metabolic sinks. Thus, each part of a stored vegetable product plays a role in ageing, providing or creating a demand for metabolites.

It is widely accepted that specific aspects of growth and differentiation depend on certain quantitative relationships between the levels of interacting growth substances (Wareing, 1978). Such relationships have been demonstrated in dormant seeds, tubers and winter buds (Samish, 1954; Wareing and Villiers, 1961) and similar control mechanisms involving growth promoters and inhibitors have been proposed for resting and storage organs of vegetable crops (Isenberg *et al.*, 1974b). The involvement of at least five major types of hormones in abscission, senescence and ripening phenomena is also well established (Thomas, 1977a).

The diversity of vegetable species entering store precludes the possibility of presenting a general scheme of hormonal control. In addition, the storage performance of any product is rarely predictable being affected not only by storage treatment but also by pre-harvest factors such as environment during growth, field management and time of harvest. Each of these factors can alter the balance of endogenous hormones and this is likely to modify the storage capacity of the crops (Isenberg *et al.*, 1974b). Nevertheless, stored vegetables can be divided into certain classes depending on the nature of the harvested organ and it is convenient to consider hormonal control under the following four crop headings: leafy vegetables, bulb crops, storage root and tuber crops, and fruit classed as vegetable produce.

CONTROL OF SENESCENCE AND REGROWTH OF LEAFY VEGETABLES

Most stored, leafy vegetables do not exhibit dormancy in the true sense of the word in that they are harvested during periods of active growth and continue to grow in store. Nevertheless they can be kept for various lengths of time after harvest depending on the species and the type of storage. Brassica crops show a considerable diversity in the type of commodity to be stored and in the maximum possible duration of storage. Thus, winter white cabbage is suitable for prolonged storage of up to seven months whereas Brussels sprouts, broccoli and cauliflower can be kept for only short periods of from 2 to 6 weeks, even at low temperatures.

With Brussels sprouts and broccoli the main problems encountered during storage are dessication, yellowing and abscission of bracts and leaves. These morphological changes are associated with high rates of respiration, reduction in protein and nucleic acids, and loss of chlorophyll. The involvement of endogenous hormones in leaf senescence is complicated and application of the same growth regulator can invoke different responses in different species. Thus, auxins are very effective in preventing senescence of detached leaves of woody species but have little effect on herbaceous species, whereas gibberellic acid (GA) treatment enhances senescence of Brussels sprout leaf discs but delays senescence in *Taraxacum officinale* (Thomas, 1977a). Thus, it seems likely that leaf senescence might be controlled by the *balance* of endogenous hormones rather than by the actual amounts of any particular hormone and that different hormones might be "limiting factors" in different species.

Measurement of hormonal changes in the green, outer leaves of Brussels sprout buttons during storage for one week in the dark at 25°C indicated that senescence is associated with decreased auxin and cytokinin levels and increased gibberellin and inhibitor levels (Table I). When mature, green leaf discs were incubated in the dark for three days, those treated with N^6-benzyladenine (BA) (2×10^{-5}M) lost 51% less chlorophyll than the untreated controls whereas GA (2×10^{-7}M)-treated discs lost 36% more. These results suggested that the maintenance of leaf chlorophyll levels in Brussels sprouts is at least partly controlled by an interaction between endogenous gibberellins and cytokinins. Moreover, analysis of hormone levels in leaves from different parts of Brussels sprout buttons (Table II) and chlorophyll determinations of etiolated inner leaves given different hormone treatments indicated that chlorophyll development is also controlled

TABLE I

Hormone and chlorophyll changes in detached Brussels sprout outer leaves during senescence at 25°C in the dark

Days stored	Leaf content				
	Chlorophyll (% of Day 1)	Gibberellin[a] (μg/kg)	Cytokinin[a] (μg/kg)	Auxin[a] (μg/kg)	Inhibitors[b] (% control)
1	100	0·14	4·1	100	82·4
2	51	1·24	0·2	104	67·4
4	42	2·00	0·1	118	74·4
5	25	1·00	0·2	60	60·4

[a]Equivalent of gibberellic acid, N^6-benzyladenine and indolylacetic acid (IAA) per kg fresh weight. [b]Percentage of control wheat coleptile extension (from Thomas, 1977a).

TABLE II

Hormone and chlorophyll content of leaves from different positions within Brussels sprout buds

Leaf number	Leaf content				
	Chlorophyll (% of leaf 1)	Gibberellin[a] (μg/kg)	Cytokinin[a] (μg/kg)	Auxin[a] (μg/kg)	Inhibitors[b] (% control)
1	100	0·14	4·1	100	82·6
3	46	0·14	0·8	80	60·8
6	45	0·07	0·1	100	60·6
9	26	2·48	0·2	136	40·5
12	19	1·20	0·2	164	52·4

[a b] For key, see Table I (from Thomas, 1977a).

by these hormones (Thomas, 1977a). Senescence of both Brussels sprout buttons and broccoli spears can be inhibited by the application of cytokinins or cytokinin-active compounds or by treatments such as low temperature storage which reduce the loss of cytokinins (Fuller *et al.*, 1977; Thomas, 1977b).

Conversely, Brussels sprout leaf senescence can be accelerated by either abscisic acid (ABA) or ethylene, and increases in endogenous inhibitor levels have been demonstrated in senescing leaves (Thomas, 1977a). However, there is some question as to whether ABA changes are a prelude to, or occur as a result of, senescence or even whether they may be due to other associated phenomena such as wilting or environmental stresses. The involvement of ethylene is more definite and ethylene-generating compounds such as ethephon have been used commercially on a number of crops to stimulate abscission and senescence of leaves and fruit. Wang (1977) showed that compounds such as rhizobitoxin analogues which inhibit ethylene synthesis reduced respiration and inhibited senescence of broccoli spears kept at 20°C. The opposite effects of ethylene and cytokinins in chlorophyll maintenance and senescence have been documented (Fletcher and Adedipe, 1970; Abeles, 1972) and it has been shown that the effects of ethephon on Brussels sprouts can be overcome by pre-treatment with cytokinin-like compounds (Thomas, 1977a). Thus the control of senescence in stored, leafy crops appears to involve a complex pattern of hormonal interaction which can be considerably modified by post-harvest storage treatments and conditions.

In cauliflowers, curd quality is affected both by leaf senescence and continued growth of the curd resulting in "riciness" and "bractiness". Low temperature storage will inhibit this peduncle elongation to some extent but excessive low temperatures can cause damage to the florets. The results of Leshem and Steiner (1968) suggest that curd extension growth is gibberellin-mediated and this is consistent with the general effects of gibberellins on cell elongation. In a study on cauliflower deterioration during storage, W.G. Tucker (personal communication) examined the effects of a number of growth inhibitors, including inhibitors of gibberellin synthesis, on curd growth and found that daminozide, chlormequat chloride, simazine, 4-hydroxycoumarin and maleic hydrazide all frequently inhibited curd growth. However, no satisfactory conclusion could be reached because of the wide variation in curd-to-curd behaviour in store. Nevertheless, further studies on growth retarding and anti-senescent compounds could result in valuable contributions to post-harvest research.

Studies on storage of winter white cabbage have identified two major physiological problems, these being senescence of the outer wrapper leaves and regrowth of the main apex and lateral buds. The hormonal changes occurring in cabbage heads during conventional storage and controlled atmosphere (CA) storage (2·5% O_2: 5% CO_2) both at 1°C are shown in Fig. 1 and indicate that metabolism and growth in store is controlled by an interaction between cytokinins, gibberellins and growth inhibitors. Soon after harvest, the cytokinin levels in cabbage heads are high but as the storage period progresses and chlorophyll is lost, cytokinin content drops

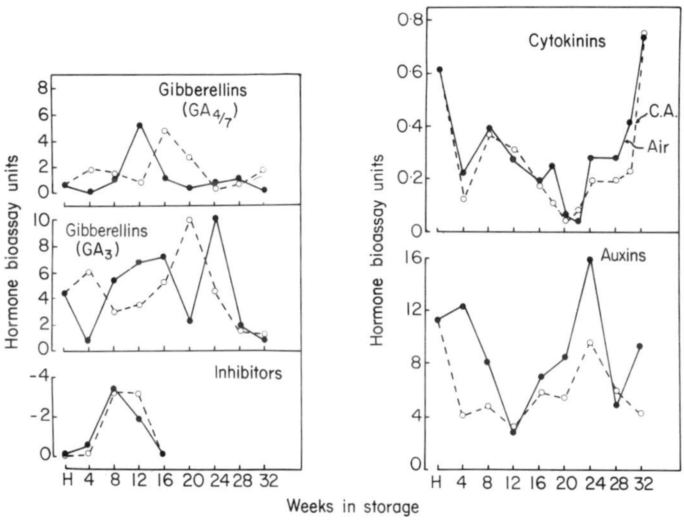

Fig. 1. Changes in endogenous growth regulators in cv. Green Winter cabbage during conventional (●——●) and controlled atmosphere (○——○) storage. Activity was determined by the lettuce hypocotyl test for gibberellins, wheat coleoptile assay for auxins and growth inhibitors and the *Amaranthus* betacyanin bioassay for cytokinins (redrawn from Isenberg *et al.*, 1974b).

until around the time of renewed meristematic activity (Fig. 2). The possibility that different cytokinins are involved during these different phases of development was suggested by Thomas *et al.* (1975) on the basis of the differences in cytokinin chromatographic profiles from cabbage cultivars of different storage capabilities. Further evidence for this hypothesis is

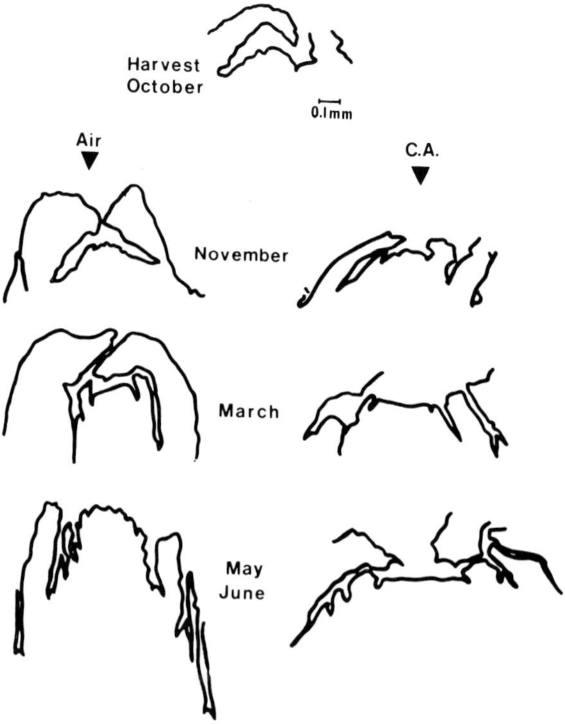

Fig. 2. Outlines of meristems sectioned from cv. Green Winter cabbage at harvest and during storage at 1°C in air or under controlled atmosphere (2·5% O_2; 5% CO_2) (unpublished data of A.M. Pendergrass, F.M.R. Isenberg and T.H. Thomas, 1974).

provided by the data shown in Fig. 3 of cytokinin profiles of cv. Green Winter before and after storage. The relevance of these cytokinin changes to dormancy and regrowth are not clear but are consistent with other data showing alterations in cytokinin and other hormone patterns during various stages of plant development (e.g. Abdel-Rahman *et al.*, 1975).

The effect of imposing CA-storage conditions on regrowth of cabbage is shown in Fig. 2 and this is reflected in a higher level of growth promoters in air-stored than in CA-stored cabbage. When heads are removed from storage after seven months and rooted in compost, those stored in air bolt and flower after four weeks whereas those stored in CA show very little

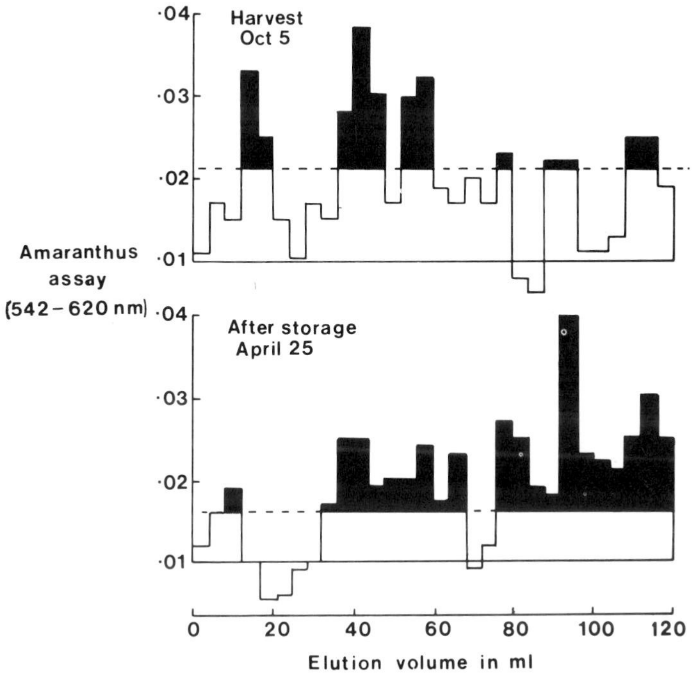

Fig. 3. Cytokinins in extracts of cv. Green Winter cabbage before and after storage at 1°C in air. Extracts were separated on an insoluble polyvinylpyrrolidone column and activity determined with the *Amaranthus* betacyanin bioassay. Shaded areas of the histograms indicate activity significant at $P = 0{\cdot}05$ (unpublished data of T.H. Thomas and F.M.R. Isenberg, 1974).

extension growth and no flowering. This difference in development might well be related to the appearance of two peaks of activity chromatographically similar to GA_3 in air-stored cabbage as compared with only one in CA, since the involvement of gibberellins in flowering processes is well established. The role of ethylene in senescence and regrowth of Danish cabbage has been discussed by Pendergrass *et al.* (1976) who showed that ethylene-induced deterioration in store was accompanied by large changes in auxins, gibberellins and ABA (Table III).

These few examples of hormonal changes in brassica crops serve to demonstrate that during storage of leafy vegetables a complex, dynamic system of hormonal control operates to control senescence, abscission and regrowth. They also show that similar hormones can operate at different phases to mobilize nutrients and control certain complicated developmental processes and this can have important practical implications. Thus, although high cytokinin levels are desirable to prevent senescence during early storage of cabbage, depression of cytokinin levels during the later storage period

TABLE III

Plant growth hormone concentrations around the apical meristem of cabbage heads before and after storage in atmospheres containing a range of ethylene (C_2H_4) concentrations for 5 weeks at 1°C

Treatment	No. of heads sampled	Growth substances μg equiv./100 g fresh wt.		
		IAA	ABA	$GA_{4/7}$
Pre-storage	2	0·460	5·50	0·292
Post-storage in:				
Air control (0·04 ppm C_2H_4)	2	0·320	3·00	0·058
Air + 1 ppm C_2H_4	2	0·104	2·80	0·088
Air + 10 ppm C_2H_4	2	0·494	[a]	0·046
Air + 100 ppm C_2H_4	2	21·104	[a]	0·720

[a] Below sensitivity level of wheat coleoptile bioassay (from Pendergrass *et al.*, 1976).

will suppress regrowth and its associated biochemical changes leading to deterioration of produce.

CONTROL OF SPROUTING IN BULB CROPS

In addition to being the only major stored vegetable bulb crop the onion is one of the more economically important vegetables in the UK. Much of the physiology including that of dormancy and sprouting has been reviewed by Brewster (1977a,b). Following harvest, the bulb undergoes a sequence of biochemical and anatomical changes (Isenberg *et al.*, 1974a) which result in the establishment of its biennial flowering phase. During storage, respiration rates increase progressively towards sprouting (Ward and Tucker, 1976), and associated changes in various enzymes, organic acids, flavour components and carbohydrates have been reported (Kato, 1966a; Pendergrass, 1969; Freeman and Whenham, 1974; Gavrilova and Lukovnikova, 1976; Darbyshire, 1978). The carbohydrate chemistry of onions and its relation to dormancy is complicated by the presence of complex fructans rather than starch as storage reserves in the outer scales but there is some evidence to suggest that sprouting is associated with increases in available fructose in the inner scales and growing point (Darbyshire, 1978; Darbyshire and Henry, 1978).

The involvement of endogenous hormones in the control of onion bulb dormancy has been reported by a number of researchers (Kato, 1966a,b; Thomas, 1969; Tsukamoto *et al.*, 1969; Isenberg *et al.*, 1974a,b; Stow, 1976) and an example based on four different cultivars of the hormonal changes occurring during storage is presented in Fig. 4. It seems clear that dormancy and sprouting are controlled by an interaction between promoters and inhibitors though the exact nature and function of these components of the hormonal complex is not clear. The results of Abdel-Rahman and Isenberg (1974a) and Mahotiere *et al.* (1976) suggest that the main operative promoters are cytokinins since neither gibberellins nor auxins

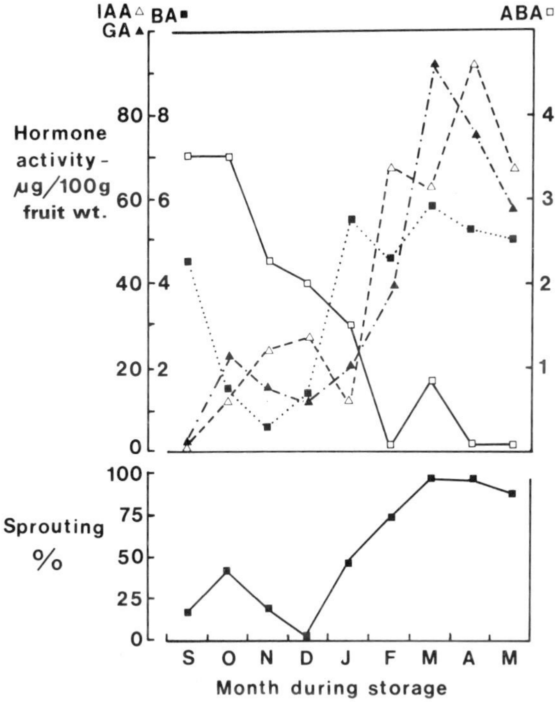

Fig. 4. Sprouting and hormonal changes in onion bulbs during storage at 5-8°C based on experiments over 3 years. Key: □, growth inhibitors; ■, cytokinins; △, auxins; ▲, gibberellins (redrawn from Isenberg *et al.*, 1974b).

but only cytokinins will induce sprouting of dormant bulbs. Experiments at the National Vegetable Research Station (NVRS) with the highly dormant New Zealand cv. Pukekohe Longkeeper have confirmed these results, untreated bulbs showing 30% sprouting as compared with 100% from BA-treated bulbs when kept at 22°C. It has been suggested that some of the gibberellin activity in stored onion bulbs is associated with vernalization and flower induction (Isenberg *et al.*, 1974a). This is consistent with the reports of gibberellin involvement in flowering of iris (Rodrigues Pereira, 1962, 1964) and tulip (Aung *et al.*, 1969) and the fact that at NVRS, GA treatment of potted, stored bulbs induced 50% more flowering than that achieved by untreated bulbs (Table IV).

TABLE IV

Effect of 25 h growth regulator soaks on the growth of onion bulbs taken from store in January and rooted in compost

Growth regulator treatment	Bulb measurements after 3 months		
	No. of shoots /bulb	% bulbs flowering	Wt. of shoots in g
Untreated control	2·0	40	64·6
NAA (10 mg/1)	3·5	40	102·9
GA (1 g/l)	1·6	90	58·0
Daminozide (4 g/l)	2·2	20	75·4
TIBA (800 mg/l)	3·0	50	98·4

Extractions of foliage and bulbs at different stages of development indicated that the growth inhibitors are produced in the leaves and translocated into the bulbs during ripening (Isenberg *et al.*, 1974b). Thus it is not surprising that premature defoliation or desiccation (Thomas and Isenberg, 1972; Stow, 1976; Rickard and Wickens, 1977) can adversely affect the performance of the bulb during storage. The nature of these inhibitors has not been elucidated but part of the activity has been ascribed to abscisic acid (Stow, 1976) and an allyl sulphide (Kato, 1966c). At present, extended storage of onion bulbs in the UK is achieved by pre-harvest spraying with maleic hydrazide, a chemical which prolongs dormancy via its effects on nucleic acid metabolism and cell division rather than on endogenous hormonal control. Since this chemical does not have universal clearance and has been regarded with some suspicion, alternative methods of sprouting control possibly based on physiological observations could be of great economical benefit.

CONTROL OF SPROUTING IN TUBERS AND STORAGE ROOTS

The main factors limiting the storage life of tubers and storage roots are the growth of sprouts and associated depletion and wilting of the storage organ. Prior to these obvious superficial changes, there is rapid conversion of storage reserves such as starch and complex sugars into the usable sugars required to support regrowth of the resting buds (Burton, 1977; Rutherford, 1977). In many respects the relationships between endogenous hormones and dormancy are similar to those demonstrated by onion bulbs. Thus in potato, cytokinins, auxins, gibberellins, abscisic acid and other inhibitors have all been implicated in the dormancy control mechanisms

(Smith and Rappaport, 1961; Hemberg, 1967; Tsukamoto and Yazawa, 1972).

Few growth regulator studies have been made with root crops. However, the storage problems are similar to those of potato and it has been demonstrated in carrots that reduction in carbohydrate conversion and therefore improvement in quality of stored roots can be achieved by low temperature and CA storage (Hansen and Rumpf, 1974; Weichmann and Ammerseder, 1974). As with onions, regrowth can be restricted by maleic hydrazide treatment but again this is not universally approved on root crops. The involvement of hormones in shoot regrowth is suggested by the experiments of Abdel-Rahman and Isenberg (1974b) who showed that auxin application stimulated lateral root development and inhibited shoot growth whereas both BA and GA stimulated shoot regrowth. However, as far as is known no critical examination of endogenous hormone changes in relation to storage of root crops other than potato has been made.

CONTROL OF RIPENING IN FRUIT CLASSED AS VEGETABLES

Although the remit of this paper does not include fruits there are certain stored fruits such as tomatoes and cucurbits which are often classed as vegetables and so should be mentioned briefly. Tomatoes are normally stored for only short periods of up to 3 or 4 weeks but recently it has been shown that certain cultivars are suitable for much longer-term storage (Thomas *et al.*, 1977). This introduces the possibility of home production of tomatoes in the UK for markets which have previously depended on imported crops.

As with most other fruit, the main problems associated with long-term storage of tomatoes are preventing shrivelling of the fruit and calyx, delaying ripening until the fruit is removed from store and subsequently ripening the fruit quickly for market. The control of ripening is highly dependent on ethylene produced by the fruit itself and so removal of this gas from the store by scrubbing, together with low temperatures or controlled atmospheres to reduce metabolism are the main methods of extending storage. There is much literature available on the control of tomato fruit ripening by ethylene and the identification of several non-ripening mutants has provided a new tool for examining the ripening processes (Tigchelaar *et al.*, 1978). Other aspects of hormonal involvement in tomato ripening have received little attention though both Dostal and Leopold (1967) and Khudairi (1972) showed that application of GA to green fruits inhibited lycopene synthesis and chlorophyll breakdown. Conversely, there is evidence (Abdel-Rahman *et al.*, 1975) that ABA plays a dominant role during the maturation and ripening process. Abdel-Rahman (1977) has suggested that ABA induces the synthesis of lycopene and various enzymes and stimulates ethylene evolution and is therefore a key factor controlling colour development, flesh softening and abscission of tomato fruits.

CONCLUSION

It is apparent that the performance of vegetable crops in store is controlled by the interactions of endogenous growth promoters and inhibitors and that the balance of these hormones can be altered both by pre-harvest treatments and the storage environment. It is envisaged that studies such as those described here will provide the information necessary for deducing the ideal storage treatments for particular crops. In addition, the breeding and selection of better storing cultivars might be facilitated by utilizing this information, for example, selection of onion lines having a high natural growth inhibitor content. However, in addition to increasing the exploration of gross hormone changes more concise studies are required to elucidate the way in which hormones are involved in the basic biochemical mechanisms controlling growth and development.

REFERENCES

Abdel-Rahman, M. (1977). *Physiologia Pl.* **39**, 115-118.
Abdel-Rahman, M. and Isenberg, F.M.R. (1974a). *J. agric. Sci. Camb.* **82**, 113-116.
Abdel-Rahman, M. and Isenberg, F.M.R. (1974b). *J. agric. Sci. Camb.* **82**, 245-249.
Abdel-Rahman, M., Thomas, T.H., Doss, G.J. and Howell, L. (1975). *Physiologia Pl.* **34**, 39-43.
Abeles, F.B. (1972). *A. Rev. Pl. Physiol.* **23**, 259-292.
Aung, L.H., de Hertogh, A.A. and Staby, G. (1969). *Pl. Physiol.* **44**, 403-406.
Brewster, J.L. (1977a). *Hort. Abstr.* **47**, 17-23.
Brewster, J.L. (1977b). *Hort. Abstr.* **47**, 103-112.
Burton, W.G. (1977). *Ann. appl. Biol.* **85**, 433-436.
Darbyshire, B. (1978). *J. hort. Sci.* **53**, 195-201.
Darbyshire, B. and Henry, R.J. (1978). *New Phytol.* **81**, 29-34.
Dostal, H.C. and Leopold, A.C. (1967). *Science, N.Y.* **158**, 1579-1580.
Fletcher, R.A. and Adedipe, N.O. (1970). *Proc. 7th int Conf. Pl. Growth Subst.* (D.J. Carr, ed.). Springer-Verlag, Berlin.
Freeman, G.G. and Whenham, R.J. (1974). *J. Sci. Fd Agric.* **25**, 517-520.
Fuller, G., Kuhnke, J.A., Corse, J.W. and Mackay, B.E. (1977). *J. Am. Soc. hort. Sci.* **102**, 480-484.
Gavrilova, V.P. and Lukovnikova, G.A. (1976). *Byull. vses Inst. Rasteniev N.I. Vavilova* **64**, 55-59.
Hansen, H. and Rumpf, G. (1974). *Acta Hort.* **38**, 321-326.
Hemberg, T. (1967). *Wiss. Z. Univ. Rostock* **16**, 661-666.
Isenberg, F.M.R., Thomas, T.H., Pendergrass, A.M., Abdel-Rahman, M. (1974a). *Acta Hort.* **38**, 95-125.

Isenberg, F.M.R., Thomas, T.H., Abdel-Rahman, M., Pendergrass, A.M., Carroll, J.C. and Howell, L. (1974b). *Proc. 19th int. hort. Congr.* **2**, 129-138.
Kato, T. (1966a). *J. Jap. Soc. hort. Sci.* **35**, 142-151.
Kato, T. (1966b). *J. Jap. Soc. hort. Sci.* **35**, 297-303.
Kato, T. (1966c). *J. Jap. Soc. hort. Sci.* **35**, 395-399.
Khudairi, K.A. (1972). *Am. Scient.* **60**, 696-707.
Leshem, Y. and Steiner, S. (1968). *Israel J. agric. Res.* **18**, 133-134.
Mahotiere, S., Herner, R.C. and Dennis, F.G. (1976). *J. Am. Soc. hort. Sci.* **101**, 211-213.
Pendergrass, A.M. (1969). Influence of maleic hydrazide on the metabolism of onion bulbs (*Allium cepa* L.) during storage. Ph.D. thesis, Cornell University, NY, USA.
Pendergrass, A.M., Isenberg, F.M.R., Howell, L.L. and Carroll, J.E. (1976). *Can. J. Pl. Sci.* **56**, 319-324.
Rickard, P.C. and Wickens, R. (1977). *Expl Hort.* **29**, 52-57.
Rodrigues Pereira, A.S. (1962). *Acta bot. neerl.* **11**, 98-138.
Rodrigues Pereira, A.S. (1964). *Acta bot. neerl.* **13**, 302-321.
Rutherford, P.P. (1977). *Ann. appl. Biol.* **85**, 440-444.
Salunkhe, D.K. (1974). "Storage, Processing and Nutritional Quality of Fruits and Vegetables". CRC Press, Inc., Cleveland, Ohio.
Samish, R.M. (1954). *A. Rev. Pl. Physiol.* **5**, 183-204.
Smith, O.E. and Rappaport, L. (1961). *Adv. Chem. Ser.* **28**, 42-48.
Stow, J.R. (1976). *Ann. appl. Biol.* **84**, 71-79.
Thomas, T.H. (1969). *J. exp. Bot.* **20**, 124-137.
Thomas, T.H. (1977a). *Ann. appl. Biol.* **85**, 421-424.
Thomas, T.H. (1977b). *Acta Hort.* **62**, 295-300.
Thomas, T.H. and Isenberg, F.M.R. (1972). *Expl Hort.* **23**, 48-51.
Thomas, T.H., Carroll, J.E., Isenberg, F.M.R., Pendergrass, A.M. and Howell, L. (1975). *Physiologia Pl.* **33**, 83-86.
Thomas, T.H., Gray, D. and Drew, R.L.K. (1977). *Ann. appl. Biol.* **87**, 117-121.
Tigchelaar, E.C., McGlasson, W.B. and Buescher, R.W. (1978). *HortScience* **13**, 508-513.
Tsukamoto, Y. and Yazawa, S. (1972). *Mem. Res. Inst. Fd Sci., Kyoto Univ.* **33**, 11-19.
Tsukamoto, Y., Fujita, M., Inaba, T., Asahira, T. (1969). *Mem. Res. Inst. Fd Sci., Kyoto Univ.* **30**, 24-37.
Wang, C.Y. (1977). *HortScience* **12**, 54-56.
Ward, C.M. and Tucker, W.G. (1976). *Ann. appl. Biol.* **82**, 135-141.
Wareing, P.F. (1978). *In* "The effect of Interactions between Plant Growth Regulators on Plant Growth and Yield", 1-9. British Plant Growth Regulator Group, Monograph No. 2.
Wareing, P.F. and Villiers, T.A. (1961). *In* "Plant Growth Regulation" (R.M. Klein, ed.), 95-107. *Iowa State Univ. Press, Ames, Iowa.*
Weichmann, J. and Ammerseder, E. (1974). *Acta Hort.* **38**, 339-344.

IV.3 Taints arising from Use of Crop Chemicals

D.H. LYON

The Campden Food Preservation Research Association, Chipping Campden, Glos. GL55 6LD, England

INTRODUCTION

I believe that quality is assessed from the relative values of several characteristics, which considered together will determine the acceptability of the product to the buyer and ultimately the consumer. In fruit and vegetable crops, either fresh or processed, the yield, size, colour, flavour, texture, nutritional value and absence of defects are the major characteristics that are used to determine the quality of the product. Kramer and Twigg (1970) have classified these quality characteristics into three groups: quantitative (which includes yield); sensory (which includes colour, flavour, texture, size and absence of defects); and hidden characteristics (which includes the nutritional value).

Although quality is made up of a number of parameters each one does not contribute equally to the general picture and there is always a bias towards the more important. Looking more closely at sensory characteristics, tenderness in peas is more important than either defects or colour, and texture of raspberries is the most important item of quality for that canned fruit. However, if any one aspect of quality is particularly poor, then the general quality of the product will be poor.

Growers and plant breeders realize that a consumer will only purchase what appears to be an attractive and acceptable product and that selection for an attractive appearance will be to the neglect of other factors. For example, a most attractive scarlet coloured strawberry may be devoid of flavour, or alternatively, a cultivar bred to withstand long distance transportation may have fruits that are tough and unpalatable. Flavour, because it is a hidden sensory characteristic, is more liable to be neglected than any other quality characteristics.

Flavour, and the chemical characteristics of the materials which contribute to flavour, is very complex (Arthey, 1974, 1975). Strong or weak flavours in processed foods may offend the average consumer and, hence restrict sales. Therefore, the flavour characteristics of each product have been designed to meet the requirements of the population as a whole, and for this reason, the flavour of many products tends to be bland. In general, what concerns processors most, apart from the strength or weakness of a flavour, is whether off-flavours are present or not in a product.

OFF-FLAVOURS AND TAINTS IN PROCESSED PRODUCTS

An off-flavour is an aroma, odour or taste which is atypical to one food, although it may occur normally in other foods. For instance, the aldehydes found as normal constituents of apple flavour, can be produced in butter as a result of microbial action and are then definitely off-flavours. A taint, on the other hand, is an unnatural or foreign flavour or odour which is not normally associated with food, such as a metallic taste or oily taste in canned fruit and vegetables.

What are the factors that can affect the flavour of a product? The cultivar itself is an obvious factor. With apples, for instance, different flavours are associated with cider, culinary or dessert cultivars. Maturity is another important factor affecting flavour. For example in pea crops, the peas become less sweet and more starchy or mealy as they mature and this is partly the reason peas are harvested at an immature stage when the flavour is still sweet.

Other factors which have been associated with flavour changes in food include the soil type on which the crop is grown, the fertilizer application, the application of agricultural chemicals and processing (taints can arise as a result of corrosion within the metal can or from chlorine from cooling water).

When a taint does occur in a manufactured or preserved food, the cause is immediately sought and remedied. The food processing industry is very conscious of the importance of quality in its products and errors in processing which could give rise to taints seldom occur, and safeguards are taken to ensure that such occurrences are kept to a minimum. Probably because of the noxious nature of agricultural chemicals, these materials are invariably placed on the suspect list when a taint is found.

CROP CHEMICALS AS CAUSES OF TAINTS

Crops grown for processing must be of high quality and with the decreasing availability of agricultural labour and the need to increase efficiency and quality, the grower has to rely more and more on agricultural chemicals (with the advantages that the use of these materials can bring). Because of the wide range of chemicals available, with different modes of action, there are numerous rates, and times and methods of application, e.g. from direct application to the seed before sowing, to application near or after harvest if the crop is to be stored. Double dosing may also occur if insufficient care is taken during spraying. At the time of processing the raw material, therefore, the chemical may be either:

(i) present as a residue in the original state at the time of application;
(ii) changed by reaction with the plant constituents; or
(iii) changed by environmental factors, e.g. solar radiation.

The nature of taints is so subtle that it is almost impossible to relate chemical residues to taints in canned products. The Campden Food Preservation Research Association through the MAFF Advisory Committee on Taint, worked for many years on this problem. Attempts were made to assess the taint following the addition of known quantities of thiram to canned strawberries, but whereas in some years taints arose at certain levels of residue, similar residues in other years produced no taint. Inconsistent results are an indication of the problems involved and it was only by accident, after studying cans which had been stored for 18 months, that a possible solution to the problem was found. It appeared that thiram, which is a sulphur-based compound, reacted with the metal container causing rubbery, sulphur taints to appear in the strawberries. Unfortunately, this work did not continue, but the testing of new agricultural chemicals to assess their taint potential when used on crops for processing continued to be developed.

Work on taint testing was initiated at Campden in 1945 with the introduction of the new organochlorine compounds, and since that time 431 chemicals have been examined in over 4800 tests (Table I). This work keeps

TABLE I

Number of chemicals examined in taint test at Chipping Campden, 1945-1977

	Chemicals	Tests
Fungicides	97	1288
Insecticides	101	1575
Herbicides	173	1391
Miscellaneous	60	547
Total	431	4801

processors informed of new materials and any adverse effects these chemicals may have on their products, and also assists chemical manufacturers to sell new materials for use on crops for processing. The information also enables chemical manufacturers to submit flavour data to the Agricultural Chemicals Approval Scheme when applying for approval for new products. In fact, the Scheme now requires taint data for materials which have a use on crops for processing.

At Chipping Campden we use the triangle test, which is the most simple and effective method of assessing the effects of pesticides on the flavour of processed foods. A panel of trained tasters, made up of Research Association employees, backed by a full-time taste panel of six people, are presented with three coded samples. Two of these samples are of material from an area treated with an agricultural chemical and the third is a control, or vice versa. Accurate processing is essential because unequal quantities of raw material in treated and control samples could lead to stronger or weaker flavours, which may be unfairly attributed to the chemical application. Each taster is asked:

(i) Can you detect an odd sample in flavour and if so which sample?
(ii) How does it differ in flavour from the other two?
(iii) Is there a taint present in any of the samples and if so, in which sample or samples?
(iv) Describe the taint and the intensity.

To assist the tasters in their judgements, lists of typical flavours and taints are provided. Examples of flavour differences include "sweeter", "stronger", "sharper", etc, whilst taints are described as "smoky", "rubbery", "burnt", etc.

Of the 431 chemicals tested for taints at Chipping Campden, 36 have been found to be associated with taints in one or more processed fruits and vegetables (Table II). With many of these chemicals, the results are not conclusive because insufficient tests have been done to establish whether the taint arose from causes other than the chemical treatment. This doubt may especially apply where taint has occurred in one year only. Nevertheless, a chemical even slightly associated with taint must remain suspect pending further tests.

A "Standard Procedure for Taint Tests" has been prepared by the MAFF Advisory Committee on Taint, based on the method used at Chipping Campden, in which the minimum number of tests required to establish whether a chemical is likely to be responsible for causing taints in processed foods is laid down. These seven tests, conducted over a three-year period and preferably on three sites, should use major processing crops treated in the most appropriate manner for that fruit or vegetable, e.g. canning for strawberries, quick freezing for Brussels sprouts. Secondary

TABLE II

Chemicals associated with taints in tests conducted at Chipping Campden, 1945 to 1977

Fungicides		Insecticides	
Captan	Nabam	Aldrin	Fenitrothion
Captafol	Quinomethionate	HCH	Formothion
Chlorothalonil	Streptomycin	Carbaryl	Malathion
Dicloran	Thiram	Chlorbenside	Phorate
Ferbam	Zineb	Demeton-S-methyl	Pirimiphos-ethyl
Griseofulvin	Zineb-ethylenebis (thiuram disulphide)	Diazinon	Propoxur
Lime-sulphur	Ziram	Ethoate-methyl	
Herbicides		**Miscellaneous**	
Chloroxuron	Simazine + prometryne	Aldicarb	Ethoxyquin + Thiophanate-methyl
MCPB		Dichloropropene	PP 073
		Ethoxyquin	Thiophanate-methyl

tests may also be conducted simultaneously on the same crop in other processed forms.

Only a few pesticide compounds have been shown positively to cause taints in processed foods. Probably the best example is still HCH, a chemical that became notorious as a tainting compound on crops such as carrots, potatoes and black currants, when root crops planted in land treated up to 10 years previously developed taints. It was these effects of this otherwise efficient insecticide that caused the taint work at Chipping Campden to begin.

Thiram and captan are also blacklisted because they cause taints and most processors prohibit their use on strawberries for canning or freezing. This led to problems in the past when growers had no suitable alternative to these compounds for the control of grey mould, but today other fungicides, e.g. dichlofluanid, that are effective against grey mould and produce no taints on processed strawberries are available.

Some years ago, the food industry experienced problems with "catty" taints. A powerful, feline odour that was immediately apparent on piercing a can, was so unmistakable that the product needed no tasting. The material that caused the offensive odours was eventually identified as mesityl oxide, which reacts with hydrogen sulphide to give the catty odour. Similar catty taints were detected in canned strawberries and the insecticide malathion was found to be the source of the taint. Whereas malathion of recent manufacture caused no taints, the chemical from containers which had been opened and left for a year did. Thereupon a MAFF circular was issued to warn against the use of old stocks of malathion on soft fruit for processing.

The rare occurrence of taints in processed foods during recent years therefore is perhaps an indication of the efficiency with which the chemicals are developed and tested. During 1977, of the 80 or so new chemicals or new formulations of chemicals examined in taint tests at Chipping Campden, not one led to the development of taints in processed fruit and vegetables. We can only hope that this trend will continue in years to come, now that manufacturers, processors and growers are aware of the problems of taints.

REFERENCES

Arthey, V.D. (1974). *Pestic. Sci.* **5**, 113-115.

Arthey, V.D. (1975). "Quality of Horticultural Products". Butterworths, London.

Kramer, A. and Twigg, B.A. (1970). "Quality Control for the Food Industry" (3rd edn), Vol. 2. Avi Publishing Co. Inc., Westport, Connecticut, USA.

Section IV
Post-harvest Changes—Diseases

IV.4 Effect of Spoilage Fungi on the Quality of Canned and Sulphited Fruit

C. DENNIS AND J.E. HARRIS

ARC Food Research Institute, Colney Lane, Norwich, NR4 7UA, England

INTRODUCTION

When properly controlled, canning (or bottling) and sulphiting are methods of preserving fruit which enable the product to be stored for several months after processing without loss of quality. Canning provides a food suitable for immediate consumption whereas sulphiting preserves fruit prior to processing into jams, candied and glacé fruits and products such as maraschino cherries.

Fungi can affect the quality of canned and sulphited fruit due to their ability to survive the processing conditions and grow in the processed product. Alternatively, they secrete extracellular metabolites during growth in the raw material prior to preservation, which are not destroyed by heating or by sulphite and can subsequently result in a change in quality during storage of the preserved fruits.

The paper reviews the importance of different fungi in the spoilage of canned (or bottled) and sulphited fruit and methods whereby such spoilage can be reduced or eliminated.

Limited groups of fungi are involved in spoilage of canned or sulphited fruits. In relation to canning and bottling, some Ascomycetes survive the processing by means of their heat-resistant ascospores. Ascomycetous yeasts also appear to have the potential to act as primary spoilers of sulphited fruits. Recently, pectolytic enzymes produced by members of the Zygomycetes and Ascomycetes have been shown to be particularly tolerant of heat and/or the acidity of sulphite liquors and are therefore capable of playing an important role in the softening and disintegration of the preserved fruits.

CANNED OR BOTTLED FRUITS

SPOILAGE DUE TO GROWTH OF FUNGI IN CANNED OR BOTTLED FRUIT

The original isolation of *Byssochlamys fulva* from spoiled cans and bottles of various fruits (Olliver and Smith, 1933) initiated a number of investigations on the spoilage aspects of this fungus (Olliver and Rendle, 1934; Gillespy, 1936-37, 1938, 1940, 1946; Hull, 1939; Beaven and Brown, 1949). This fungus is able to cause spoilage of canned or bottled fruit due to the ascospores having an abnormally high heat-resistance (Gillespy, 1936, 1938, 1940 and 1946), its ability to grow at low oxygen tension (Hull, 1939; Yates *et al.*, 1967) and to produce pectolytic enzymes (Olliver and Rendle, 1934; Beaven and Brown, 1949; Reid, 1952; Yates and Mooney, 1968). Rice *et al.*, (1977) have also shown that the maturity of canned peaches affects the rate of softening caused by *B. fulva.*

According to Yates (1973) spoilage of canned or bottled fruits by *Byssochlamys* species appears to be increasing or is at least more frequently recognized. Recently *B. nivea* has been implicated in the spoilage of bottled fruits in the Netherlands and was shown to produce ascospores with a similar heat-resistance to those of *B. fulva* (Put and Kruiswijk, 1964), which is contrary to the findings of Gillespy (1946). Put (1964) also found *B. nivea* more frequently than *B. fulva* in Dutch soils while Gillespy and Thorpe (1962) found a predominance of *B. nivea* in English soils. The latter authors also observed an outbreak of spoilage of bottled fruits to be due almost exclusively to *B. nivea* (see Yates *et al.*, 1968).

Byssochlamys spp. were originally considered to be restricted to Britain (Olliver and Rendle, 1934; Raper *et al.*, 1949) but have since been reported in Switzerland (Lüthi and Hochstrasser, 1952), Yugoslavia (Michovski, 1961-62), Canada (Yates and Ferguson, 1963; Yates, 1974), Holland (Put, 1964), Australia (Spurgin, 1964; Richardson, 1965) and the USA (Michener *et al.*, 1966). Splittstoesser *et al.* (1971, 1974) consider that contamination of fruit by soil, at or before harvest, is the main source of inoculum on the fruit, although the level of contamination is generally low (Splittstoesser *et al.*, 1971; Yates, 1974). Thorough washing of the fruits prior to processing will therefore help in reducing the inoculum and where possible the heat treatment should destroy the ascospores. The latter may be difficult to achieve completely since increasing the severity of the thermal treatment may reduce product quality. Furthermore, different isolates of *Byssochlamys* produce ascospores showing different heat-resistances (Splittstoesser *et al.*, 1974). Beuchat and Toledo (1977) showed that the syrup and fruit content affected subsequent germination of the ascospores as well as their heat-resistance.

Although *B. fulva* and *B. nivea* have also been reported to produce mycotoxins, byssochlamic acid (Raistrick and Smith, 1933), patulin (Ciegler *et al.*, 1971; Scurti *et al.*, 1973; Escoula, 1975; Percebois *et al.*, 1975) and byssotoxin A (Kramer *et al.*, 1976), there is as yet no evidence that such compounds are present in cans or bottles of fruit spoiled by *Byssochlamys* species.

Few other fungi have been reported to cause spoilage of heat processed fruit. A *Penicillium* specie was reported by Williams *et al.* (1941) to spoil canned blueberries. The fungus survived the canning by means of heat-resistant sclerotia and to prevent such spoilage Ruyle *et al.* (1946) suggested elimination of oxygen from the can by substituting with nitrogen or other inert gases. Although Wilson *et al.* (1978) reported *Saccharomyces cerevisiae* to have been isolated from a spoiled can of raspberries, *Saccharomyces* spp., despite their relatively heat-resistant ascospores (Put *et al.*, 1976) do not generally affect the quality of canned or bottled fruit.

SPOILAGE DUE TO GROWTH OF FUNGI IN FRUIT PRIOR TO PROCESSING

Breakdown of canned apricots was first reported to be a commercial problem in Australia in 1958 (see Harper and Beattie, 1971). Breakdown occurred sporadically and unpredictably in subsequent seasons and was of considerable economic significance because large proportions of a year's production could become unsaleable in less than 9 months. No evidence of contamination of the cans by *Byssochlamys* was obtained. A similar disorder was subsequently reported from California (Luh *et al.*, 1974; Claypool, 1974).

The initial work on this problem in Australia (Harper and Beattie, 1971) showed that addition of exudate from apricots infected with *R. stolonifer* (≡ *R. nigricans*; von Arx, 1970) to healthy apricots before canning caused disintegration of the whole can of fruit. Maceration of the fruit occurred since pectolytic activity survived the canning process (100°C for 12 min), and still retained some of its activity after boiling for 16 min (Harper, 1971; Harper *et al.*, 1972, 1973). This was the first report of a heat-stable pectolytic enzyme of fungal origin, although heat-stable polygalacturonases had been reported from cranberries (Arakji and Yang, 1969) and tomatoes (McCulloch and Kertesz, 1948). Archer and Fielding (1975) have recently reported a highly heat-stable polygalacturonase produced by *Sclerotinia fructigena*.

Later work with apricots in California showed that the presence of exudate from apricots infected with *Rhizopus arrhizus* and *Gilbertella persicaria*, as well as *R. stolonifer*, contained heat-tolerant pectolytic enzymes (Luh *et al.*, 1974; Ogawa *et al.*, 1974; Sommer *et al.*, 1977). In canning trials isolates of *R. arrhizus* were found to be considerably more active than *R. stolonifer*. Ogawa *et al.* (1974) also reported *R. arrhizus* to be the predominant *Rhizopus* specie in Californian apricot orchards.

More recent work by Sommer *et al.* (1977) has demonstrated that extremely small amounts of juice (< 1 ml in a No. 303 can) from fruit infected with *R. arrhizus* and *R. stolonifer* was sufficient to soften the fruits within 7½ months. It is considered that more rigorous heat treatment is not possible because of the tendency of apricots to disintegrate with excessive cooking. Furthermore, some pectolytic activity was still detectable after autoclaving (121°C for 20 min) (Sommer *et al.*, 1977). Luh *et al.* (1978) suggested that sugars and other cellular compounds may protect the polygalacturonase produced in apricots infected with *R. arrhizus* from heat denaturation. Heat-stability studies on polygalacturonase activity by these workers showed that heat denaturation occurred in two steps. An initial rapid inactivation was followed by a slower inactivation which may indicate the presence of polygalacturonase isoenzymes possessing different heat-stabilities (see also Archer and Fielding, 1975, 1979).

In theory the problem of fungal enzyme mediated breakdown can be reduced by controlling pre- and post-harvest infection of fruit by the causal fungi or by the removal of any infected fruits prior to canning. The time between harvest and canning should be reduced to a minimum so that further mould development does not occur after harvest, especially on mechanically damaged fruits.

Since extremely small amounts of exudate from infected fruits can result in softening of the canned product, inactivation of pectolytic enzymes prior to canning would be a more effective method of control. In this context Sommer *et al.* (1977) showed that treatment of infected fruit with sodium hydroxide (0·5-1 N for 0·5-2 min) followed by washing, inactivated pectolytic enzymes in apricot halves infected with *R. arrhizus* and consequently prevented softening of the canned fruit. Thus Sommer *et al.* (1977) suggest that a satisfactory treatment would consist of a combination of a vigorous water-spray wash (to open infected areas) followed by a dip or spray of sodium hydroxide. The success of such a cold lye treatment will depend on maintaining the correct concentration of lye and contact time, followed by a thorough rinsing of treated fruit.

Strand *et al.* (1977) reported that the pectolytic activities of fungi important in causing progressive softening of canned apricots have bimodal heat stability curves with minimal stabilities at 70°-80°C (see also Archer and Fielding, 1975). A 6-minute treatment at 75°C before canning eliminated detectable enzyme activity and significantly reduced subsequent softening in No. 2½ cans of apricots to which single fruit halves with 5 mm lesions had been added (Strand *et al.*, 1977). Such data suggest that pre-treatment at about 75°C could help to alleviate the breakdown problem of canned apricots.

SULPHITED FRUIT

The food processing industry requires an all year round supply of many fruits which are harvested for only a few weeks of the year in most countries. Sulphite preservation adequately provides this facility and when properly controlled is a relatively cheap and efficient method of preservation. Fruits preserved in this way (strawberries, raspberries, blackberries, blackcurrants, apricots, citrus, cherries), can be used for the manufacture of jams, marmalades, candied and glacé fruits and maraschino cherries.

Sulphite liquor consists of sulphurous acid and calcium bisulphite solutions each containing equal concentrations of sulphur dioxide (SO_2), although the exact composition of the liquor can be varied according to the fruits concerned and their condition at harvest (Doesburg, 1965; Watters, 1975). Sulphite preservation of fruits often aids the manufacturing process. The formation of calcium pectate gels in the fruit and liquor helps to firm the fruit and provides a matrix which prevents physical disruption of the tissue during transport and handling prior to manufacture. In the case of sulphited sweet cherries, bleaching by the sulphur dioxide to a light straw colour facilitates subsequent dyeing to the desired shade for brightly coloured products such as maraschino cherries.

SPOILAGE DUE TO GROWTH OF FUNGI IN SULPHITED FRUIT

When used at the recommended concentrations, there is sufficient sulphur dioxide present in the sulphite liquor to completely inhibit fungal growth.

Dennis (1978), however, reported spoilage of sulphited strawberries as a result of fungal growth. This appears to be the only report of such spoilage and it is of interest that similar spoilage of other sulphited fruits does not appear to occur. Spoilage of sulphited strawberries results in the fruit turning pink, having a fruity, winey smell and this renders it unusable. Only the top 3-4 inches of fruit in the barrel is affected with the remainder being completely bleached and suitable for the manufacture of good quality jam. *Saccharomyces* species (*S. bailii, S. florentinus, S. cerevisiae, S. kluyveri, S. uvarum*) as well as film yeasts, *Hansenulla subpelliculosa* and *Pichia membranefaciens*, have been isolated from spoiled samples of fruit. In addition *Penicillium* species have also been observed to develop on the surface of severely spoiled barrels of fruit.

Saccharomyces bailii has been widely reported as a preservative-resistant yeast (Pitt and Richardson, 1973; Pitt, 1974) growing in the presence of 500 ppm SO_2 (Reed and Peppler, 1973). In contrast, Hammond and Carr (1976) reported that SO_2 rapidly kills yeasts. They showed that less than 25 ppm SO_2 (pH 3·4) markedly reduced the viable count of *S. cerevisiae* after 8 h at 25°C. Similarly Reed and Peppler (1973) report *Pichia membranefaciens* to be sensitive to SO_2.

Since SO_2-sensitive yeasts have been isolated from the spoiled fruit, it would appear that spoilage results from inadequate mixing of liquor and fruit when sufficient inoculum of appropriate yeasts is present. More information is required on the incidence of the above yeasts in relation to fruit ripeness since Buhagiar and Barnett (1971) only very rarely isolated the relevant yeasts from strawberries. The sensitivity of these yeasts to SO_2 in relation to their growth on sulphited fruit also needs to be investigated.

SPOILAGE DUE TO GROWTH OF FUNGI IN FRUIT PRIOR TO SULPHITING

Despite the presence of calcium in the sulphite liquor, softening or disintegration of the sulphited fruit can occur during storage prior to secondary processing. This type of spoilage occurs in a whole range of sulphited fruits and can result in a serious loss of quality of the fruit as a raw material for the manufacturing process, especially where whole or recognizable parts of the original fruits are required in the manufactured product (e.g. cherries for the production of maraschino or glacé fruit; strawberries and apricots for the production of good retail jams; citrus for the production of marmalade and candied fruits). In other cases, however, (e.g. raspberries, blackberries) disintegrated fruit is still suitable for the manufacture of good quality jams.

Investigations have shown that pectolytic enzymes of fungal origin cause disintegration of sulphited cherries and strawberries (see below), and personal observations have also indicated that similar enzymes cause disintegration of sulphited orange peel, apricots and raspberries.

Cherries

Softening of cherries preserved in calcium bisulphite brines was first noticed in about 1947 (see Wiegand, 1957). The extensive softening recorded in Oregon and Washington in 1957 when 11 800 barrels, valued at more than £½ million, were affected (Wiegand, 1957), stimulated investigation into the cause of the softening.

McCready and McComb (1954) related softening of brined cherries to the presence of pectin-degrading enzymes. Steele and Yang (1960) demonstrated the presence of large concentrations of polygalacturonase in cherries showing the "cats claw" disease and showed that such cherries (10% by number) caused rapid deterioration of a sulphited sample. Thus the polygalacturonase was considered to be of microbial origin. Later work by Lewis *et al.* (1963) showed that cherries infected with *Aspergillus niger, Cytospora leucostoma* and *Penicillium expansum* caused softening of sound sulphited cherries. Rapid softening was correlated with the presence of a sulphite-stable polygalacturonase. Surveys of naturally-rotted cherries indicated the involvement of *C. leucostoma* in the softening of commercially-brined cherries in the Pacific North-west (Lewis *et al.*, 1963).

The demonstration by Steele and Yang (1960) that polygalacturonase activity and softening of sulphited cherries could be inhibited by alkyl aryl sulphonates in model systems, and a similar effect in commercial samples of sulphited fruit (Yang *et al.*, 1960), has provided a satisfactory method for the elimination of the problem. The addition of 0·025% alkyl aryl sulphonate (e.g. Nacconol) as part of the sulphite liquor did not affect the manufacture or subsequent quality of maraschino cherries (Yang *et al.*, 1960). Brekke *et al.* (1963) provided evidence that heating the liquor and cherries to about 75°C or lowering the pH of the liquor by addition of calcium chloride prevented softening. According to Yang *et al.* (1960) and Wiegand (1957), however, the addition of calcium chloride causes the cherries to crack and their skins to injure easily.

Strawberries

The softening and complete breakdown of sulphited strawberries has occurred as a sporadic commercial problem in the UK and the Netherlands since the 1950s and resulted in a number of investigations (Pandhi, 1953; Hinton and Pandhi, 1958; Staden and Doesburg, 1960, 1961, 1962, 1963; Staden, 1964; Dakin and Tampion, 1965). The possible relationship between fruit infected with various spoilage fungi and breakdown was first reported by Dakin and Tampion (1968). These workers showed that the inclusion of 25% by weight of fruit infected with *Mucor* spp., *Rhizopus stolonifer* and *R. sexualis* resulted in complete disintegration of samples of sulphited strawberries. Fruit infected with a single strain of *Botrytis cinerea* caused no softening. They concluded from laboratory experiments that *R. stolonifer* was the most active fungus in causing breakdown. Recent work by Dennis and Harris (1979) confirmed the potential for berries infected with *R. stolonifer* and *R. sexualis* to cause breakdown and also illustrated that *Mucor piriformis* was the most likely *Mucor* species involved. As little as 1-2% of infected tissue for each of the three species consistently caused breakdown, while with *R. sexualis* even 0·1% resulted in some softening and breakdown of the fruit. The inability of berries infected with *B. cinerea* to cause breakdown was confirmed in experiments with artificially infected fruit (6 strains) and naturally infected fruit (10 samples) during 12 months' storage in sulphite. Dennis and Harris (1979) also showed that culture filtrates of the *Mucor* and *Rhizopus* species possessing pectolytic activity, together with those of other pectolytic fungi associated with strawberries (*Aureobasidium pullulans, Trichosporon pullulans, Cryptococcus albidus* var. *albidus*), also possessed the potential to cause breakdown. The destruction of pectolytic activity in culture filtrates and infected tissue by boiling was coincident with a loss in the ability to cause breakdown, and provided evidence that pectolytic enzymes of fungal origin are the primary cause of breakdown (Dennis and Harris, 1979). Archer (1979) further demonstrated that the disintegration of strawberries in sulphite liquor was associated with pectin degradation since a loss of liquor viscosity, a reduction in degree of polymerization of pectic substances and the appearance of galacturonic acid in the liquor coincided with the presence of fungal endopolygalacturonase. The endopolygalacturonases from *R. sexualis* and *R. stolonifer* persisted longer in sulphite liquor than the enzymes from *Mucor piriformis* (Harris and Dennis, 1979). The biphasic loss of polygalacturonase activity for the *Rhizopus* species suggested the possible existence of polygalacturonases showing different rates of inactivation in sulphite liquor. This was confirmed in electrofocusing experiments by Archer and Fielding (1979), who demonstrated the production of an endopolygalacturonase isoenzyme (pI 7·2) by *Rhizopus* species which was particularly stable at low pH. It is considered that this isoenzyme is mainly responsible for the breakdown of sulphited strawberries.

Observations on commercial samples of fruit by Dennis *et al.* (1979) indicated that *R. sexualis* was the most important cause of breakdown of sulphited fruit (cf. Dakin and Tampion, 1968). *Rhizopus stolonifer* occurred very rarely on strawberries from any of the fruit growing areas studied (Wisbech area, East Norfolk; County Wexford, Southern Ireland) and thus was not considered to be important. A greater proportion of the sulphited fruit from the Wisbech area showed softening and disintegration which corresponded to both a higher incidence of *R. sexualis* and of sulphite liquors showing detectable pectolytic activity.

In order to achieve maximum control of breakdown Dennis *et al.* (1979) showed that the interval between harvesting and sulphiting should be kept to a minimum. In addition they suggest that due to the absence of fungicides effective against *Mucor* and *Rhizopus* species, good plantation hygiene and regular picking would help to control the problem by reducing infection in the field. Despite these measures it is unlikely that breakdown of sulphited fruit would be completely prevented since only a very small amount of infected tissue causes softening and breakdown (Dennis, 1978; Dennis and Harris, 1979). To eliminate breakdown it would be necessary to inhibit all polygalacturonase activity in the sulphite liquor. Archer (1979) demonstrated that surfactants (0·01%), similar to those used by Steele and Yang (1960) and Yang *et al.* (1960) to prevent softening of sulphited cherries, were inhibitory to the polygalacturonases from *Mucor* and *Rhizopus.* However, the incorporation of the sulphite liquor into the manufactured product prevents their use with sulphited strawberries.

Our recent work has shown that treatment of infected strawberries, prior to sulphiting, with alkali (1·0-0·1 N for 0·5-5 min) failed to reduce pectolytic activity in the liquor (cf. Sommer *et al.*, 1977). Even treatment of homo-

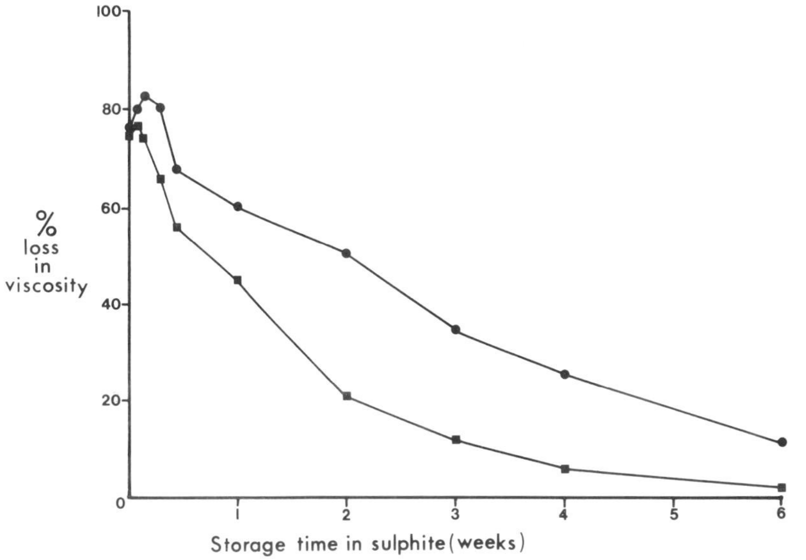

Fig. 1. Pectolytic activity in liquors of sulphited strawberries. The pectolytic activity of dialysed liquors was measured viscometrically using sodium polypectate as substrate. ●——●, liquor containing 10% ($^{w}/_{w}$) strawberry tissue infected with *Rhizopus sexualis*; ■——■ liquor containing 50% ($^{w}/_{w}$) strawberry tissue infected with *Botrytis cinerea* and 10% ($^{w}/_{w}$) tissue infected with *R. sexualis.* The sulphited fruit was stored at 15°C.

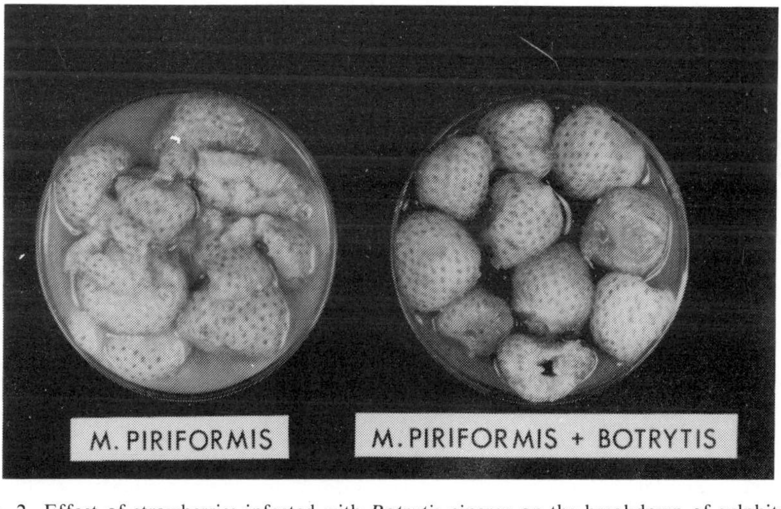

Fig. 2. Effect of strawberries infected with *Botrytis cinerea* on the breakdown of sulphited strawberries caused by *Mucor piriformis.* The sulphited fruit was stored at 15°C for 16 weeks.

genized infected tissue (*R. sexualis*) with 1 N NaOH (pH 12) for 1 min did not completely destroy pectolytic activity. Thus, such alkali-treatment would not give the same measure of control of breakdown as achieved for canned apricots (Sommer *et al.*, 1977).

It was mentioned earlier that the presence of strawberries infected with *B. cinerea* did not cause softening of sulphited fruit. Further work showed that pectolytic activity could not be detected in extracts of berries infected with *B. cinerea*, and that the presence of such berries also accelerates the loss of pectolytic activity in samples of sulphited fruit containing 10% tissue infected with *R. sexualis, R. stolonifer* and *M. piriformis* (Fig. 1). Although breakdown of the healthy fruit was not prevented in the case of *R. sexualis* and *R. stolonifer*, the berries infected with *M. piriformis* remained whole (Fig. 2). This data indicates that depending on the relative incidence of infected fruit, the presence of fruit infected by *B. cinerea* could prevent enzymic breakdown of sulphited fruit caused by *Mucor* and *Rhizopus* species. Further work is in progress on the fractionation and identification of the pectolytic enzyme inhibitor present in strawberries infected with *B. cinerea.*

REFERENCES

Arakji, O.A. and Yang, H.Y. (1969). *J. Fd Sci.* **34**, 340-342.
Archer, S.A. (1979). *J. Sci. Fd Agric.* **30**, 692-703.
Archer, S.A. and Fielding, A.H. (1975). *J. Fd Sci.* **40**, 423-424.
Archer, S.A. and Fielding, A.H. (1979). *J. Sci. Fd Agric.* **30**, 711-723.
Arx, J.A.von (1970). *In* "The Genera of Fungi Sporulating in Pure Culture". J. Cramer, Vaduz.
Beuchat, L.R. and Toledo, R.T. (1977). *Trans. Br. mycol. Soc.* **68**, 65-71.
Beaven, G.H. and Brown, F. (1949). *Biochem. J.* **45**, 221-224.
Brekke, J.E., Watters, G.C. and Powers, M.J. (1963). *Cann. Pack.* **132**, (8), 29.
Buhagiar, R.W.M. and Barnett, J.A. (1971). *J. appl. Bact.* **34**, 727-739.
Ciegler, A., Detroy, R.W. and Lillehoj, E.B. (1971). *In* "Microbial Toxins" (A. Ciegler, S. Kadis and S.J. Ajl, eds), Vol. VI, pp. 409-434. Academic Press, New York and London.
Claypool, L.L. (1974). *Calif. Agric.* **28**, (7), 4-6.
Dakin, J.C. and Tampion, J. (1965). *Tech. Circ. Br. Fd Mfg Ind. Res. Ass. No. 280.* 1-9.
Dakin, J.C. and Tampion, J. (1968). *J. Fd Technol.* **3**, 39-48.
Dennis, C. (1978). *ARC Res. Rev.* **4**, 38-42.
Dennis, C. and Harris, J.E. (1979). *J. Sci. Fd Agric.* **30**, 687-691.
Dennis, C., Davis, R.P., Harris, J.E., Calcutt, L.W. and Cross, D. (1979). *J. Sci. Fd Agric.* **30**, 959-973.
Doesburg, J.J. (1965). *In* "Pectic Substances in Fresh and Preserved Fruits and Vegetables", I.B.V.T. Communication No. 25. Wageningen, Netherlands.
Escoula, L. (1975). *Ann. Rech. vété.* **6**, 303-310.
Gillespy, T.G. (1936-37). *Rep. Fruit Veg. Preserv. Res. Stn* 68-75.

Gillespy, T.G. (1938). *Rep. Fruit Veg. Preserv. Res. Stn* 68-78.
Gillespy, T.G. (1940). *Rep. Fruit Veg. Preserv. Res. Stn* 54-61.
Gillespy, T.G. (1946). *Rep. Fruit Veg. Preserv. Res. Stn* 31-39.
Gillespy, T.G. and Thorpe, R.H. (1962). *Tech. Memo No. 44. Fruit Veg. Cann. Quick Freez. Res. Ass.*
Hammond, S.M. and Carr, J.G. (1976). *In* "Inhibition and Inactivation of Vegetative Microbes" (F.A. Skinner and W.B. Hugo, eds), pp. 89-110. Academic Press, London and New York.
Harper, K.A. (1971). *Chemy Ind.* 462.
Harper, K.A. and Beattie, B.B. (1971). *Fd Res. Q.* **31**, 33-36.
Harper, K.A., Beattie, B.B., Pitt, J.I. and Best, D.J. (1972). *J. Sci. Fd Agric.* **23**, 311-320.
Harper, K.A., Beattie, B.B. and Best, D.J. (1973). *J. Sci. Fd Agric.* **24**, 527-531.
Harris, J.E. and Dennis, C. (1979). *J. Sci. Fd Agric.* **30**, 704-710.
Hinton, C.L. and Pandhi, P.N. (1958). *Scient. tech. Surv. Br. Fd Mfg Ind. Res. Ass. No. 32*, 2-16.
Hull, R. (1939). *Ann. appl. Biol.* **26**, 800-822.
Kramer, R.K., Davis, N.D. and Diener, U.L. (1976). *Appl. Environ. Microbiol.* **31**, 249-253.
Lewis, J.C., Pierson, C.F. and Powers, M.J. (1963). *Appl. Microbiol.* **11**, 93-99.
Luh, B.S., Peupier, L.Y. and Liu, Y.K. (1974). *Calif. Agric.* **28**, (7), 4-6.
Luh, B.S., Ozbilgin, S. and Liu, Y.K. (1978). *J. Fd Sci.* **43**, 713-716.
Lüthi, H. and Hochstrasser, R. (1952). *Schweiz. Z. Obstr.-u. Weinb.* **61**, 301-358.
McColloch, R.J. and Kertesz, Z.I. (1948). *Arch. Biochem.* **17**, 197-199.
McCready, R.M. and McComb, E.A. (1954). *West. Cann. Pack.* **46**, (13) 17-19.
Michener, H.D., King, A.D., Jr., and Ito, K.A. (1966). *Bact. Proc.* (A 15), 3.
Mickovski, M. *Goaišen Zb. zenijod.-šum. Fak. Univ. Skopje* 1961-62, 190-191.
Ogawa, J.M., Rumsey, J., Manji, B.T., Tate, G., Toyoda, J., Bose, E. and Dugger, L. (1974). *Calif. Agric.* **28**, (7) 6-7.
Olliver, M. and Rendle, T. (1934). *J. Soc. chem. Ind.* (*Lond.*), **53**, 166-172.
Olliver, M. and Smith, G. (1933). *J. Bot.* (*Lond.*), **71**, 196-197.
Pandhi, P.N. (1953). *Scient. tech. Surv. Br. Fd Mfg Ind. Res. Ass. No. 22.* 2-22.
Percebois, G., Basile, A.M. and Schwertz, A. (1975). *Mycopath.* **57**, 109-111.
Pitt, J.I. (1974). *Fd Technol. Aust.* **26**, 238-241.
Pitt, J.I. and Richardson, K.C. (1973). *C.S.I.R.O. Fd Res. Q.* **33**, 80-85.
Put, H.M.C. (1964). *J. appl. Bact.* **27**, 59-64.
Put, H.M.C. and Kruiswijk, J. Th. (1964). *J. appl. Bact.* **27**, 53-58.
Put, H.M.C., De Jong, J., Sand, F.E.M.J. and Van Grinsven, A.M. (1976). *J. appl. Bact.* **40**, 135-152.
Raistrick, H. and Smith, H. (1933). *Biochem. J.* **27**, 1814-1819.
Raper, K.B., Thom, C. and Fennel, D.I. (1949). *In* "A Manual of the *Penicillia*". Williams and Wilkins Co., Baltimore.
Reed, G. and Peppler, H. (1973). *In* "Yeast Technology". Avi Publishing Co. Inc., Westport, Connecticut, USA.
Reid, W.W. (1952). *Biochem. J.* **50**, 289-292.
Rice, S.L., Beuchat, L.R. and Heaton, E.K. (1977). *J. Fd Sci.* **42**, 1562-1565.
Richardson, K.C. (1965). *Qd J. agric. Sci.* **22**, 347-350.
Ruyle, E.H., Pearce, W.E. and Hays, G.L. (1946). *Fd Res.* **11**, 274-279.
Scurti, J.C., Codignola, A., Nobili, G. and Caputo, O. (1973). *Allionia* **19**, 39-42.
Sommer, N.F., Buchanan, J.R. and Fortlage, R.G. (1974). *Calif. Agric.* **28**, (7) 8.

Sommer, N.F., Buchanan, J.R., Fortlage, R.J. and Mitchell, F.G. (1977). *Calif. Agric.* **31**, (4) 16-17.

Splittstoesser, D.F., Kuss, F.R., Harrison, W. and Prest, D.B. (1971). *Appl. Microbiol.* **21**, 335-337.

Splittstoesser, D.F., Downing, D.L., Roger, N.F. and Murdock, D.I. (1974). *J. Milk Fd Technol.* **37**, 445-447.

Spurgin, M.W. (1964). *Qd J. agric. Sci.* **21**, 247-250.

Staden, O.L. (1964). *Rep. Inst. Res. Stor. Process. hort. Prod. Wageningen,* 42.

Staden, O.L. and Doesburg, J.J. (1960). *Rep. Inst. Res. Stor. Process. hort. Prod. Wageningen*, 53-54.

Staden, O.L. and Doesburg, J.J. (1961). *Rep. Inst. Res. Stor. Process. hort. Prod. Wageningen*, 60-62.

Staden, O.L. and Doesburg, J.J. (1962). *Rep. Inst. Res. Stor. Process. hort. Prod. Wageningen*, 65-67.

Staden, O.L. and Doesburg, J.J. (1963). *Rep. Inst. Res. Stor. Process. hort. Prod. Wageningen*, 47.

Steele, W.F. and Yang, H.Y. (1960). *Fd Technol., Champaign,* **14**, 121-126.

Strand, L.L., Ogawa, J.M., Bose, E. and Rumsey, J.W. (1977). *Proc. Am. phytopath. Soc.* **4**, 102.

Watters, G.G. (1975). *In* "Commercial Fruit Processing" (J.G. Woodroof and B.S. Luh, eds), pp. 430-446. Avi Publishing Co. Inc., Westport, Connecticut, USA.

Weigand, E.H. (1957). *Proc. Oregon Hort. Soc.* **36**, 47-49.

Williams, C.C., Cameron, E.J. and Williams, O.B. (1941). *Fd Res.* **6**, 69-73.

Wilson, J.M., Wood, J.M. and Jarvis, B. (1978). *Res. Rep. Br. Fd Mfg Ind. Res. Ass. No. 275*, pp. 17.

Yang, H.Y., Steele, W.F. and Graham, D.J. (1960). *Fd Technol. Champaign,* **14**, 644-647.

Yates, A.R. (1973). *Can. Inst. Fd Sci. Technol. J.* **6**, 244-247.

Yates, A.R. (1974). *Can. Inst. Fd Sci. Technol. J.* **7**, 148-150.

Yates, A.R. and Ferguson, W.E. (1963). *Can. J. Bot.* **41**, 1599-1601.

Yates, A.R. and Mooney, D.B. (1968). *J. Inst. Can. Technol. Aliment.* **1**, 106-109.

Yates, A.R., Seaman, A. and Woodbine, M. (1967). *Can. J. Microbiol.* **13**, 1120-1123.

Yates, A.R., Seaman, A. and Woodbine, M. (1968). *Can. J. Microbiol.* **14**, 319-325.

IV.5 The Effect of Bacteria on Post-harvest Quality of Vegetables

BARBARA M. LUND

*ARC Food Research Institute,
Colney Lane, Norwich,
NR4 7UA, England*

INTRODUCTION

Bacteria can affect the post-harvest quality of vegetables by causing blemishes or decay and consequent wastage. The extent of post-harvest losses in fresh fruit and vegetables due to all causes has been discussed in recent reviews (Eckert, 1975; Harvey, 1978) and may be particularly great in less-developed, tropical countries. While some of these losses are attributable to mechanical damage and to physiological causes, a substantial part is due to microbial spoilage. In the case of vegetables and salad crops bacteria can be responsible for a considerable part of this wastage (Ramsey *et al.*, 1959; Ramsey and Smith, 1961; Smith *et al.*, 1966; McCollock *et al.*, 1968; Ceponis, 1970; Ceponis *et al.*, 1970; Coursey and Booth, 1972; Ceponis and Butterfield, 1974a,b; Beraha and Kwolek 1975; Smith and Wilson, 1978); the proportion will depend on environmental conditions and information regarding the extent of bacterial spoilage is sparse.

Spoilage of vegetables by bacteria can be divided into two categories (Lund, 1971):

(i) That due to bacteria which cause extensive soft rot; vegetables may be attacked in the field (Chupp and Sherf, 1960) or spoilage initiated in the field or post-harvest may progress during transit or storage.

(ii) That due to field infection by bacteria causing blemishes or surface defects which detract from the value of the crop; this spoilage may progress after harvest and be followed by decay due to soft rot bacteria. Examples of this second category are described in the US Department of Agriculture Handbooks cited above and include blemishes on tomato fruit caused by bacterial canker (*Corynebacterium michiganense*), bacterial speck (*Pseudomonas tomato*) (Upstone, 1971; Wilkie and Dye, 1974) and bacterial spot (*Xanthomonas vesicatoria*), on beans caused by halo-blight (*Pseudomonas phaseolicola*), on cabbage (Smith and Ramsey, 1956) and lettuce (Grogan *et al.*, 1977) by *Pseudomonas cichorii*, and on potatoes and other root vegetables by scab (*Streptomyces scabies*).

This paper is mainly concerned with the first category of spoilage, and with the influence of post-harvest factors.

When considering the effect of bacteria on the quality of vegetables it should be noted that salad crops, in particular, can act as a vehicle for transfer of bacteria and other micro-organisms which cause human disease. Increasing demands on water supplies for irrigation and changes in methods used during transit, storage and marketing of produce could make this a factor of increasing importance. Further discussion of this topic is outside the scope of this paper; it has been considered recently by Geldreich and Bordner (1971), Nicholls *et al.* (1971), Ercolani (1976), Kaferstein (1976), Bryan (1977), Maxcy (1978), Robinson and Adams (1978), Tamminga *et al.* (1978).

BACTERIA WHICH CAUSE EXTENSIVE SOFT ROT OF VEGETABLES

The most important bacteria involved are *Erwinia carotovora* (Lelliott, 1974) and fluorescent pseudomonads including *Pseudomonas marginalis* and similar organisms. [In the classification of Doudoroff and Palleroni (1974), *P. marginalis* was incorporated into *P. fluorescens* biotype II. In this paper the former name is used for reasons stated by Dye *et al.* (1975).] *Erwinia carotovora* and *P. marginalis* can cause spoilage of a very wide range of vegetables in the field (Chupp and Sherf, 1960) and after harvest (Ryall and Lipton, 1972); this effect on plant tissue results from their ability to produce active pectic enzymes which include pectinesterase, pectate lyase and polygalacturonase (Nasuno and Starr, 1966; Rombouts *et al.*, 1978; refs. cited in Lund, 1979). *Erwinia carotovora* var. *atroseptica* and var. *carotovora* are usually the main cause of bacterial soft rot in potatoes but pectolytic *Clostridium* spp. are often also involved (Lund, 1979), while *P. marginalis* and pectolytic *Bacillus* spp. may sometimes be implicated.

BACTERIA ASSOCIATED WITH FRESH VEGETABLES

Large numbers of bacteria may be present on the surface of fresh vegetables. Viable counts of 8×10^4-2×10^6 bacteria g^{-1}, 1×10^3-5×10^3 pectate-degrading bacteria g^{-1} and 80-1×10^3 fluorescent pseudomonads g^{-1} were reported on celery at harvest (Robinson and Adams, 1978). On the outer leaves of freshly harvested winter white cabbage viable counts of 5×10^2 - 10^5 bacteria cm^{-2} of leaf surface were found, including 2-$1{\cdot}2 \times 10^4$ pectolytic bacteria cm^{-2} about 80% of which were fluorescent pseudomonads similar to *P. marginalis* (Geeson, 1979). Freshly harvested cauliflower curd may have bacterial counts of 10^5 g^{-1} including 10^4 - 10^5 pectolytic bacterial g^{-1} tissue; a high proportion of these pectolytic bacteria are usually *P. marginalis* (Tables I and II). With the methods used in this work *E. carotovora* was seldom detected on fresh curd, but the presence of this bacterium was revealed by its occurrence in high numbers on curds which subsequently showed decay during storage. Rombouts (1972) also reported that the majority of pectolytic bacteria from five fresh vegetables were *Pseudomonas* spp, 75% of which were fluorescent strains.

Because of the importance of soft rot bacteria in decay of lettuce in the field (Burkholder, 1954), in bottom rot of lettuce (in conjunction with *Rhizoctonia solani*, Pieczarka and Lorbeer, 1975a), and of *P. marginalis* in

TABLE I

Numbers of micro-organisms[a] ($\log_{10}$ count g^{-1} curd) on fresh cauliflower (Lund and Wyatt, unpublished)

Cultivar	Barrier Reef	Nevada	South Pacific	Barrier Reef
Date	2-16/10/67[b]	29/7/74[b]	19/9/74[c]	17/10/74[c]
Site[d]	1	2	3	3
Number of heads	9	5	20	5
Total bacteria	5·1	5·3	5·3	5·8
Fluorescent pseudomonads[e]	(40%)[f]	4·9	5·0	5·7
Pectolytic bacteria	(12%)[f]	4·7[g]	4·5[h]	5·0[h]
E. carotovora	ND[i]	ND	2·8[j]	ND
Yeasts	3·6	3·6	3·9	4·1
Moulds	2·3	2·8	2·6	2·3

[a]Total bacteria counted on Plate Count agar (Oxoid); fluorescent pseudomonads on medium B of King *et al.* (1954); pectolytic bacteria including *E. carotovora* on Paton's (1959) medium; yeasts and moulds on yeast-malt agar pH 3·4. Counts expressed as geometric means. [b]Counts made on day of harvest. [c]Counts made after storing heads overnight at 2·5°C. [d]Site 1 = Cambridge, 2 = Kirton, Lincs 3 = Luddington, Warwickshire. [e]The majority of strains were *P. fluorescens*-*P. putida* group and included *P. marginalis*. [f]See Table 2. [g]The majority of pectolytic colonies were probable *Flavobacteria* (Lund, 1969). [h]Between 10% and 100% of pectolytic colonies were *P. marginalis*. [i]ND = not detected. [j]Mean count based on 4 heads in which this bacterium was detected.

TABLE II

Bacteria[a] isolated from freshly harvested, sound cauliflowers cv. Barrier Reef, 2/10/67-16/10/67 (Lund, unpublished)

Total isolates	346
Fluorescent *Pseudomonas* spp.	138 (40%)
Pectolytic bacteria[b]	40 (12%)
P. marginalis	13
P. viridiflava	2
Flavobacterium spp.[c]	16
Cytophaga spp.	4
Unidentified	5

[a]Viable counts of bacteria were made on Heart Infusion Agar. From these count plates, 20 representative strains were isolated from counts on both the curd and leaves of each of nine cauliflowers. [b]Tested on Paton's (1959) medium. [c]Lund (1969).

pink rib (Hall *et al.*, 1972) and in soft rot in transit, storage and market conditions, a study was made of the incidence of soft rot bacteria on lettuce plants during growth (Pieczarka and Lorbeer, 1975b). On the lowest healthy green leaves of eight-week-old plants, the number of pectolytic bacteria reached 10^3 cm^{-2} leaf surface in non-ridged fields and approximately 60 cm^{-2} in ridged fields. The pectolytic bacteria were not identified and not all were able to cause decay of lettuce. The presence of lower numbers of bacteria on leaves from ridged than from non-ridged plants was attributed to the drier conditions and better circulation of air in the former case.

These examples show that *P. marginalis* can usually be detected readily on the surface of fresh, sound vegetables where it forms a relatively high proportion of the bacteria present. There appear to be few data relating to the presence of *E. carotovora* on sound, leafy vegetables possibly because, if present, it usually forms a relatively low proportion of the total bacteria. An indication of its occurrence is given by the report that it can be detected on the leaves of potatoes during wet weather at the end of the growing season (Pérombelon and Lowe, 1976), and dissemination of this bacterium by insects and in aerosols is being investigated (Graham and Harrison, 1975; Pérombelon and Lowe, 1976; Harrison *et al.*, 1977). *Erwinia carotovora* survives on stored potato tubers (Pérombelon, 1972, 1973) and there is also evidence of its occurrence in the rhizosphere of cruciferous plants (Kikumoto, 1974; Mew *et al.*, 1976; refs. cited in Lund, 1979).

During harvesting and transport a considerable increase in bacterial counts on vegetables may result from contamination of harvesting equipment (e.g. pea viners) and multiplication of bacteria in warm, humid conditions before produce is cooled (Splittstoesser, 1970).

Bacteria may also be present within healthy plant tissue (Hayward, 1974). Careful studies have shown the presence of bacteria, particularly pseudomonads and *Enterobacteriaceae* within cucumbers and tomatoes (Samish and Dimant, 1959; Samish *et al.*, 1961; Samish and Etinger-Tulczynska, 1963; Samish *et al.*, 1963; Meneley and Stanghellini, 1972, 1974). In tomatoes bacteria were found most frequently in the stem scar and underlying central core; evidence from inoculating *Serratia* sp. on to sepals of young fruit indicated that bacteria could enter through connective tissue at the stem end. In contrast, bacteria were more numerous in the outer tissues of cucumbers than in the centre, possibly indicating entry of bacteria through stomata or broken hairs. The marked differences found between the numbers of bacteria in tomatoes from different fields were attributed to differences in varietal characteristics, cultural practices and climatic influences, and overhead irrigation was associated with the occurrence of higher numbers of bacteria within fruit. In healthy tissue, growth of bacteria was probably limited, but under conditions in which the structure of tissue was stressed rapid multiplication could occur. During preparation of fermented, brined tomatoes and cucumbers by the action of lactic acid bacteria, growth of gram-negative organisms within the fruit resulted in undesirable changes such as gas formation in the internal cavities (''bloaters'', Samish *et al.*, 1963).

Multiplication of endophytic bacteria in tomatoes has been implicated as a cause of ''graywall'' in field conditions in Florida (Stall and Hall, 1969; Hall *et al.*, 1970). Sound watermelons have also been reported to contain bacteria which, under certain conditions, can multiply and cause the symptom of rind necrosis (Hopkins and Elmstrom, 1977).

FACTORS AFFECTING BACTERIAL SPOILAGE OF VEGETABLES

INITIAL QUALITY AND DAMAGE

If wastage is to be minimized, vegetables should be harvested at the optimum stage of maturity and should be free from incipient infections. In some cases bacterial spoilage can probably result from entry of bacteria via stomata, or via lenticels in potatoes (Lund, 1979), but mechanical damage during harvesting and handling is a major factor leading to bacterial spoilage. Bacterial soft rot of field-grown tomatoes in the United States is a problem in crops which show damage due to soil-scarring or to mechanical harvesting (Bartz and Crill, 1974). Careful harvesting and handling of cauliflowers resulted in maintenance of better quality of heads stored in retail conditions than normal, commercial handling (Table III). Damage to curds resulted in discolouration associated with increased numbers of micro-organisms, particularly pectolytic bacteria (Table IV), and soft rot was likely if leaf cover or packing of the heads provided humid conditions around the curd.

TABLE III

The effect of method of harvesting and handling on quality of cauliflowers during shelf-life. Heads were kept at 20°C and 70% r.h. for 3-4 days (Lund and Wyatt, unpublished)

Date of harvest	Cultivar	Treatment[a]	% of heads in class after storage[b] Class[c] 1	2	3	4
17/9/74	South	a	100	0	0	0
	Pacific	b	51	35	11	3
17/10/74	Barrier	a	50	39	11	0
	Reef	b	25	45	23	7
18/2/75	Saint	a	100	0	0	0
	Keverne	c	80	12	8	0
23/6/76	Delso	a	98	2	0	0
		b	67	24	9	0

[a] Treatment a, heads harvested and handled individually, transported in face-packs. Treatment b, heads harvested by commercial method, transported in crates of 16 heads, held overnight in 20°C room before unloading into single-layer trays. Treatment c, heads harvested by commercial method, transported in crates of 16 heads, transferred to single-layer trays on the same day, before placing at 20°C. [b] Assessed after 3 days except in the case of cv. Barrier Reef (4 days). [c] Class of head was defined according to the extent of surface blemishes:
Class 1, total blemished area ≤ half a flower cluster.
Class 2, total blemished area ≤ one flower cluster.
Class 3, total blemished area ≤ 1/8th of the surface area of the curd.
Class 4, total blemished area > 1/8th of the surface area of the curd.

TEMPERATURE

The risk of bacterial spoilage can be greatly reduced by rapid cooling and refrigerated storage of vegetables. Where this is not practicable the risk of spoilage is invariably greater, as in the case of potatoes held at 10-15°C for a curing period (about 14 days) followed by storage at 7-10°C to minimize low temperature sweetening, and in tomatoes, cucumbers and peppers stored at temperatures above 7·5°C to avoid chilling injury. At low temperatures *P. marginalis* may be a more important cause of spoilage than *E. carotovora*. Decay of whitloof chicory transported in refrigerated conditions from Belgium to New York was due to *P. marginalis* (Friedman, 1951a). After inoculation into chicory leaves, *E. carotovora* caused more extensive rotting than *P. marginalis* at 24°C whereas at temperatures of <2°C-21°C, *P. marginalis* caused the greater damage. The minimum growth temperatures reported for *E. carotovora* (vars. *atroseptica* and *carotovora*) vary between 1°C and 6°C (Burkholder and Smith, 1949; Elliott, 1951; Chupp and Sherf, 1960) but there appears to be little information on the rates of growth of these bacteria at low temperatures. Strains of *Pseudomonas* frequently grow at 4°C and below (Doudoroff and Palleroni, 1974) and for strains of *P. fluorescens* generation times of 26-30 h at 0°C, 6·7 h at 0·5°C, and 4·2-8·2 h at 4°C have been reported (Olsen and Jezeski, 1963; Morita, 1975).

TABLE IV

Numbers of micro-organisms ($\log_{10}$ count g^{-1} curd) on Class 1 cauliflowers and on cauliflowers showing spoilage (Class 2 or Class 3) after storage for 3 days at 20°C and 70% r.h.[a] cv. South Pacific, harvested 17/9/74 (Lund and Wyatt, unpublished).

	Class 1 heads	Class 2 and Class 3 heads
Total bacteria	5·3	7·9
Fluorescent pseudomonads	4·9	7·8
Pectolytic bacteria	2·1	>7·0
E. carotovora	ND[c]	5·9[b]
Yeasts	3·7	4·2
Moulds	2·5	2·5

[a] Media for counts as in Table I. Counts are geometric means of results for 10 heads. [b] Based on 8 heads on which this bacterium was detected. [c] ND = not detected.

Spoilage caused by *P. marginalis* occurred in mechanically harvested celery stored at 0·4-1·0°C in an ice-bank cooler for 6 weeks at the Food Research Institute. When strains of *P. marginalis* isolated from the decayed celery were inoculated on to wounded celery petioles subsequently stored at 0·4-1·0°C, soft rot occurred after 6-10 days, but no decay was caused on unwounded petioles. Two of the isolates of *P. marginalis* grew with a generation time of 15-17 h at 0·2°C in a nutrient medium (Brocklehurst and Lund, 1981).

RELATIVE HUMIDITY

A relative humidity of about 95% has generally been recommended for storage of vegetables (Ryall and Lipton, 1972). Higher relative humidities (98-100%) can result in a decrease in weight loss and wilting without incurring an increase in decay provided the temperature is maintained as low as 0-4°C (Ryall and Lipton, 1972; Eckert, 1975; Harvey, 1978; Grierson and Wardowski, 1978; Van den Berg and Lentz, 1978). The availability of water, expressed as water potential, for growth of spoilage bacteria will be somewhat between that of the plant tissue in contact with the bacterium and that of the surrounding atmosphere. In potatoes and certain fruits, and presumably within vegetables, the water potential is high enough to provide ideal conditions for bacterial growth. An atmosphere with a relative humidity of 95% at 25°C equivalent to a water potential of −70 bars is low enough to prevent growth of bacteria such as *Pseudomonas* spp. (Cook and Papendick, 1978). In these conditions the site occupied by spoilage bacteria, for example, whether they are in contact with damaged tissue, enclosed in a film of water between leaves, protected by close-packing of produce or by the presence of a wrapping film, will have a determining effect on the incidence of spoilage.

When cauliflowers are stored under supermarket conditions (20°C, 70% r.h.) water is lost rapidly (Robinson *et al.*, 1975), bacterial spoilage is prevented by drying of the heads, and shelf-life is limited mainly by wilting. Packaging in a polyvinyl chloride film of suitable permeability to water vapour, O_2 and CO_2 can reduce the loss of water, and hence wilting, and extend the shelf-life of the produce. Heads packed in Resinite VP-50 pvc film (Borden (UK) Ltd.) (Table V) and stored at 20°C and 70% r.h. decreased in weight, due mainly to loss of water, by less than 4% in 6 days compared with a decrease in weight of 16% in 4 days for unwrapped heads. The unwrapped heads became unacceptable because of wilting in 3-4 days.

TABLE V

Permeability of Resinite VP-50 pvc film (Data from manufacturer, Borden (UK) Ltd.)

Property	Result	Test method
O_2 permeability minimum	16 000 cm³/m²/24 h/bar	BS 2782/514A
CO_2 permeability minimum	100 000 cm³/m²/24 h/bar	BS 2782/514A
Water vapour transmission	150 g/m²/24 h at 25°C/75% r.h.	BS 2782/5132
Nominal thickness	12·5 microns	

In seven experiments an average of 68% of wrapped heads were graded as Class 1 and 21% as Class 2 (according to the presence of surface blemishes) after storage for 6 days and were still in a good marketable condition. In some cases, particularly when heads were harvested in wet autumn conditions or soon after irrigation, film-wrapping led to an increase in bacterial soft rot. This developed on the curd and the butt in 3 out of 7 experiments affecting the butt in 17%, 24% and 70% of heads. From soft rots of the curd both *P. marginalis* and *E. carotovora* were isolated, whereas *E. carotovora* was the main isolate from affected butts (Lund and Wyatt, unpublished data). Soft rot of the cut stem in wrapped heads cannot be readily removed by trimming. Frequent cleaning and disinfecting of knives in order to prevent build-up and spread of bacteria during trimming would probably help to reduce this problem.

Storage of cauliflower in an atmosphere containing 5% CO_2 at 5°C can lead to off-flavour and off-odour after cooking (Lipton *et al.*, 1967). In the experiments described above, tests by a taste panel failed to detect such off-odours or flavours and measurements in some of the experiments showed that the CO_2 concentration in packs usually remained $< 6\%$. With any less-permeable film, accumulation of CO_2 in the pack could have a deleterious effect.

Free water on the surface of vegetables tends to encourage bacterial spoilage. Treatment of winter white cabbage with post-harvest fungicide drenches to control *Botrytis cinerea* increased the incidence of soft rot caused by *P marginalis* during refrigerated storage at 1°C (Kear and Symonds, 1973; Bobbin and Geeson, 1977). The presence of a film of water on potatoes, causing a decrease in the rate of gaseous diffusion through lenticels, is a major factor leading to an increased risk of bacterial soft rot (Lund, 1979).

CONTROLLED ATMOSPHERES

Controlled atmospheres can be used to extend the storage life of vegetables, the precise requirements depending on the commodity (Ryall and Lipton, 1972; Lipton, 1975). In some cases such atmospheres have been reported to reduce bacterial decay. Soft rot of asparagus stored at 3°C could be reduced by storing in air enriched with 7% CO_2 (Lipton, 1965; 1975). When lettuce was stored at 5°C for one month, soft rot was reduced by 50% in an atmosphere containing 3% O_2 and was negligible in 3% O_2 + 4% CO_2 (Lipton, 1971). An atmosphere of 3% O_2 + 5% CO_2 at 13°C reduced decay of pink or red tomatoes inoculated with *E. carotovora*, but not of green or breaking fruits (Parsons and Spalding, 1972).

The improvement in quality obtained by the use of controlled atmospheres may be due in part to the effect on the physiological condition of the vegetable and in part to inhibition of spoilage organisms. In nutrient medium at 20°C the growth of six strains of *E. carotovora* var. *atroseptica, E. carotovora* var. *carotovora* and *P. fluorescens* decreased with decreasing concentrations of O_2 in the presence of 3% CO_2; in an atmosphere containing 3% O_2 + 3% CO_2 growth was 54-64% of that in air, whereas in 0% O_2 + 3% CO_2 growth was 3-12·5% of that in air (Wells, 1974). Growth of these bacteria was retarded to various extents in the absence of CO_2. In the presence of 3% O_2 or 21% O_2 concentrations of CO_2 greater than 10% were required to inhibit growth, and in the presence of 30% CO_2 growth was 15-34% of that in air.

Reduction of the oxygen concentration to 2% had a negligible effect on growth of two strains of *P. fluorescens* in a complex medium at 22°C (Clark and Burki, 1972). Oxygen concentrations between 2% and 0·5% caused an increase in the lag period before growth; below 0·5% O_2 the generation time increased and the yield decreased, and no growth occurred in the absence of O_2. Growth of a strain of *Pseudomonas* at 5°C was only slightly retarded in 0·8% O_2 compared with growth in air, was reduced by 70% in 0·2% O_2 and inhibited in the absence of O_2 (Shaw and Nichol, 1969).

These limited experiments indicate that, in the conditions most frequently proposed for controlled atmosphere storage of vegetables, the decreased concentrations of O_2 and increased concentrations of CO_2 are unlikely to have a marked effect on the rate of growth of spoilage bacteria. However, in situations where nutrients, pH, water potential and temperature are sub-optimal, the effect of near-limiting concentrations of CO_2 and O_2 might be greater.

ANTI-BACTERIAL TREATMENTS

Bacterial spoilage of vegetables cannot be reliably controlled by the chemical treatments at present available for this purpose (Eckert, 1977).

Hypochlorite in the form of free, available chlorine is rapidly lethal to bacteria suspended in water and can be a useful means of preventing the build-up and dissemination of soft rot bacteria in water used for washing and hydrocooling vegetables. Although less than 10 ppm free, available chlorine at neutral pH will rapidly kill most vegetative bacteria, concentrations of 50-100 ppm available chlorine at pH 7-8·5 may be used commercially where wash waters are heavily contaminated with soil and organic matter. Because of the inactivation of chlorine by organic material, maintenance of an adequate concentration of free, available chlorine is essential if the treatment is to be effective (Eckert, 1977).

The addition of hypochlorite to wash water has helped to reduce the incidence of bacterial black spot of radish (Segall and Smoot, 1962), bacterial soft rot and bacterial necrosis in field-washed tomatoes (Segall, 1968), and bacterial soft rot in pre-packed carrots (Segall and Dow, 1973). Although addition of hypochlorite reduced the numbers of bacteria in water used to wash spinach and on the leaves, it failed to prevent decay of the pre-packed vegetable (Friedman, 1951b).

Dipping potatoes in solutions containing 600-1200 ppm free, available chlorine (pH approximately 9·4) reduced the incidence of bacterial soft rot starting from lenticels but failed to give complete control (Lund and Wyatt, 1979). Experiments in which tubers were mechanically damaged showed that although treatment with 2000 ppm available chlorine reduced the bacterial soft rot developing in pre-packed potatoes in 2 days, it resulted in increased rotting after 16 days, presumably due to a toxic effect on surface cells of the tubers (Scholey *et al.*, 1968). The control of bacterial soft rot of potatoes is an application for which antibacterials are particularly required (Harris, 1978; Lund and Wyatt, 1979).

Hot water treatment has been used commercially to reduce the incidence of bacterial soft rot caused by *E. carotovora* in peppers. The treatment consisted of immersion for 1·5 min in water at 126-128°F (52-53°C), and time and temperature were critical. The control achieved by treatment with hot water was completely nullified by subsequent hydrocooling with water alone or containing 300 ppm chlorine; moisture added by the hydrocooling was considered to be responsible for this effect (Johnson, 1966) presumably by promoting growth of bacteria which survived the hot water treatment.

CONCLUSIONS

Spoilage bacteria are usually present on crops and are a potential cause of wastage of all vegetables. Control is probably best achieved by refrigeration where this is possible; antibacterial compounds at present available do not prevent bacterial spoilage of vegetables. Measures aimed at reducing the inoculum level of soft rot bacteria in the environment can probably help to decrease wastage, but an increase in mechanical harvesting and consequent damage to produce could intensify post-harvest disease problems in general (Eckert, 1977) and bacterial spoilage in particular.

REFERENCES

Bartz, J.A. and Crill, J.P. (1974). *Proc. Fla St. hort. Soc.* **86**, 153-156.
Beraha, L. and Kwolek, W.F. (1975). *Pl. Dis. Reptr.* **59**, 1001-1004.
Bobbin, P. and Geeson, J.D. (1977). A.D.A.S., Experiments and Development in the Eastern Region 1976, 151-154.
Brocklehurst, T.F. and Lund, B.M. (1981). *J. appl. Bact.* **50**, 259-266.
Bryan, F.L. (1977). *J. Fd Protection* **40**, 45-56.
Burkholder, W.H. (1954). *Phytopathology* **44**, 592-596.
Burkholder, W.H. and Smith, W.L. (1949). *Phytopathology* **39**, 887-897.
Ceponis, M.J. (1970). *Pl. Dis. Reptr.* **54**, 964-966.

Ceponis, M.J. and Butterfield, J.E. (1974a). *Pl. Dis. Reptr.* **58**, 367-369.
Ceponis, M.J. and Butterfield, J.E. (1974b). *Pl. Dis. Reptr.* **58**, 558-560.
Ceponis, M.J., Kaufman, J. and Butterfield, J.E. (1970). *Pl. Dis. Reptr.* **54**, 263-265.
Chupp, C. and Sherf, A.F. (1960). "Vegetable Diseases and their Control". The Ronald Press Co., New York.
Clark, D.S. and Burki, T. (1972). *Can. J. Microbiol.* **18**, 321-326.
Cook, R.J. and Papendick, R.I. (1978). *HortScience* **13**, 559-564.
Coursey, D.G. and Booth, R.H. (1972). *Rev. Pl. Pathol.* **51**, 751-765.
Doudoroff, M. and Palleroni, N.J. (1974). *Pseudomonas. In* "Bergey's Manual of Determinative Bacteriology" (8th edn), (R.E. Buchanen and N.E. Gibbons, eds), pp. 217-243. Williams and Wilkins Co., Baltimore.
Dye, D.W., Bradbury, J.F., Dickey, R.S., Goto, M., Hale, C.N., Hayward, A.C., Kelman, A., Lelliott, R.A., Patel, P.N., Sands, D.C., Schroth, M.N., Watson, D.R.W. and Young, J.M. (1975). *Int. J. Syst. Bact.* **25**, 252-257.
Eckert, J.W. (1975). *In* "Postharvest Biology and Handling of Fruits and Vegetables" (N.F. Haard and D.K. Salunkhe, eds), pp. 81-117. The Avi Publishing Co. Inc., Westport, Connecticut, USA.
Eckert, J.W. (1977). *In* "Antifungal Compounds" (M.R. Siegel and H.D. Sisler, eds), Vol. 1, pp. 269-352. Marcel Dekker Inc., New York.
Elliott, C. (1951). "Manual of Bacterial Plant Pathogens" (2nd edn). Chronica Botanica Co., Waltham, Mass.
Ercolani, G.L. (1976). *Appl. Environ. Microbiol.* **31**, 847-852.
Friedman, B.A. (1951a). *Phytopathology* **41**, 880-8.
Friedman, B.A. (1951b). *Phytopathology* **41**, 709-713.
Geeson, J.D. (1979). *J. appl. Bacteriol.* **46**, 189-193.
Geldreich, E.E. and Bordner, R.H. (1971). *J. Milk Fd Technol.* **34**, 184-195.
Graham, D.C. and Harrison, M.D. (1975). *Phytopathology* **65**, 739-741.
Grierson, W. and Wardowski, W.F. (1978). *HortScience* **13**, 570-574.
Grogan, R.G., Misaghi, I.J., Kimble, K.A., Greathead, A.S., Ririe, D. and Bardin, R. (1977). *Phytopathology* **67**, 957-960.
Hall, C.B., Stall, R.E. and Burdine, R.E. (1972). *Proc. Fla St. hort. Soc.* **84**, 163-165.
Hall, C.B., Stall, R.E. and Strobel, J.W. (1970). *Proc. Fla St. hort. Soc.* **83**, 179-182.
Harris, R.I. (1978). *Potato Res.* **21**, 231-233.
Harrison, M.D., Quinn, C.E., Sells, I.A. and Graham, D.C. (1977). *Potato Res.* **20**, 37-52.
Harvey, J.M. (1978). *A. Rev. Phytopathol.* **16**, 321-341.
Hayward, A.C. (1974). *A. Rev. Phytopathol.* **12**, 87-97.
Hopkins, D.L. and Elmstrom, G.W. (1977). *Phytopathology* **67**, 961-964.
Johnson, H.B. (1966). Bacterial soft rot in Bell peppers. Cause and Commercial Control. U.S. Dep. Agr. Marketing Res. Rept. 738, 1-6.
Kaferstein, F.K. (1976). *J. Milk Fd Technol.* **39**, 837-840.
Kear, R.W. and Symonds, J.P. (1973). *Mededel. Fac. Landbouwwetenscha. Gent.* **38**, 1549-1560.
Kikumoto, T. (1974). *Bull. Inst. agric. Res. Tohuku Univ.* **25**, 125-137.
King, E.O., Ward, M.K. and Raney, D.E. (1954). *J. Lab. clin. Med.* **44**, 301-307.
Lelliott, R.A. (1974). *Erwinia. In* "Bergey's Manual of Determinative Bacteriology" (8th edn), (R.E. Buchanen and N.E. Gibbons, eds) pp. 332-338. Williams and Wilkins Co., Baltimore.
Lipton, W.J. (1965). *Proc. Am. Soc. hort. Sci.* **86**, 347-356.

Lipton, W.J. (1971). Controlled atmosphere effects on lettuce quality in simulated export shipments. U.S. Dept. Agr. Agr. Res. Serv. Rept. 51-45.

Lipton, W.J. (1975). *In* "Postharvest Biology and Handling of Fruits and Vegetables" (N.F. Haard and D.K. Salunkhe, eds), pp. 130-143. The Avi Publishing Co. Inc., Westport, Connecticut, USA.

Lipton, W.J., Harris, C.M. and Couey, H.M. (1967). *Proc. Am. Soc. hort. Sci.* **91**, 852-859.

Lund, B.M. (1969). *J. appl. Bact.* **32**, 60-67.

Lund, B.M. (1971). *J. appl. Bact.* **34**, 9-20.

Lund, B.M. (1979). *In* "Plant Pathogens" (D.W. Lovelock, ed.), Soc. Appl. Bact. Tech. Ser. No. 12, pp. 19-49. Academic Press, London and New York.

Lund, B.M. and Wyatt, G.M. (1979). *Potato Res.* **22**, 191-202.

Maxcy, R.B. (1978). *J. Fd Protection* **41**, 435-438.

McColloch, L.P., Cook, H.T. and Wright, W.R. (1968). Market Diseases of Tomatoes, Peppers and Eggplants. U.S. Dep. Agr. Agr. Handbook 28, pp. 74.

Meneley, J.C. and Stanghellini, M.E. (1972). *Phytopathology* **62**, 778 (Abstr).

Meneley, J.C. and Stanghellini, M.E. (1974). *J. Fd Sci.* **39**, 1267-1268.

Mew, T.W., Ho, W.C. and Chu, L. (1976). *Phytopathology* **66**, 1325-1327.

Morita, R.Y. (1975). *Bact. Rev.* **39**, 144-167.

Nasuno, S. and Starr, M.P. (1966). *Phytopathology* **56**, 1414-1415.

Nicholls, A.A., Davies, P.A., King, K.P., Winter, E.J. and Blackwall, F.L.C. (1971). *J. hort. Sci.* **46**, 425-433.

Olsen, R.H. and Jezeski, J.J. (1963). *J. Bact.* **86**, 429-433.

Parsons, C.S. and Spalding, D.H. (1972). *J. Am. Soc. hort. Sci.* **97**, 297-299.

Paton, A.M. (1959). *Nature, Lond.* **183**, 1812-1813.

Pérombelon, M.C.M. (1972). *Ann. appl. Biol.* **71**, 111-117.

Pérombelon, M.C.M. (1973). *Ann. appl. Biol.* **74**, 59-65.

Pérombelon, M.C.M. and Lowe, R. (1976). *Rep. Scot. hort. Res. Inst.* 66-67.

Pieczarka, D.J. and Lorbeer, J.W. (1975a). *Phytopathology* **65**, 16-21.

Pieczarka, D.J. and Lorbeer, J.W. (1975b). *Phytopathology* **65**, 509-513.

Ramsey, G.B. and Smith, M.A. (1961). Market Diseases of Cabbage, Cauliflower, Turnips, Cucumber, Melons and Related Crops. U.S. Dep. Agr. Agr. Handbook 184, pp. 49.

Ramsey, G.B., Friedman, B.A. and Smith, M.A. (1959). Market Diseases of Beets, Chicory, Endive, Escarole, Globe Artichokes, Lettuce, Rhubarb, Spinach and Sweet Potatoes. U.S. Dep. Agr. Agr. Handbook 155, pp. 44.

Robinson, I. and Adams, R.P. (1978). *J. appl. Bact.* **45**, 83-90.

Robinson, J.E., Browne, K.M. and Burton, W.G. (1975). *Ann. appl. Biol.* **81**, 399-408.

Rombouts, F.M. (1972). "Occurrence and properties of bacterial pectate lyases". Doctoral Thesis, Agricultural University, Wageningen, The Netherlands, pp. 33-39.

Rombouts, F.M., Spaansen, C.H., Visser, J. and Pilnik, W. (1978). *J. Fd. Biochem.* **2**, 1-22.

Ryall, A.L. and Lipton, W.J. (1972). "Handling, Transportation and Storage of Fruits and Vegetables". Vol. 1, Vegetables and Melons. The Avi Publishing Co. Inc., Westport, Connecticut, USA.

Samish, Z. and Dimant, D. (1959). *Fd Mf.* **34**, 17-20.

Samish, Z. and Etinger-Tulczynska, R. (1963). *Appl. Microbiol.* **11**, 7-10.

Samish, Z., Etinger-Tulczynska, R. and Bick, M. (1961). *Appl. Microbiol.* **9**, 20-25.

Samish, Z., Etinger-Tulczynska, R. and Bick, M. (1963). *J. Fd Sci.* **28**, 259-266.

Scholey, J., Marshall, C. and Whitbread, R. (1968). *Pl. Path.* **17**, 135-139.
Segall, R.H. (1968). *Proc. Fla St. hort. Soc.* **81**, 212-214.
Segall, R.H. and Dow, A.T. (1973). *Pl. Dis. Reptr.* **57**, 896-899.
Segall. R.H. and Smoot, J.J. (1962). *Phytopathology* **52**, 970-973.
Shaw, M.K. and Nichol, D.J. (1969). Proc. 15th Eur. Meet. Meat Research Workers. Institute of Meat Technology, University of Helsinki, Helsinki, Finland. Paper B16, 226-232.
Smith, M.A. and Ramsey, G.B. (1956). *Phytopathology* **46**, 210-213.
Smith, M.A., McColloch, L.P. and Friedman, B.A. (1966). Market Diseases of Asparagus, Onions, Beans, Peas, Carrots, Celery and Related Vegetables. U.S. Dep. Agr. Agr. Handbook 303, pp. 65.
Smith, W.L. Jr., and Wilson, J.B. (1978). Market Diseases of Potatoes. U.S. Dep. Agr. Agr. Handbook 479, pp. 99.
Splittstoesser, D.F. (1970). *J. Milk Fd Technol.* **33**, 500-505.
Stall, R.E. and Hall, C.B. (1969). *Phytopathology* **59**, 1650-1653.
Tamminga, S.K., Beumer, R.R. and Kampelmacher, E.H. (1978). *J. Hyg., Camb.* **80**, 143-154.
Upstone, M.E. (1971). *Proc. 6th Br. Insectic. Fungic. Conf.* **1**, 211-214.
Wells, J.M. (1974). *Phytopathology* **64**, 1012-1015.
Wilkie, J.P. and Dye, D.W. (1974). *N.Z. Jl. agric. Res.* **17**, 131-135.
Van den Berg, L. and Lentz, C.P. (1978). *HortScience* **13**, 565-569.

IV.6 The Effect of Field Nitrogen and Post-harvest Fungicide Dips on Storage of Dutch White Cabbage

S. J. WALE* AND H. A. S. EPTON

Cryptogamic Botany Laboratories,
University of Manchester, Manchester, M13 9PL, England

INTRODUCTION

Burton (1973) has stated that the main forms of deterioration affecting long-term storage of Dutch white cabbage are water loss, rotting caused by pathogens, leaf senescence and regrowth. Storage under the optimum practical conditions (1-2°C and 95% r.h.) minimizes losses attributable to each of these forms of deterioration (Robinson *et al.*, 1975). However, even in optimal environments rotting by pathogens can still be important.

Botrytis cinerea Pers. ex Fr. is the most prevalent fungus found on decaying stored cabbage in many commercial stores (Yoder and Whalen, 1975) and the main cause of wastage (Geeson, 1978). Brown *et al.* (1975) considered *B. cinerea* infection to be the most important factor limiting successful long-term storage. Infection of cabbage by *B. cinerea* is more common during the latter part of the storage period, after 3-4 months (Adair, 1971; Yoder and Whalen, 1975), and has been found mainly on chlorotic tissue (Pendergrass and Isenberg, 1974).

Attempts to control the incidence of infection by *B. cinerea* and other fungal pathogens on stored cabbage have been made using a wide range of fungicides, applied before harvest as sprays or after harvest as dips or dusts (Anon, 1973a; Kear and Symons, 1973; Shipway, 1975; Brown *et al.*, 1975; Davies, 1976; Bobbin and Geeson, 1976, 1977; Kear *et al.*, 1977). In a number of storage trials benomyl, a broad-spectrum systemic fungicide, was used successfully to control *B. cinerea*. However, on two occasions

*Present address: Plant Pathology Section, North of Scotland College of Agriculture, 581 King St., Aberdeen, AB9 1UD, Scotland.

a benomyl dip was found to aggravate symptoms described as black speck or speckling (Anon, 1973a, Kear and Symons, 1973). Turnip mosaic virus has since been found to be the cause of a disease which has the same symptoms (Walkey and Webb, 1977). Symptoms of another necrotic disorder, pepper spotting, which develops as small superficial spots around stomata, have been found to develop more rapidly in benomyl-treated cabbages (Bobbin and Geeson, 1977). At the end of storage, however, these workers found no differences in pepper spotting between the benomyl-treated and the untreated cabbages.

Benzimidazole fungicides, including benomyl, have some cytokinin-like properties (Skene, 1972; Thomas, 1974) such as delaying the onset of senescence. However, a delay in senescence of stored white cabbage has not been reported following dipping in benzimidazole fungicides. If this effect occurs it could prolong storage life.

Two studies of the effect of field nitrogen on the storage potential of Dutch white cabbage have reached different conclusions. Suhonen (1969) found that high levels of nitrogen, four to six times the normal dose (34·2 kg nitrogen ha^{-1}), resulted in a reduction in marketable weight. This reduction only became apparent at the end of a seven-month storage period and was not found after four months' storage. At Stockbridge House Experimental Horticulture Station, however, where up to 300 kg ha^{-1} of nitrogen was applied, no effect on the storage quality was noted (Anon, 1972).

The present investigation examines the effect of field nitrogen on the storage potential of Dutch white cabbage, particularly in relation to disease incidence and severity, and the efficacy of a benomyl post-harvest dip in the control of rotting by *B. cinerea.*

METHODS

CULTIVATION

Three cultivars of Dutch white cabbage—Jupiter, Holland Winter White and Holland Winter Export— were sown in cold frames and transplanted after four weeks into a sandy-loam soil on 19 May, 1975. The average total nitrogen in the soil was 4·0 mg g^{-1} dry weight of soil; the nutrient index for nitrogen was therefore, considered to be between 3 and 4 for cultivation of Brussels sprouts or cabbage (Anon, 1973b). Three plots of each cultivar received ammonium sulphate in the form of three equal top dressings such that the amounts of nitrogen applied were 113, 69 and 25 kg ha^{-1}. Plots were laid out in a graeco-latin square design. Potassium and phosphorus were applied with the first top dressing at the rates of 226 and 125 kg ha^{-1} respectively.

At transplanting, seedlings were treated with calomel dust (4% mercuric chloride). Diazinon (Basudin 5G), an organophosphorus insecticide against cabbage root fly, was applied on the same day at a rate of 450 g/500 plants. A post-emergence herbicide, aziprotryne (Brasoran 50 WP), was applied seven weeks after transplanting at a rate of 3·92 kg ha^{-1}. One application of dimethoate (Dimethoate 40) at a rate of 0·84 l ha^{-1} was used to control cabbage aphis ten weeks after transplanting.

HARVESTING AND STORAGE

Cabbages were harvested by hand on 11 November, 1975, trimmed to remove all but two wrapper leaves and placed in polyethylene crates. Only heads conforming to EEC standards (Anon, 1974) were used in the storage trial, for which four crates of each cultivar at each nitrogen level were used. Crates were placed in a cold store maintained at 1 ±0·5°C and 90-95% relative humidity throughout the six months' storage period. Before storage two of each set of four crates were dipped in a 500 mg active ingredient l^{-1} suspension of benomyl (Benlate) for one minute and allowed to drain for two hours at room temperature.

MEASUREMENTS AND ASSESSMENTS

Immediately after harvest, the total nitrogen content of the outermost 3-4 leaves from a number of heads from each cultivar/nitrogen level treatment was determined by Kjeldahl analysis, and leaf dry matter was also measured. At the beginning and end of storage, individual heads were assessed visually for symptoms of pepper spotting and black speck as described by Geeson and Robinson (1975) and Geeson (1978). The assessment at the end of storage took place after trimming. Heads were then examined externally and, after being cut in half, internally.

Pepper spotting was assessed in three categories: (i) no pepper spotting; (ii) slight pepper spotting, which was insufficient to warrant discarding the head; and (iii) severe spotting, rendering the head unmarketable.

Senescence and fungal coverage were assessed every four weeks during storage. Senescence was considered to be synonymous with yellowing (cf. Eberhardt, 1955). On each head, the percentages of the surface area covered by yellow tissue and fungal rotting were assessed and divided into seven categories, as shown in Table I. Each category was given a score and by totalling scores and dividing by the number of heads, a mean score for senescence and fungal coverage was determined for each treatment. The contents of each crate were weighed before and after the storage period to determine the weight loss due to evaporation of water and metabolic activity. Heads were then carefully trimmed to a marketable quality and reweighed to find the marketable weight. The trimmings from the heads in each crate were weighed and divided into rotted tissue and uninfected tissue (waste). The weight of the rotted tissue was taken to represent the loss due to fungal rotting.

TABLE I

Categories used for assessment of senescence and surface fungal coverage of Dutch white cabbage during storage

	% Surface area affected						
	0	<1	1-3	4-9	10-24	25-50	>50
Score	0	1	2	3	4	5	6

RESULTS

YIELDS, TOTAL NITROGEN AND DRY MATTER CONTENT OF LEAF TISSUE

An analysis of variance of yield, expressed as the mean head weight, showed significant differences between nitrogen levels but not between cultivars. An increase in the level of nitrogen resulted in an increase in head weight, in some instances a significant increase (Table II). There was, however, no consistent response to nitrogen fertilizer by the three cultivars.

TABLE II

Mean head weights (kg) of three cultivars of Dutch white cabbage grown at each of three nitrogen fertilizer levels

Nitrogen level (kg ha^{-1})	Jupiter	Holland Winter White	Holland Winter Export
113	1·97	2·20	1·67
69	1·45	1·75	1·38
25	1·20	1·01	1·36

Least significant difference between nitrogen levels for each cultivar $= 0{\cdot}292$ at $P \leq 0{\cdot}05$.

The total nitrogen content of leaf tissue increased in response to the application of the higher levels of nitrogen, except for cv. Holland Winter White (HWW) at 113 kg ha^{-1} (Table III). In contrast, the dry matter content of leaf tissue declined with increasing nitrogen supply, overall significant differences between cultivars and between nitrogen levels being detected by an analysis of variance.

TABLE III

Total nitrogen and dry matter of leaf tissue of three cultivars of Dutch white cabbage grown at each of three nitrogen fertilizer levels, immediately after harvest

	Total nitrogen content ($mg\ g^{-1}$ dry wt. of leaf tissue)[a]			Dry matter content (% by wt.)[b]		
Nitrogen level ($kg\ ha^{-1}$)	Jupiter	Holland Winter White	Holland Winter Export	Jupiter	Holland Winter White	Holland Winter Export
113	36·3	28·8	36·5	11·5	9·4	10·0
69	31·6	30·8	30·6	12·2	9·8	10·1
25	26·7	27·2	26·2	12·9	11·0	10·3

[a] For total nitrogen, least significant difference between nitrogen levels for each cultivar = 6·36 at $P \leq 0{\cdot}05$.
[b] For dry matter, least significant difference between nitrogen levels for each cultivar = 1·13 at $P \leq 0{\cdot}05$.

CHANGES AFTER SIX MONTHS' STORAGE

Details of measurements made at the end of the storage period are given in Table IV a-e. Storage cabbage (cv. Decema) was determined by Robinson *et al.* (1975) to be unmarketable when water loss due to evaporation and metabolism had reached 10% of the original weight. The results (Table IVa) show that benomyl dipping had a significant effect on weight loss ($\leq 10\%$) compared with the majority of undipped controls which lost $> 10\%$ of their original weight. No particular effect of nitrogen level on weight loss was evident as there was considerable variation within cultivars.

As for total trimming losses, there was a significant overall difference between cultivars (Table IVb). The cv. Jupiter required much less trimming than HWW whilst Holland Winter Export (HWE) was intermediate. The benomyl dip reduced total trimming losses in all cultivars compared to undipped controls, this reduction being significant for the cvs HWW and HWE.

The majority of rotting occurring during storage was fungal in origin. *Botrytis cinerea* was the principal pathogen causing rotting, but many firm penetrating rots, typified by that of *Alternaria brassicicola*, were also present. Very little bacterial rotting occurred.

Since most of the trimming was necessitated by the presence of fungal invasion, we assessed the amount of rotted tissue to obtain an indication of relative susceptibility to fungal attack. The percentage by weight of fungal-rotted tissue in the cv. Jupiter was significantly less than HWE, which in turn was significantly less than in HWW (Table IVc). The correlation between the weight of fungal-rotted tissue and uninfected "waste" trimming was only moderate (correlation coefficient = 0·539, 95% confidence limits 0·450 and 0·616). Therefore, the amount of total trimming loss could not be used reliably to predict the amount of fungal rotting, i.e. the relative susceptibilities of different cultivars.

In every instance, the benomyl dip treatment reduced the amount of diseased tissue compared with the untreated controls and was generally significant in the cultivars HWW and HWE. In the untreated controls of these two cultivars the amount of disease was considerable, reaching 20% by weight in one instance. There was no consistent relationship between losses due to fungal rotting and the nitrogen levels.

When the marketable product was considered (Table IVd) an analysis of variance showed significant differences between cultivars and between the untreated controls and benomyl dip treatments. The cv. Jupiter had a significantly larger marketable product than HWE which, in turn, had significantly more marketable product than HWW. The benomyl dip treatments, in every instance, increased the marketable product by 2-25% over the controls.

Considering the effect of nitrogen level on the undipped controls, the two cultivars with the greatest losses, HWW and HWE, showed the least loss at the highest nitrogen level. However, the benomyl dip negated the differences between the various nitrogen levels (Table IVd).

The effectiveness of the benomyl dip in reducing fungal infection was also shown when the percentage of heads with basal lesions was considered (Table IVe). Infection of leaf bases occurred only after 3-4 months' storage and mainly began at the point of abscission of the leaf. Lesions initiated here often developed around the head and caused extensive rotting. The percentage of heads showing basal lesions was generally reduced by dipping in benomyl, particularly in the cv. Jupiter. No consistent effect of nitrogen level was apparent.

The incidence of black speck was very low both at the beginning and end of storage. No conclusion could be drawn with respect to the effect of nitrogen level or benomyl dipping on the incidence of black speck.

Pepper spotting was commonly found, although was rarely severe enough to render a head unsuitable for marketing. Dipping in benomyl increased the incidence but not the severity of pepper spotting both on the surface exposed after trimming and within the cabbage head. However, no consistent effect of nitrogen level could be detected that either affected the incidence or severity of pepper spotting. Before storage the incidence of pepper spotting on cv. Jupiter was lower than that on the other two cultivars. By the end of storage this difference had almost disappeared.

SENESCENCE AND SURFACE FUNGAL COVERAGE

With each treatment combination some senescence was recorded after only one month of storage, and in every instance senescent areas covered more than 40% (often >50%) of the total surface area after six months' storage. Considering individual cultivars, those heads grown at the lowest nitrogen level and dipped in benomyl consistently showed the slowest rate of senescence, but the remaining treatment combinations had varying effects depending on the cultivar. Dipping generally reduced the rate of senescence within each treatment combination, particularly with cv. Jupiter (Fig. 1).

Fungal lesions did not usually develop on the surface of the heads until after one month's storage, and lesions rarely enlarged to cover more than 50% of the surface area by the end of storage. There were large differences in fungal coverage between different treatment combinations within each cultivar. At each nitrogen level fungal coverage developed at a slower rate in the benomyl dip treatment compared to the untreated controls. The benomyl dipped heads at the lowest nitrogen level exhibited the slowest rate of lesion spread, whilst all dip treatments generally had the least surface

TABLE IV

Effect of nitrogen fertilizer level and a post-harvest benomyl dip on various losses, damage and marketable product of three cultivars of Dutch white cabbage after storage for six months at 1 °C and 90-95% relative humidity

Nitrogen level (kg ha^{-1})	Fungicide treatment	(a) Weight loss (mean % by wt.)[a]			(b) Total trimming losses (mean % by wt.)[a]		
		Jupiter	Holland Winter White	Holland Winter Export	Jupiter	Holland Winter White	Holland Winter Export
113	Control	11·2	11·6	11·3	20·2	32·5	26·1
	Benomyl-dipped	10·1	10·3	8·8	18·8	20·1	18·4
69	Control	11·4	13·4	11·3	20·2	34·9	23·5
	Benomyl-dipped	9·5	10·4	10·3	16·1	19·7	15·0
25	Control	9·4	13·5	13·3	17·7	36·3	29·5
	Benomyl-dipped	9·6	9·7	8·8	14·2	18·1	19·3

For (a) least significant difference between variables within cultivars = 2·81 at $P \leq 0{\cdot}05$. For (b) least significant difference between variables within cultivars = 5·81 at $P \leq 0{\cdot}05$. [a]Percentages are relative to weight before storage.

TABLE IV (contd.)

Nitrogen level (kg ha^{-1})	Fungicide treatment	(c) Losses due to fungal rotting (mean % wt.)[a]			(d) Marketable product (mean % by wt.)[a]		
		Jupiter	Holland Winter White	Holland Winter Export	Jupiter	Holland Winter White	Holland Winter Export
113	Control	5·7	14·2	12·5	64·9	56·0	62·6
	Benomyl-dipped	4·5	8·5	5·9	67·0	62·6	72·8
69	Control	6·3	17·3	9·3	54·5	41·7	55·4
	Benomyl-dipped	3·1	7·1	4·7	70·7	66·9	70·9
25	Control	4·5	20·5	13·9	72·9	48·9	52·5
	Benomyl-dipped	2·5	5·5	4·8	76·2	68·6	71·9

For (c) least significant difference between variables within cultivars = 5·32 at $P \leq 0{\cdot}05$. For (d) least significant difference between variables within cultivars = 8·18 at $P \leq 0{\cdot}05$. [a] Percentages are relative to weight before storage.

TABLE IV (contd.)

Nitrogen level (kg ha^{-1})	Fungicide treatment	(e) Basal lesions (% of heads showing basal lesions) Jupiter	Holland Winter White	Holland Winter Export
113	Control	63	44	65
	Benomyl-dipped	38	58	60
69	Control	44	60	35
	Benomyl-dipped	17	50	50
25	Control	37	71	64
	Benomyl-dipped	16	56	50

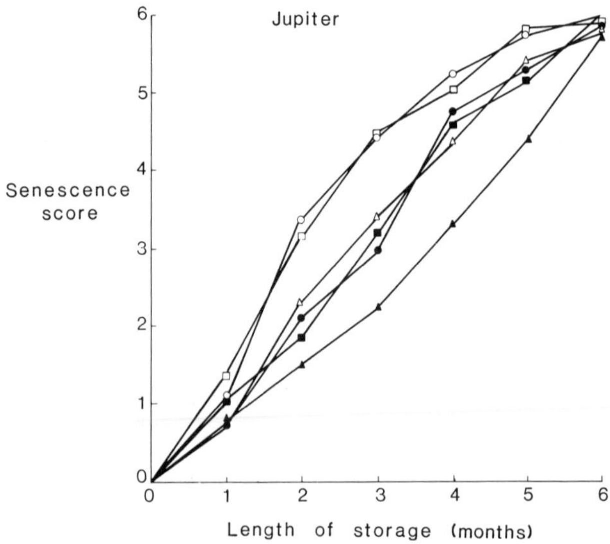

Fig. 1. Effect of nitrogen fertilizer level and a post-harvest benomyl dip on the rate of senescence of cv. Jupiter during storage for six months. Key to treatments. ○——○ 113 kg N ha^{-1} and undipped; ●——● 113 kg ha^{-1} plus benomyl dip; □——□ 69 kg ha^{-1} and undipped; ■——■ 69 kg ha^{-1} plus benomyl dip; △——△ 25 kg ha^{-1} and undipped; and ▲——▲ 25 kg ha^{-1} plus benomyl dip. For assessment of senescence scores see Table I.

area affected every month. In those months (months 3-4) where the technique of assessment allowed critical comparisons to be made, the differences between the benomyl dip treatments and the undipped controls were usually significant. No consistent relationship between nitrogen level and the rate of fungal coverage could be established when all cultivars were considered.

DISCUSSION

Data concerning the effect of high levels of nitrogen are worthy of close examination. Differences between the cultivars make it difficult to reach general conclusions on the effect of nitrogen levels on the undipped controls. Considering cv. Jupiter, variation in the nitrogen level rarely had a significant effect, but the general trend was often towards an increase in the various types of loss with increasing levels of nitrogen. This trend was generally reversed for cvs HWW and HWE and the improvements in storage quality resulting from the use of higher levels of nitrogen were frequently significant. However, dipping in benomyl had a far greater effect in reducing losses and tended to even out any trend resulting from variation in the nitrogen levels, when percentage changes were considered.

In terms of total yield, and consequently marketable weight, the benefits obtained by using increased nitrogen greatly outweighed any deleterious effect. There was no firm evidence that the resulting increases in leaf-nitrogen and associated decreases in dry matter content increased the susceptibility of the heads to fungal invasion.

Benomyl dipping reduced losses resulting from evaporation and metabolic activity, disease, trimming, and in some instances reduced the rate of senescence. As might be expected, the cultivars in which losses were greatest in the untreated controls (cvs HWW and HWE) showed the greatest benefits from dipping. In commercial practice it would seem that cultivars known to have a poor storage potential could benefit significantly from a fungicidal dip. Benefits may not outweigh the costs for cultivars such as Jupiter, which may be considered to have a good storage potential.

The occurrence of lesions penetrating the head through a number of leaf layers, typically caused by *A. brassicicola*, was more frequent following benomyl dipping than when heads were not dipped. This observation may be explained by the reduction in competition from other fungi which would result from the use of a broad-spectrum fungicide. Such deeply penetrating lesions increase the amount of trimming loss disproportionately to their externally visible area. This may explain why the correlation between the weight of rotted tissue and uninfected "waste" trimmings is only moderate.

It appeared that in some instances benomyl had a marked effect in delaying senescence. Senescence of tissue was well advanced by the time fungal infection of the surface of heads began, after 3-4 months. Except where damage had occurred, infection by *B. cinerea* usually commenced in senescing tissue. Once established the further growth of lesions appeared, usually, to be unhindered by green "healthy" tissue. Delaying senescence may delay invasion and growth of *B. cinerea* and ultimately reduce trimming losses. Studies on the pre-harvest application of cytokinins, which could delay senescence, may be useful.

REFERENCES

Adair, C.N. (1971). *Pl. Dis. Reptr.* **55**, 864-868.

Anon. (1972). *Rep. Stockbridge House expl Hort. Stn for 1972*, 33-34.

Anon. (1973a). *Rep. Stockbridge House expl Hort. Stn for 1973*, 20-22.

Anon. (1973b). "Fertiliser Recommendations" Ministry of Agriculture, Fisheries and Food, Bulletin No. 209, pp. 102. HMSO.

Anon. (1974). "EEC Standards for Cabbage (round headed)." Ministry of Agriculture, Fisheries and Food. Department of Agriculture and Fisheries for Scotland. Department of Agriculture for Northern Ireland.

Bobbin, P. and Geeson, J.D. (1976). ADAS, Experiments and Development in the Eastern Region 1975, 284-286.

Bobbin, P. and Geeson, J.D. (1977). ADAS, Experiments and Development in the Eastern Region 1976, 151-154.

Brown, A.C., Kear, R.W. and Symons, J.P. (1975). *Proc. 8th Br. Insectic. Fungic. Conf.* **1**, 339-346.

Burton, W.G. (1973). *Proc. 7th Br. Insectic. Fungic. Conf.* **3**, 1037-1055.

Davies, A.C.W. (1976). *Rep. Luddington expl Hort. Stn for 1976*, 97.

Eberhart, F. (1955). *Planta* **45**, 57-67.

Geeson, J.D. (1978). *Grower* **89**, 27-31.

Geeson, J.D. and Robinson, J.E. (1975). *Comml Grow.* No. 4147, 1245-1246.

Kear, R.W. and Symons, J.P. (1973). *Meded. Fac. Landbouwwet. Rijksuniv. Gent*, **38**, 1549-1560.

Kear, R.W., Williams, D.J. and Stevens, C.C. (1977). *Proc. 9th Br. Insectic. Fungic. Conf.* **1**, 189-195.

Pendergrass, A.M. and Isenberg, F.M.R. (1974). *HortScience* **9**, 226-227.

Robinson, J.E., Browne, K.M. and Burton, W.G. (1975). *Ann. appl. Biol.* **81**, 399-408.

Shipway, M.R. (1975). *Rep. Kirton expl Hort. Stn for 1975*, 35-36.

Skene, K.G.M. (1972). *J. hort. Sci.* **47**, 179-182.

Suhonen, I. (1969). *Acta Agric. scand.* **19**, 18-32.

Thomas, T.H. (1974). *Ann. appl. Biol.* **76**, 237-241.

Walkey, D.G.A. and Webb, M.J.W. (1977). *Rep. natn. Veg. Res. Stn for 1976*, 109.

Yoder, O.C. and Whalen, M.L. (1975). *Can. J. Bot.* **53**, 691-699.

Section IV
Post-harvest Changes—Control of Post-harvest Changes

IV.7 Cold Stores for Fresh Vegetables

R. T. LINDSAY AND M. A. NEALE

National Institute of Agricultural Engineering, Wrest Park, Silsoe, Bedford MK45 4HS, England

INTRODUCTION

It is likely that the quantity of vegetables stored for a day or two on farms or at the packhouses of marketing co-operatives will increase as once-over harvesting becomes more prevalent and as more vegetables are sold through supermarkets or to processors.

The demand for vegetables is irregular and a breakdown of trading figures over a week in a supermarket has shown that 56% of the weekly turnover occurred on Thursday and Friday with a further 19% on Saturday. The remaining three days, Monday to Wednesday, accounted for only 25% of the turnover (Love, 1977). Furthermore there is a price advantage for the producer roughly coinciding with peak sales (Hinton, 1975). On the other hand, factories which process vegetables require an even supply throughout the week. However, in a packhouse it is easier to organize labour and transport and reduce overtime payments by packing regularly and evenly throughout the week especially as inclement weather or non-working days break the flow from producer to processor.

The supply of produce can be matched to demand by using buffer storage for fresh vegetables. Cool stores can also be used speculatively to lengthen the period of supply by short- or long-term storage. A vegetable store will be successful only if it will keep the produce in good condition with minimum deterioration as inevitably the stored produce is compared with freshly harvested produce by the customer.

The storage and shelf-lives of most vegetables are extended if they are cooled rapidly (Tomkins, 1959), say within 10 h to below 5°C, and stored just above freezing. In addition all vegetables should be stored in air saturated with water vapour, as their quality rapidly deteriorates because of loss of moisture to the surrounding air when it has a water vapour pressure deficit (Burton, 1972). Wilted produce may also be more susceptible to attack by micro-organisms (Derbyshire and Shipway, 1978).

COMMERCIAL VEGETABLE STORES

In a survey of 12 cool stores on farms, the cooling rates of different vegetables in various containers were recorded on 25 occasions (Gibbon, 1972). The rate of cooling varied with the type of vegetable and container, but the mean cooling rate was slightly less than 0·5°C an hour, a rate that would be inadequate for vegetables harvested in the summer at say 20-25°C. This cooling rate might be acceptable for vegetables harvested in winter but it is not likely to be achieved because of poor performance due to low temperature differences and bad air distribution. Furthermore, a crop like Brussels sprouts, whether in bulk or nets on pallet bases, and cool on arrival at the store, can heat up even when the cooling equipment is working.

In the stores surveyed, the air from the cooler was directed at the boxes of vegetables standing in the store. No attempt was made to channel the air into the containers and it was noticeable that the cooling rate was fastest for crops and containers with an open structure. For example, leeks and spinach, each in open-sided crates, cooled at a rate of 1·2°C and 0·4°C an hour respectively, and celery in crates cooled at 0·5°C an hour, whereas in cardboard boxes the rate was only 0·2°C an hour.

The mean water vapour pressure deficit was about 1 mbar while the humidity in some stores was so low (Gibbon and Boyce, 1971), that some leaf vegetables could have lost as much as 1% of their moisture content in 3¼ hours (Robinson *et al.*, 1975).

At the NIAE, cooling rates were improved and weight losses reduced by positive ventilation of the vegetables with air conditioning by a cooling tower using chilled water from an ice bank.

POSITIVE VENTILATION

VENTILATING SIMULATED VEGETABLES

In early experiments to study the value of channeling cooling air through bins of produce, straw was packed into pallet-based bulk bins 1·05 × 1·00 × 0·65 m high at such a density that its resistance to airflow was similar to that of Brussels sprouts (Neale and Messer, 1978). The rate of cooling at the centre of the bin was measured by thermocouples in bottles holding 0·55 kg of water starting at about 28°C and buried in the straw. The cooling rate was expressed as "time to half-temperature": that is, the time needed to halve the difference between the produce temperature at the start of cooling and 1°C—the temperature of the cooling medium.

Slatted bins were positively ventilated by stacking them in a store and blocking the space between them and the walls and ceiling, so that most of the ventilating air had to pass through the bins instead of round them. The pallet bases of other bins were used as air ducts and air directed into them could escape only through the produce. In some cases this was via an upturned cauliflower crate placed over a hole in the centre of solid floors (Fig. 1) and in others via the slatted floors of solid-sided bins (Fig. 2). As a control, standard slatted bins were stacked in the store with no way of directing air through them and all were cooled with air at around 1 °C and 98% r.h.

Fig. 1. Model of bulk bins and slotted plenum wall.

In six comparative runs with groups of 9, 18, 20 or 27 bins the mean time taken to cool the centre of free-standing bins was between 37 and 84% longer than for the positively ventilated ones (Lindsay *et al.*, 1975b). In other runs the rate of cooling within bins ventilated through slatted floors was similar to that of those with solid floors and a centre inlet, but because in these bins the cauliflower crate reduced the storage capacity of the bin by about 10% it was not considered to be a practical design.

VENTILATING VEGETABLES

The vegetables were cooled in solid-sided bins using the pallet bases as ducts to distribute air to all the boxes in the store (Fig. 2). Cool air entered the

Fig. 2. Solid sided bin with slatted floor.

ducts from slots in a plenum chamber wall and, after passing through the slatted floor of each box and between the vegetables, returned to the cooler which fed the plenum chamber (cf. Fig. 5). In this way we showed that the cooling time for carrots was reduced by 88%, Brussels sprouts by 85%, cauliflower 75%, spring cabbage 60%, beetroot 46% and lettuce by 38% compared with the time taken when they were not positively ventilated. For example, the half-cooling time of cauliflower was reduced from 24 to 6 h by positive ventilation. Furthermore, hot spots in the bins of vegetables were eliminated so preventing a small proportion of the crop spoiling the whole sample. In experiments even the slowest cooling parts of positively ventilated bins were cooled up to ten times quicker than the slowest parts of bins cooled in the normal way. Air was supplied to the bins according to the weight of material in them and rates between 0·05 and 0·22 $m^3\ s^{-1}\ tonne^{-1}$ (0·25 and 1·0 kg of air $h^{-1}\ kg^{-1}$ of produce) gave satisfactory cooling rates.

Vegetables stored in small boxes were also positively ventilated at an air pressure of less than 5 mm water gauge. In one method the boxes were stacked on pallet bases in the form of hollow squares with a lid so that the air passed from the duct in each base into the central space and so outwards through holes in the sides of the boxes (Fig. 3). Solid stacks on pallet bases were enclosed in polythene sleeves or a number of stacks pushed together and the pallet exits closed so that air (at rates up to 0·3 $m^3\ s^{-1}\ tonne^{-1}$) had to pass upwards through the stacked produce. These methods reduced the

Fig. 3. Lettuce in cardboard boxes stacked in hollow square.

cooling time by 69-84%. For example, 9 pallet loads of lettuce in cardboard boxes ventilated upwards from the pallet base cooled to half-temperature in 6 h whereas a smaller block of 6 pallets took 43 h when ventilated in the usual way. Vegetables in nets, trays or crates can also be cooled rapidly using similar methods of positive ventilation (Lindsay and Neale, 1977).

COOLING RATES WITH POSITIVE VENTILATION

Following the first comparative experiments with bins of vegetables, ten species of fresh vegetables were positively ventilated with air at rates ranging from 0·1 to 2·5 kg h^{-1} kg^{-1} of produce. The time taken to reach half-temperature at the top of the bins varied from 2¼ to just over 7 h depending on the size and structure of the vegetable. The speed of the "time to half-temperature" front for most vegetables was constant at any airflow rate, although it increased linearly with increasing airflow (Lindsay and Neale, 1978). Table I lists the constants to calculate the speed of the front and linear regressions for each vegetable. However, the speed of the "half-temperature" fronts of vegetables with different physical properties were similar and it was possible to place the vegetables into three groups; root crops, potatoes and Brussels sprouts; lettuce, spinach, spring cabbage and cauliflowers; and hard-centred cabbage. The linear regression lines for each group are given in Table I (see also Fig. 4), and so it is possible to determine the time to half-temperature of vegetables in a bin if the airflow rate is known or alternatively select an airflow rate to give the desired time to half-temperature. Furthermore, the cooling rate can be calculated whatever the depth of bin or length of the air path using:

$Z = \frac{D}{S}$ where Z = time to half-temperature (h)
D = depth of bin (mm)
S = speed of half-cooling front (mm h^{-1})
and $S = mV + b$ where V = airflow (kg h^{-1} per kg of produce)
and m and b are constants (listed in Table I).

TABLE I

Speed of half-cooling front constants

Vegetable	Group	Speed of front: $S = mV + b$ m	b	Correlation coefficient
Beetroot	A	293·5	45·1	0·96
Carrots	A	244·0	80·0	0·73
Potatoes	A	148·3	57·6	0·95
Brussels sprouts	A	314·7	45·0	0·99
Lettuce	B	80·9	108·1	0·93
Spinach	B	167·6	51·4	0·92
Spring cabbage	B	111·0	64·4	0·96
Cauliflower	B	119·1	46·5	0·83
Cabbage—Primo	C	48·3	48·1	0·91
Cabbage—Winningstadt	C	61·0	44·0	0·77
	A	251·7	49·9	0·85
	B	115·6	65·8	0·87
	C	48·4	49·2	0·90

In hollow stacks of boxes or nets the relationship between airflow and cooling rate takes the form:

$$Z = \left(\frac{K_1}{V - K_2}\right)^{0\cdot5} \tag{1}$$

where: K_1 and K_2 = constants (listed in Table II).

Similarly soft fruit in open trays cool quickly when the air is blown horizontally over the fruit which can be cooled to below 5°C in 4-6 hours. When the ratio of airflow approach velocity to mass of fruit in each tray is used for V in formula (1), $K_1 = 188{\cdot}4$ and $K_2 = 8{\cdot}646$.

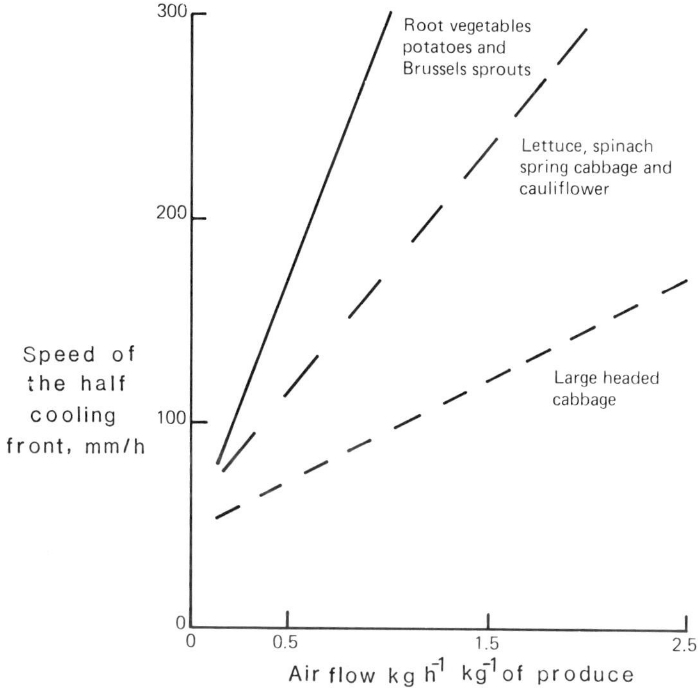

Fig. 4. Speed of cooling front/airflow rate for vegetables.

TABLE II

Cooling rate of vegetables in hollow stacks with positive ventilation via the pallet base

Vegetable	Container	Time to half-temp: $Z=\left(\frac{K_1}{V-K_2}\right)^{0\cdot5}$		Root mean square of deviation	Time to ½ temp. at airflow (V) of 1 kg h^{-1} kg^{-1}
		K_1	K_2	(h)	(h)[a]
Brussels sprouts	Nets	7·0	0·01	0·67	2·7
Calabrese	Cardboard boxes	9·8	−0·29	0·83	2·8
Lettuce	Cardboard boxes	33·7	−0·36	0·45	5·0

[a]Double the amount of air to allow for losses in full-scale stores; 1 kg h^{-1} kg^{-1} = 0·22 m^3 s^{-1} t^{-1}

COOLING THE AIR

ICE BANK COOLER AND COOLING TOWER

Although the conventional refrigeration plant can be used to cool the air in a positively ventilated store, the rate at which heat is removed from the vegetables by positive ventilation is so fast that the overall cooling load is up to 15 times greater during the cooling period than when produce has been cooled in the normal way. Further, stores are filled and emptied irregularly with differing quantities of produce so that cooling loads can be extremely variable. Thus, a relatively large refrigeration plant is needed and there will be long periods when it is underused in just maintaining the store temperature. An alternative is to use an ice bank cooler (Fig. 5) with a comparatively small compressor employing its spare capacity in periods of low or no cooling load to build ice in an insulated tank of water. The water

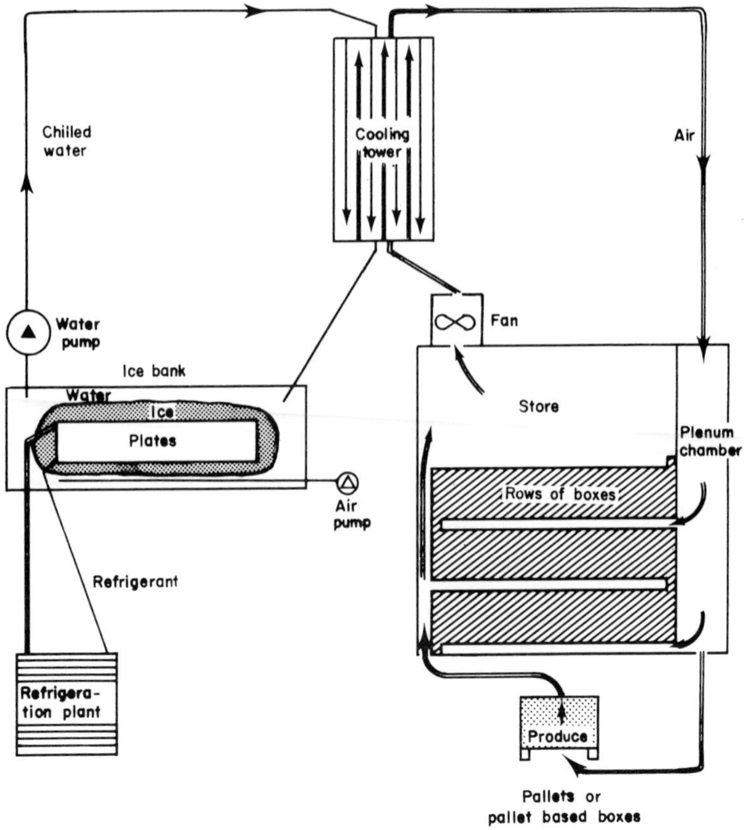

Fig. 5. Diagram of ice bank cooling using positive ventilation.

surrounding the ice is then used to cool the ventilating air and ice is melted to achieve rates of heat extraction much greater than the capacity of the refrigeration plant. At first an experimental two-stage refrigeration system was used where the ice was built on coils carrying chilled brine (Lindsay *et al.* 1975a) but it was inefficient. Instead a direct system with plate evaporators cooled by refrigerant is now used and the plates immersed in the tank of water.

Chilled water from the ice bank can be used in a spray heat exchanger but a packed tower is more efficient (Lindsay *et al.*, 1975c). In the cooling tower the air, and the water cooled by the ice, flow in opposite directions over a packing which greatly increases the area of contact between them so that large quantities of heat can be removed from the air. Warmed water from the tower is returned to the ice bank where the water is maintained at about 0·5°C by agitation with air even at high heat extraction rates.

HUMIDITY

The larger the difference between the temperature of the air and the coolant in a heat exchanger the greater is the amount of moisture removed from the air and hence from the stored vegetables. In conventional stores the temperature difference is large and water is removed from the air by freezing on the evaporators so that its water vapour deficit increases and it dehydrates the produce. By increasing the area of contact between the air and the coolant as in a cooling tower the temperature difference between them and hence the quantity of moisture removed from the air can be reduced and in addition the cooled air returning to the store is saturated with a little free moisture, so that dehydration of the vegetables is kept to a minimum. Furthermore, because the coolant is water the vegetables cannot be accidently frozen. Conventional cooling has to be interrupted by a defrosting cycle and vegetables cannot be cooled to the temperature of the cooling air because it is mixed in the store with air already warmed by the produce and other heat sources before it reaches the produce. On the other hand, the cooling air in positively ventilated stores removes heat from the produce first and then passes into the store before mixing with any heat coming through the structure and leaks at the door. Thus, although store air temperatures are generally higher in positively ventilated stores, the heat gain through the walls are reduced as is the load on the cooler. Stores ventilated in this way may also be able to have larger doors for better access provided they are reasonably airtight.

WEIGHT LOSS

The weight loss during cooling of vegetables in bulk bins rarely exceeded 1% whilst using positive ventilation with ice bank cooling. During storage, summer cauliflower lost 0·4% per week and winter cabbage 0·2% whereas similar vegetables in a conventional store lost up to twice as much, that is 0·8% and 0·3% respectively (Shipway, 1977a). We have shown that the weight loss of vegetables in cardboard boxes and nets is small and for soft fruit is about 1%; the weight loss of vegetables stored for more than 2 months was beetroot 0·18%, washed carrots 0·32% and white cabbage 0·13% per week (Farrimond *et al.*, 1979).

FULL-SCALE EXPERIMENTAL STORES

NIAE EXPERIMENTAL STORE

With a capital grant from the Agricultural Research Council, the NIAE in co-operation with Engineering Department at the ARC Food Research Institute and ADAS Land Service, Cambridge, built a full-scale experimental store adjoining the packhouse of Messrs Bedfordshire Growers Ltd., at Biggleswade in Bedfordshire. The store (Fig. 6) is about 11 × 6 × 4·5 m to the eaves and is of timber frame construction supported on a dwarf brick wall. It is insulated with 100 mm of expanded polystyrene with a vapour barrier on the outside and clad externally with steel sheeting. The store has two cooling rooms separated by a space for the refrigeration plant, ice bank, cooling tower and ventilating fans. Each cooling room holds 24 pallet-based

Fig. 6. The NIAE full-scale experimental vegetable store.

bins (1·35 m^3) stacked four deep, two wide and three high, plus one row of smaller bins or pallet loads. A plenum chamber at the back of each cooling room has a wall of reinforced pvc fabric with slots in it to correspond with the pallet bases of the bins stacked against it. The slots can be closed when not in use. For access, the rooms have large double sliding doors the full height and width of the room so that the bins can be stacked easily in rows at right angles to the plenum chamber walls.

The ice is made by two 2·25 kW refrigeration plants on six evaporator plates in a tank 5 × 2·5 × 1 m deep. The chilled water is pumped from round the ice to the top of the cooling tower where it overflows out of plastic gutters and so is distributed evenly over plastic rings randomly packed into the tower. At the bottom, the warmed water is collected and passed via an air trap to the tank where it is distributed over two-thirds of its surface area so as to prevent uneven ice formation. Air from a small compressor is distributed through drilled pipes laid on the tank floor (Hoyle, personal communication) to agitate the water and to make sure that the ice is melted quickly enough to maintain the supply of chilled water.

After passing up through the tower, the cooling air enters the store through the plenum chamber in each cooling room at a temperature of 1°C and 99% r.h. Under commercial conditions the dimensions of the same type of bins varied slightly so that there were small gaps between adjacent ones and about half of the ventilating air escaped from the pallet bases and did not pass through the produce. Thus, the rate of ventilation had to be increased to twice that suggested by the laboratory experiments. Otherwise full-scale experimental stores performed as predicted by the research results and it was possible to demonstrate that cooling vegetables by positive ventilation with ice bank cooling was faster and less weight was lost than when they were cooled by a conventional cooling plant with positive ventilation and very much better than one without it (Table III).

KIRTON EHS STORE

Working in co-operation with NIAE, staff at Kirton EHS converted an existing cold room 5 × 5 × 3 m high into a positively ventilated store with an ice bank cooler. The conversion cost £1200 but did not include the cost of a refrigeration plant as one of the two 3 hp condensing units previously used for cooling the store was adequate for the ice bank. Brussels sprouts, cauliflower and cabbage have been successfully cooled and stored with times to half-temperature of 8, 3 and 5 hours respectively and weight losses during the storage period of about 0·8, 0·7 and 0·8% per week. Further the possible storage period for cauliflowers has been extended from 7 days to 14-21 days without significant loss of quality or shelf-life (Shipway, 1977b).

TABLE III

Comparison of positive ventilation and ice bank cooling with conventional methods

Crop	Ventilation	Cooling	Airflow rate ($kg\ h^{-1} kg^{-1}$)	Start temperature (°C)	Mean time to half temp. (h)	Weight loss (%)
Brussels sprouts	Positive	Ice bank	0·7	12·6	4·7	0·80
	Positive	Conventional	2·5	17·5	7·7	1·25
	Not positive	Conventional	4·0	13·0	39·0	0·50
Cos lettuce	Positive	Ice bank	2·5	16·1	4·0	1·03
	Positive	Conventional	10·0	12·7	6·4	1·14
	Not positive	Conventional	30·0	21·1	23·4	1·75

FOOD RESEARCH INSTITUTE STORE

A cold store at the Food Research Institute, Norwich, was converted to positive ventilation and an ice bank cooler installed, the ice bank being housed in the existing underfloor ducts. Experiments on the long-term storage of carrots and preliminary experiments with soft fruit have been encouraging.

COMMERCIAL STORES

We now have sufficient information to allow the size of the various parts of an ice bank cooler and tower heat exchanger to be specified, given the desired half-cooling time and the heat load of the produce together with the heat gains from the building and equipment. A computer program for the purpose has been written and used to design the equipment for two commercial stores in Lincolnshire with positive ventilation, an ice bank cooler and a tower heat exchanger. A commercial firm has also developed an ice bank cooler unit consisting of a 4 hp compressor, water/ice tank and cooling tower which can be used in new or existing stores. These units are now under trial in commercial stores and elsewhere.

CONCLUSIONS

(i) Positive ventilation whereby air is directed between produce in bulk bins or small containers, reduces cooling times by as much as 88%.

(ii) A cooling tower and ice bank cooler provides the cooling capacity needed to cool vegetables quickly.

(iii) The weight losses of vegetables in stores with this equipment are small, rarely exceeding 1% when cooling leafy vegetables and 0·8% per week of the subsequent storage period.

Given a store with positive ventilation and ice bank cooling it should be possible to maintain shelf-life during storage so that there is little difference between fresh and stored produce; to extend the time between harvesting and marketing or processing of produce whilst the quality is maintained; and have greater flexibility and control over supplies for market, packhouse or processor.

REFERENCES

Burton, W.G. (1972). *Agric. Engng Lond.* **27**, 43-53.
Derbyshire, D.M. and Shipway, M.R. (1978). *Outlook Agric.* **9**, 246-252.
Farrimond, A., Lindsay, R.T. and Neale, M.A. (1979). *Int. J. Refrig.* **2**, 199-206.
Gibbon, J.M. (1972). *J. agric. Engng Res.* **17**, 332-337.

Gibbon, J.M. and Boyce, D.S. (1971). Dep. Note DN/SY/108/3040. Natn. Inst. Agric. Engng. Obtainable from the Library, National Institute of Agricultural Engineering, Wrest Park, Silsoe, Bedford MK45 4HS.

Hinton, W.L. (1975). "Marketing Vegetables from Store". Occasional papers No. 19, Agricultural Economics Unit, University of Cambridge.

Lindsay, R.T. and Neale, M.A. (1977). *In* "Vegetable Cooling and Storage Conference" pp. 5-23. National Agricultural Centre, Stoneleigh.

Linsday, R.T. and Neale, M.A. (1978). Dep. Note DN/CS/690/10005. Natn. Inst. Agric. Engng. Obtainable from the Library, National Institute of Agricultural Engineering, Wrest Park, Silsoe, Bedford MK45 4HS.

Lindsay, R.T., Neale, M.A. and Farrimond, A. (1975a). *Proc. Inst. Refr.* **72**, 23-30.

Lindsay, R.T., Neale, M.A. and Messer, H.J.M. (1975b). *J. agric. Engng Res.* **20**, 235-243.

Lindsay, R.T., Neale, M.A., Messer, H.J.M. and Hawkins, J.C. (1975c). Dep. Note DN/FB/187/3040. Natn. Inst. Agric. Engng. Obtainable from the Library, National Institute of Agricultural Engineering, Wrest Park, Silsoe, Bedford MK45 5HS.

Love, J. (1977). *In* "Vegetable Cooling and Storage Conference" pp. 1-4. National Agricultural Centre, Stoneleigh.

Neale, M.A. and Messer, H.J.M. (1978). *J. agric. Engng Res.* **23**, 67-75.

Robinson, J.M., Browne, K.M. and Burton, W.G. (1975). *Ann. appl. Biol.* **81**, 399-408.

Shipway, M.R. (1977a). *Rep. Kirton expl Hort. Stn for 1976*, 58-59.

Shipway, M.R. (1977b). *In* "Vegetable Cooling and Storage Conference" pp. 26-31. National Agricultural Centre, Stoneleigh.

Tomkins, R.G. (1959). *Br. Pakr* **21**, 92-94.

IV.8 The Effect of Storage Conditions on the Quality of Vegetables and Salad Crops

C. DENNIS

ARC Food Research Institute, Colney Lane, Norwich, NR4 7UA, England

INTRODUCTION

In assessing the quality of stored vegetables and salad crops the criteria for acceptability are ultimately based on a comparison with the characteristics of the commodity harvested at an optimum stage for immediate consumption. This stage can be ephemeral and it may be necessary to harvest before the optimum is reached, so that changes which inevitably occur during storage do not result in unacceptable produce.

The produce termed as vegetables and salad crops is very varied in nature, including stems, leaves, buds, inflorescences, fruits as well as storage organs and can thus be expected to possess very different storage potentials. Although fruits have the same basic reproductive function they are very varied in structure and hence in storage potential. Burton (1973) discussed how the nature of the commodity can influence the type of deterioration and the characteristics of the storage environment.

In some cases the storage conditions are designed to maintain acceptable quality throughout the storage period of an essentially fully developed organ, whereas in others it is necessary to control the inevitable development which occurs in crops harvested at an "immature" stage, so that they are in an acceptable condition for sale and consumption at the end of the storage period.

All stored vegetables and salad crops are subject to the same types of deterioration, physical, physiological, biochemical and pathological, although their relative importance varies considerably between commodities. Ultimately, deterioration or loss of quality will result in wastage, and much of the work on the storage of vegetables and salad crops has been concerned with reducing wastage rather than on improving the quality of visually acceptable produce. Nevertheless, conditions which reduce overall wastage may also maintain higher quality produce for a longer period of time.

Physical deterioration occurs as evaporative loss which affects the texture of the produce and results in shrivelling or withering, whereas the physiological or biochemical changes may be manifest in several ways. Such changes may have a direct influence on acceptability in terms of colour, texture and flavour, on nutritive value or safety as well as resulting in changed physiological behaviour as in the case of renewed growth of storage organs. Deterioration caused by fungal and bacterial growth may affect appearance and safety of the produce initially but usually results in wastage of whole or part of the stored plant structure.

The first aim in storing any given crop must be to reduce the most important cause of deterioration of that crop. In doing so it is quite possible that the conditions used will cause another form of deterioration, for example, a high humidity may reduce water loss but will enhance microbial growth. Thus, in such conditions a compromise has to be reached.

This paper discusses the effect of temperature, relative humidity (r.h.) and gaseous composition (CO_2, O_2) of the storage environment on maintaining and controlling the quality of stored vegetables and salad crops. Particular emphasis will be given to examples of recent and potential developments in storage methods for such crops in the UK.

TEMPERATURE

All the processes leading to loss of quality of stored produce are related to temperature. The temperature determines the amount of water vapour required to saturate the atmosphere, thus affecting evaporation loss, as well as the rate and type of physiological and biochemical changes and microbial growth. Thus, the correct storage temperature is probably the most important factor in reducing deterioration during storage. Ryall and Lipton (1972), Burton (1973), Ryall and Pentzer (1974), Robinson *et al.* (1975), and Goldenberg (1974, 1976) have published optimum temperatures for the transport and storage of fruits and vegetables. Many crops should be stored at temperatures approaching 0°C (e.g. onions, root crops, leafy vegetables) and will rapidly deteriorate at higher temperatures, including loss of nutritive value. For example, Harkett (1977) showed that the vitamin C content of certain vegetables and salad crops (Brussels sprouts, cauliflower) rapidly decreased when stored at 20°C. In the case of lettuce and broccoli, however, even storage at 2°C resulted in a 50% loss of ascorbic and dehydroascorbic acid within 7 days.

Crops of tropical or sub-tropical origin (sweet potato, tomato, aubergine, pepper) have an optimum storage temperature between 5° and 13°C. Storage at temperatures below the optimum can result in chilling injury which is discussed in more detail below.

Since temperature has such a marked influence on deterioration, the rapid removal of "field heat" after harvest is often necessary in order to maintain quality during storage and subsequent shelf-life. However, vegetables and salad crops vary in their cooling requirements. The bulb and root crops, although benefiting from storage at or near 0°C, do not need to be cooled as rapidly as produce with a relatively high metabolic rate such as some leafy vegetables, buds and inflorescences. Conversely, some crops require a short period at a high temperature to allow wound healing. Davies (1977) recommends storage of carrots at 24-26°C, 98% r.h. for 2 days prior to cold storage. Those crops sensitive to chilling benefit from rapid cooling to the optimum storage temperature.

Boa and Lindsay (1976), Lindsay and Neale (1977), Bartlett (1977) and Shipway (1977) discussed methods for cooling and storing produce at refrigerated temperatures and particularly emphasized the advantages of positive ventilation and the ice-bank cooler. This system recently developed jointly by the National Institute for Agricultural Engineering and the Food Research Institute (Hawkins *et al.*, 1978), ensures rapid and even cooling of produce to approximately 1°C (Lindsay and Neale, 1977) without the danger of freezing, since the cooling air comes into contact with water at a temperature slightly above 0°C (see also Lindsay and Neale, pp. 315-328, this Symposium).

The ability to maintain a uniform low temperature without the danger of freezing has proved particularly suitable for the storage of carrots (Dennis and Browne, unpublished). Carrots maintained at 0·5-0·7°C in an ice-bank store showed only slight regrowth after 10 months' storage as well as remaining turgid and sweet, whereas roots stored at 2-3°C showed obvious signs of shoot and root growth within 6 months.

Such a system should also be suitable for the long-term storage of crops such as beetroot and Dutch white cabbage which require a high relative humidity.

Commodities such as tomatoes, peppers, cucumbers, aubergines and beans can be severely injured by storage below 10-12°C. Chilling injury (Fidler, 1968) is a disorder induced by low but non-freezing temperatures in whole plants or susceptible tissues and limits the storage life. The extent of chilling injury is influenced by the temperature, the duration of exposure to a given temperature and the chilling sensitivity of the crop as illustrated by Tomkins (1966) and Ryall and Lipton (1972).

The symptoms of chilling injury may not be evident while the produce is held at chilling temperature but only become apparent after transfer to a higher temperature. Conspicuous symptoms are discolouration or pitting, enhanced microbial spoilage, and in the case of immature fruits the loss of the ability to ripen normally. Discolouration and pitting may only occur externally as in beans (Watada and Morris, 1966 a,b; McColloch and Vaught, 1968) and cucumbers (Wiebe 1969) but may be accompanied by internal discolouration as in aubergines (McColloch, 1966) and peppers (McColloch, 1962; Ogata *et al.*, 1968).

Rapid fungal spoilage of peppers (McColloch, 1962), aubergines (McColloch, 1966) and tomatoes (Tomkins, 1963; Dennis and Davis, unpublished data) occurs after exposure to chilling temperatures. In unchilled tomatoes, infection usually occurs via the calyx whereas, after chilling, infection occurs at points all over the surface of the fruit (Tomkins, 1963; Dennis and Davis, unpublished data).

The failure to ripen is shown most conspicuously with mature green tomatoes where chilled fruits develop a very uneven and blotchy colour when transferred to a ripening temperature and even if fungal spoilage is prevented the fruit never reaches an edible stage.

All of these symptoms result in reduced quality and often wastage. Chilling may also reduce the nutritive value of some crops. Carotene, the precursor of vitamin A, increases in sweet potatoes stored at 13°C or above, but not when stored at 10°C (Ezell and Wilcox, 1952), which is in the chilling range. Chilling also reduces the ascorbic acid (vitamin C) content of some vegetables (Ryall and Lipton, 1972) although Craft and Heinze (1954) showed that chilling temperatures had no effect on the ascorbic acid content in tomatoes during storage. Rhodes and Wooltorton (unpublished data) also showed that chilling temperatures had very little effect on the sugar/acid balance of stored tomatoes.

The effect of chilling on other nutritionally important constituents of vegetables has received little attention, most probably because the obvious symptoms of chilling injury are economically much more important than the invisible and possibly minor changes in nutritive value.

Lyons and Raison (1970) suggest that chilling injury which occurs below critical temperatures involves a phase change in the lipids of sensitive crops at this critical temperature. This has been postulated as the primary response of sensitive species to chilling temperatures and is thought to lead to changes in membrane permeability and in the activity of membrane-bound enzymes with the accumulation of toxic intermediates and hence damage or even death of the cells (Lyons, 1973). A number of cellular functions of the tomato fruit associated with membranes show marked changes at about 12°C (Lyons and Raison, 1970; Shneyour *et al.*, 1973; Nobel, 1974). Ultrastructural changes also occur in the mitochondria and plastids of tomatoes during chilling (Moline, 1976). Rhodes and Wooltorton (1978a) reported increased metabolism of chlorogenic acid in chilled tomatoes which has also been reported to occur in chilled peppers (Kozube and Ogata, 1971) and sweet potatoes (Lieberman *et al.*, 1959; Rhodes and Wooltorton, 1978b).

For some sensitive crops, intermittent warming during storage at low temperatures is effective in preventing chilling injury (Hruschka *et al.*, 1969).

RELATIVE HUMIDITY

Although the term relative humidity is usually used in describing the water vapour content of the air in a storage environment, when considering the evaporative loss from fresh produce (see Burton, 1973) a more useful concept is water vapour pressure deficit (WVPD). Although the rate of evaporation from different vegetables and salad crops varies enormously, depending on the structure of the commodity and the surface area exposed to the ambient air (Burton, 1972, 1973), the evaporation from any given commodity is directly proportional to the WVPD of its immediate environment. The amount of water loss which can occur before a given commodity becomes unacceptable ranges from about 3% for lettuce to 10% for onions (Robinson *et al.*, 1975).

In order to avoid excessive water loss during storage, many vegetables and salad crops (e.g. carrots, lettuce) require high relative humidities (low WVPD) whereas commodities which are more resistant to water loss (e.g. tomatoes) can be stored satisfactorily at a higher WVPD. Onions are exceptional vegetables in tolerating a relatively high WVPD (0·7-1·2 mb) for successful storage (Robinson *et al.*, 1975; Shipway, 1977).

The relative humidity of the storage environment also affects the activity of micro-organisms on the surface of commodities but its effect is dependent on temperature. Van den Berg and Lentz (1977) have reported that very high humidities or even the presence of droplets of condensation on the produce caused no increase in losses due to microbial growth at temperatures of 0-1°C. Earlier reports by these workers (1973a,b; 1974) using jacketed stores (Lentz and Van den Berg, 1973) at 0-1°C, 98-100% r.h. (WVPD 0-0·1 mb) had shown that the amount of wastage was reduced and a higher quality was achieved in terms of crispness with a range of vegetables including carrots, parsnips, cabbage, celery and Brussels sprouts, compared to samples stored at 90-95% r.h. (WVPD 0·2-0·5 mb). A similar environment can also be achieved by other methods such as the humi-fresh system (Meredith, 1973) and the ice-bank system (Lindsay and Neale, 1977).

Where produce has to be stored at higher temperatures, for example, to avoid chilling injury, then a greater WVPD is required to avoid excessive microbial growth. For example, a relative humidity of between 93-95% is required for successful storage of mature green tomatoes at 12·5°C for

6-8 weeks (Dennis, Browne and Adamicki, unpublished data). Higher relative humidities result in fungal growth on the surface of the fruits which may eventually result in rotting.

COMPOSITION OF THE ATMOSPHERE

The use of controlled atmospheres (CA), that is, the modification and control of the oxygen and/or carbon dioxide concentration in the atmosphere surrounding the produce, is justified only if the vegetable or salad crop is in a more desirable condition after storage in CA than after a similar period in air. The value of CAs in preserving the quality and extending the storage life of vegetables and salad crops is not as clearly demonstrated as it is with fruits (Stoll, 1974). However, reducing the oxygen tension to a level which limits the activity of terminal oxidases can be effective in maintaining quality and thus extending the storage life of commodities in which biochemical changes are an important factor in deterioration (Burton, 1974). Such changes include discolouration of cut surfaces, loss of chlorophyll during ripening and senescence or toughening in green vegetables, development of physiological disorders and the ripening process in commodities harvested unripe.

Discolourations at cut surfaces are typically caused by oxidation of phenolic substrates by phenolases, which have a low affinity for oxygen and are therefore susceptible to moderate changes in oxygen tension. Thus, browning of the cut ends of broccoli (Smith, 1938) and lettuce (Singh *et al.*, 1972a,b) is reduced or almost prevented by oxygen concentrations of 5% or below.

Ripening of tomatoes can be delayed by storage in CAs of 3-5% O_2 in combination with up to 10% CO_2 (Kidd and West, 1933; Furlong, 1946; Eaves and Lockhart, 1961; Tomkins, 1963; Parsons *et al.*, 1970). Salunkhe and Wu (1973) showed that low oxygen atmospheres inhibited the degradation of chlorophyll and starch and syntheses of lycopene, β-carotene and soluble sugars in tomato fruits.

The loss of chlorophyll which occurs during the senescence of leafy vegetables can be reduced by high carbon dioxide in combination with low oxygen. This improves the appearance of vegetables such as broccoli (Smith, 1938), spinach (McGill *et al.*, 1966), asparagus (Wang *et al.*, 1971) and cabbage (Isenberg and Sayles, 1969; Pendergrass and Isenberg, 1974; Bohling and Hansen, 1977) compared to those stored in air.

In asparagus, an increase in fibrousness and consequent toughening renders the commodity unacceptable. This toughening can be delayed or even reversed by holding asparagus in about 12% CO_2 and under high relative humidity at 4°C or below. Such treatment produces tender spears and a greener colour, both effects being enhanced by cooking (Lipton, 1960, 1968; Lougheed and Dewey, 1966; Wang *et al.*, 1971). High carbon dioxide concentrations (10%) can also improve the tenderness of broccoli after harvest due to an increase in pH of the tissue resulting from storage in such an atmosphere (Lebermann *et al.*, 1968; Lipton and Harris, 1974).

Two economically important physiological disorders which can be controlled by CA storage are "russet-spotting" in lettuce (Lipton *et al.*, 1972) and pepper spot in winter storing cabbage (Geeson and Robinson, 1975; Geeson and Browne, 1978). Lipton (1975) showed that "russet-spotting" on midribs of lettuce could be reduced by holding lettuce in 2-6% O_2, while Isenberg *et al.* (1971), Bohling and Hansen (1977) and Geeson and Browne (personal communication) showed that low oxygen combined with high carbon dioxide reduced or eliminated pepper spot on Danish and Dutch white cabbage.

Storage in CAs can affect the flavour of vegetables. For example, Hansen and Rumpf (1974) reported that an atmosphere of 1·5-3% O_2 inhibited the production of isocoumarins in carrots which, when present, may impart a bitter flavour. Isenberg and Sayles (1969), using Danish cabbage, showed that very low oxygen and carbon dioxide concentrations resulted in a sweeter taste than normal, while higher carbon dioxide concentrations produced more pungent tastes characteristic of fresh cabbage and superior to that of cabbage stored in air. Similarly, the flavour and texture of Dutch white cabbage stored for 34 weeks in 3% O_2 and 5-6% CO_2 were superior to that of cabbages stored in air (Geeson and Griffiths, unpublished data).

Controlled atmospheres may also reduce the amount of microbial spoilage either by a direct effect on the micro-organism or by affecting the resistance of the commodity to invasion by fungi and bacteria. For example, rotting of stored winter cabbage is reduced under CA conditions (Pendergrass and Isenberg, 1974; Henze, 1977; Geeson and Browne, unpublished data). Geeson and Browne (personal communication) found an atmosphere of 3% O_2 and 5-6% CO_2 particularly effective in reducing rots caused by *Botrytis cinerea*.

Apart from the beneficial effects, described above, CAs in which the oxygen is too low and/or the carbon dioxide is too high can result in visible injury of commodities which may lead to increased microbial spoilage. Such cases include the brownish-red discolouration (brown stain) of lettuce induced by high carbon dioxide concentrations and enhanced further in combination with low oxygen (Stewart and Uota, 1971). Lipton (1977a) concludes that symptoms of low oxygen or high carbon dioxide injury are invariably most prominent in white or yellow tissue of leafy vegetables but never in green tissue, and thus suggests that photosynthetic ability of the tissue influences its susceptibility to injury.

Cauliflower is very sensitive to carbon dioxide, exposure to 5% or more results in excessive softening and discolouration on cooking (Lipton *et al.*, 1967).

Severe off-odours and off-flavours may also be induced by CAs; Smith (1938) reported strong, unpleasant odours in broccoli held in 5% O_2 and 15% CO_2 while McGill *et al.*, (1966) found spinach to have an unacceptable flavour after one week's storage in 13% CO_2.

Thus, vegetables and salad crops differ widely in their reaction to exposure to CAs (see also Kader and Morris, 1977; Morris and Kader, 1977a; Lipton, 1977b; Weichmann, 1977a,b). This may be influenced by physiological age at harvest, storage temperature and duration of holding (Lipton, 1975; Morris and Kader, 1977b). Large differences in tolerance to low oxygen and high carbon dioxide occur between closely related vegetables. For example, cauliflower is injured after one week at 5°C in 10% CO_2 whereas broccoli remains in excellent condition in that atmosphere. Even cultivars of the same crop (e.g. lettuce) differ greatly in their reaction to CAs (Lipton, 1977b).

Despite the wealth of research on long-term CA storage of many crops, notably in the USA, it is only the top fruit industry which uses such a system extensively in the UK. Modified atmospheres are used during transport of various commodities, again especially in the USA (Lipton, 1975) although with the introduction of the "Tectrol" system, vegetables and salad crops are now shipped from Europe under modified atmospheres (Anon, 1979).

In the UK, as far as the author is aware, Dutch white cabbage is the only vegetable stored long-term under CA. Controlled atmosphere storage of cabbage began commercially in 1976 but still only represents less than 5% of the estimated 20 000 tonnes in refrigerated stores. Controlled atmosphere storage offers the potential for cabbage grown in the UK to be available for "fresh" and processing outlets for a greater proportion of the year.

It may also be possible to extend the UK tomato season by using CA storage for mature green tomatoes, both for glasshouse-grown cultivars and commercial outdoor crops from the Channel Islands. Recent work has shown that current commercial cultivars (Sonato, Sonatine, Eurocross BB, Hundred Fold) can be stored in 5% CO_2 with 3% or 5% O_2 for 6-10 weeks at 12·5°C, 92-95% r.h. and subsequently ripened in air at 15° or 20°C (Dennis, Browne and Adamicki, unpublished data). This successful use of high carbon dioxide is contrary to the findings of Morris and Kader (1974) where injury occurred within 7 days' storage at 12·5°C. This presumably reflects differences in susceptibility between cultivars and possibly also a difference in growing conditions. Samples of fruit stored in CAs ripen more uniformly than those stored continuously in air. Taste panel assessments indicate that fruit of certain cultivars, especially Hundred Fold, are superior in flavour and overall quality after CA storage to fruit imported from Spain and the Canary Islands into the UK between November and January. The main limitation to storage potential is fungal spoilage, especially with outdoor-grown cultivars. Although dipping the fruit in "Rovral" prior to storage does reduce the incidence of rots, at present, post-harvest use of this chemical is not permitted on tomatoes. Although further research and development is required, CA storage of tomatoes may also be used to enable the supply of fruit to be adjusted to meet the demand throughout the season and is especially attractive because of the uniform ripening achieved. Similar possibilities should also be explored for the increasing quantity of capsicums and courgettes grown in the UK.

REFERENCES

Anon. (1979). *Fruit Trades J.* **155**, 19.

Bartlett, D.I. (1977). *In* "Vegetable Cooling and Storage Conference", pp. 20-25. National Agricultural Centre, Stoneleigh.

Boa, W. and Lindsay, R.T. (1976). *ARC Res. Rev.* **2**, 86-87.

Bohling, H. and Hansen, H. (1977). *Acta Hort.* **62**, 49-54.

Burton, W.G. (1972). *Agric. Engng Lond.* **27**, 43-53.

Burton, W.G. (1973). *Proc. 7th Br. Insect. and Fung. Conf.* **3**, 1037-1055.

Burton, W.G. (1974). *Ann. appl. Biol.* **78**, 149-168.

Craft, C.C. and Heinze, P.H. (1954). *Proc. Am. Soc. hort. Sci.* **64**, 343-350.

Davies, W.P. (1977). *Ann. appl. Biol.* **85**, 163-164.

Eaves, C.A. and Lockhart, C.L. (1961). *J. hort. Sci.* **36**, 85-92.

Ezell, B.D. and Wilcox, M.S. (1952). *Pl. Physiol.* **27**, 81-94.

Fidler, J.C. (1968). *In* "The Low Temperature Biology of Foodstuffs" (J. Hawthorn and E.J. Rolfe, eds), pp. 271-283. Recent adv. Fd Sci. Vol. 4. Pergamon Press, Oxford.

Furlong, C.R. (1946). *J. Pomol.* **22**, 197-208.

Geeson, J.D. and Browne, K.M. (1978). *Grower* **89**, 27-31.

Geeson, J.D. and Robinson, J.E. (1975). *Comml Grow.* No. 4147, 1245-1246.

Goldenberg, N. (1974). *Outlook Agric.* **8**, 105-112.

Goldenberg, N. (1976). *Rep. Long Ashton Res. Stn 1975*, 184-199.

Hansen, H. and Rumpf, G. (1974). *Acta. Hort.* **38**, 321-326.

Harkett, P.J. (1977). *Ann. appl. Biol.* **85**, 159.

Hawkins, J.C., Messer, H.J.M. and Lindsay, R.T. (1978). *ARC Res. Rev.* **4**, 34-37.

Henze, J. (1977). *Acta. Hort.* **62**, 71-78.

Hruschka, H.W., Smith, W.L.Jr. and Baker, J.E. (1969). *Am. Potato J.* **46**, 38-53.

Isenberg, F.M.R. and Sayles, R.M. (1969). *J. Am. Soc. hort. Sci.* **94**, 447-449.

Isenberg, F.M.R., Oyer, E.G. and Engst, C.B. (1971). *Acta Hort.* **20**, 7-18.

Kader, A.A. and Morris, L.L. (1977). *In* "Proceedings of 2nd National Controlled Atmosphere Research Conference" (D.H. Dewey ed.), pp. 260-265. Michigan State University.

Kidd, F. and West, C. (1933). *Rep. Fd Invest. Bd. 1928-1932*, 209-221.

Kozubue, N. and Ogata, K. (1971). *J. Jap. Soc. hort. Sci.* **40**, 300-304.

Lebermann, K.W., Nelson, A.I. and Steinberg, M.P. (1968). *Fd. Tech. Champaign.* **22**, 490-493.
Lentz, C.P. and Van den Berg, L. (1973). *ASHRAE J.* **15**, 55-60.
Liebermann, M., Craft, C.C. and Wilcox, M.S. (1959). *Proc. Am. Soc. hort. Sci.* **74**, 642-648.
Lindsay, R.T. and Neale, M.A. (1977). *In* "Vegetable Cooling and Storage Conference" pp. 1-16. National Agricultural Centre, Stoneleigh.
Lipton, W.J. (1960). *Mktg Res. Rep. U.S. Dept. Agric.* 428, 17-26.
Lipton, W.J. (1968). *Mktg Res. Rep. U.S. Dept. Agric.* 817, pp. 7.
Lipton, W.J. (1975). *In* "Post-harvest Biology and Handling of Fruits and Vegetables" (N.F. Haard and D.K. Salunkhe, eds) 130-143. Avi Publishing Co. Inc., Westport, Connecticut, USA.
Lipton, W.J. (1977a). *In* "Proceedings of the 2nd National Controlled Atmosphere Research Conference" (D.M. Dewey, ed.), pp. 137-141. Michigan State University.
Lipton, W.J. (1977b). *In* "Proceedings of the 2nd National Controlled Atmosphere Research Conference" (D.M. Dewey, ed.), pp. 277-281. Michigan State University.
Lipton, W.J. and Harris, C.M. (1974). *J. Am. Soc. hort. Sci.* **99**, 200-205.
Lipton, W.J., Harris, C.M. and Couey, M.H. (1967). *Proc. Am. Soc. hort. Sci.* **91**, 852-859.
Lipton, W.J., Stewart, J.K. and Whitaker, T.W. (1972). *Mktg Res. Rep. U.S. Dept. Agric.* 950.
Lougheed, E.C. and Dewey, D.H. (1966). *Proc. Am. Soc. hort. Sci.* **89**, 336-345.
Lyons, J.M. (1973). *A. Rev. Pl. Physiol.* **24**, 445-466.
Lyons, J.M. and Raison, J.K. (1970). *Pl. Physiol.* **45**, 386-389.
McColloch, L.P. (1962). *Mktg Res. Rep. U.S. Dept. Agric.* 536, pp. 16.
McColloch, L.P. (1966). *Mktg Res. Rep. U.S. Dept. Agric.* 749, pp. 5.
McColloch, L.P. and Vaught, C. (1968). *U.S. Agr. Res. Serv. ARS.* 31-23, pp. 3.
McGill, J.N., Nelson, A.I. and Steinberg, M.P. (1966). *J. Fd Sci.* **31**, 510-517.
Meredith, M. (1973). *ASHRAE J.* CH-73-7, 29-34.
Moline, H.E. (1976). *Phytopathology* **66**, 617-624.
Morris, L.L. and Kader, A.A. (1974). *In* "Fresh Market Tomato Research", 1973, Vegetable Crops Series 166, pp. 36-48. University of California, Davis.
Morris, L.L. and Kader, A.A. (1977a). *In* "Proceedings of the 2nd National Controlled Atmosphere Research Conference" (D.H. Dewey, ed.), pp. 266-276. Michigan State University.
Morris, L.L. and Kader, A.A. (1977b). *In* "Proceedings of the 2nd National Controlled Atmosphere Research Conference" (D.H. Dewey, ed.), pp. 142-148. Michigan State University.
Nobel, P.S. (1974). *Planta* **115**, 369-372.
Ogata, K., Kozube, N. and Murata, T. (1968). *J. Jap. Soc. hort. Sci.* **37**, 249-255.
Parsons, C.S., Anderson, R.E. and Penney, R.W. (1970). *J. Am. Soc. hort. Sci.* **95**, 791-794.
Pendergrass, A.M. and Isenberg, F.M.R. (1974). *HortScience.* **9**, 226-227.
Rhodes, M.J.C. and Wooltorton, L.S.C. (1978a). *Phytochemistry* **16**, 655-659.
Rhodes, M.J.C. and Wooltorton, L.S.C. (1978b). *Phytochemistry* **17**, 1225-1229.
Robinson, J.E., Browne, J.M. and Burton, W.G. (1975). *Ann. appl. Biol.* **81**, 399-408.

Ryall, A.L. and Lipton, W.J. (1972). "Handling, Transportation and Storage of Fruits and Vegetables. Vol.1. Vegetables and Melons". Avi Publishing Co. Inc., Westport, Connecticut, USA.
Ryall, A.L. and Pentzer, W.T. (1974). "Handling, Transportation and Storage of Fruits and Vegetables. Vol.2. Fruits and Tree Nuts". Avi Publishing Co. Inc., Westport, Connecticut, USA.
Salunkhe, D.K. and Wu, M.T. (1973). *J. Am. Soc. hort. Sci.* **98**, 12-14.
Singh, B., Yang, C.C. and Salunkhe, D.K. (1972a). *J. Fd Sci.* **37**, 48-51.
Singh, B., Wang, D.J., Salunkhe, D.K. and Ramman, A.R. (1972b). *J. Fd Sci.* **37**, 52-55.
Shipway, M.R. (1977). *In* "Vegetable Cooling and Storage Conference", pp. 26-31. National Agricultural Centre, Stoneleigh.
Shneyour, A., Raison, J.K. and Smillie, E.M. (1973). *Biochim. biophys. Acta,* **292**, 152-161.
Smith, W.M. (1938). *Rep. Fd Invest. Bd.* 1937, 185-187.
Stewart, J.K. and Uota, M. (1971). *J. Am. Soc. hort. Sci.* **96**, 27-30.
Stoll, K. (1974). *Bull. int. Inst. Refrig.* **54**, 1302-1324.
Tomkins, R.G. (1963). *J. hort. Sci.* **38**, 335-347.
Tomkins, R.G. (1966). *Ditton Lab. Memoirs, No. 91*, 1-17.
Van den Berg, L. and Lentz, C.P. (1973a). *J. Am. Soc. hort. Sci.* **98**, 129-132.
Van den Berg, L. and Lentz, C.P. (1973b). *ASHRAE J.* CH-73-7, 12-18.
Van den Berg, L. and Lentz, C.P. (1974). *Can. Inst. Fd Sci. Technol.* **7**, 260-262.
Van den Berg, L. and Lentz, C.P. (1977). *Acta Hort.* **62**, 209-215.
Wang, S.S., Haard, N.F. and DiMarco, G.R. (1971). *J. Fd Sci.* **36**, 657-661.
Watada, A.E. and Morris, L.L. (1966a). *Proc. Am. Soc. hort. Sci.* **89**, 368-374.
Watada, A.E. and Morris, L.L. (1966b). *Proc. Am. Soc. hort. Sci.* **89**, 375-380.
Weichmann, J. (1977a). *In* "Proceedings of 2nd National Controlled Atmosphere Research Conference" (D.H. Dewey, ed.), pp. 122-136. Michigan State University.
Weichmann, J. (1977b). *In* "Proceedings of the 2nd National Controlled Atmosphere Research Conference" (D.H. Dewey, ed.), pp. 282-286. Michigan State University.
Wiebe, M.J. (1969). *Gemüse* **24**, 54-56.

IV.9 The Influence of Controlled Atmosphere Storage on the Ripening of Apples in Relation to Quality

M. KNEE AND R. O. SHARPLES

East Malling Research Station, Maidstone, Kent ME19 6BJ, England

DEVELOPMENT OF CONTROLLED ATMOSPHERE STORAGE

Refrigeration provides the basic means whereby the deterioration of fruit after harvest is delayed. Some fruits, for example pears, can be kept just above their freezing point at $-1\,°C$. Control of ripening and deterioration due to physiological or pathological causes at this temperature is so effective that normally refrigeration alone is adequate for most commercial requirements. However, many English cultivars of apple must be stored at around $3\,°C$, because below this temperature they suffer injury. Consequently storage life is limited; for example, Cox's Orange Pippin can be stored until only about mid-December in air at $3\,°C$. This has provided an incentive for the development of supplementary methods for extending storage life.

Kidd and West began the systematic work in this direction in Cambridge after the First World War by discovering that ripening and senescence were delayed by the carbon dioxide (CO_2) which accumulated when apples were enclosed in a gas-tight chamber. In 1921 they supervised the operation of the first "gas" store with a mixture of apple cultivars on a farm at Histon near Cambridge. They paid £114-19-0 for the fruit at harvest, and in spite of 24% wastage, it was sold for £262-10-0 in the following March (Kidd, 1923).

Subsequently these workers demonstrated that reduced oxygen (O_2) levels also prolonged the storage life of fruit (Kidd and West, 1936) but it was not until the early 1950s that control of O_2 was obtained in commercial stores, by using "scrubbers" containing lime, sodium hydroxide or other absorbants to remove excess CO_2. As the technique became more widely adopted, it was renamed controlled atmosphere (CA) storage. For many years the standard conditions for UK apples were 5% CO_2 with 3% O_2. Prolonged storage in 5% CO_2 with 3% O_2 revealed the injurious effects of high concentrations of CO_2 on Cox (manifest as core flush), and new CA regimes were developed in which CO_2 was kept below 1% and O_2 was reduced to 2%, whereby the disorder was avoided and ripening was still controlled satisfactorily.

Today, with the total production of apples and pears in the UK averaging about 520 000 tons, there is storage capacity for about 300 000 tons; this provides for about 80% of the crop of cultivars suitable for storage. Just over a quarter of the stores are equipped with scrubbers and can be used for low O_2 and high CO_2; a further 50% are suitable for storing apples in high CO_2 regimes and the remainder are suitable for air storage only (Anon., 1973). The bulk of the CA store capacity is found at co-operative pack-houses and on the farms of larger independent growers.

In the post-war years increasing yields of apples together with increasing store capacity enabled the UK industry to supply a larger proportion of consumer requirements for apples. Rising standards of living have not led to greatly increased apple consumption and the structural surplus of the European fruit industry has depressed the market so that for the individual grower there is no financial reward for storage; the price of a given cultivar of apple is remarkably constant over the storage season. Of course, prices would be much lower if the entire crop had to be sold after harvest and there is a need for continuous research to provide the technology to extend the marketing period still further.

FRUIT NUTRITION AND POST-HARVEST CHEMICAL TREATMENTS

The demand for longer storage has been met, in part, by refinement of the traditional variables, so that optimum temperatures, and CO_2 and O_2 levels can be specified for the main apple cultivars in the UK as in other fruit growing regions (Fidler, 1973).

In the last 25 years, advances have also been made in improving the storage potential of the crop by attention to the influence of growing conditions and through the introduction of various chemical treatments designed to control wastage.

Particular attention has been paid to the effects of nutritional factors on fruit quality. Recognition of the importance of calcium, in reducing susceptibility to several physiological storage disorders and rotting, has led to modifications in fertilizer usage and the development of methods of supplementing fruit calcium levels by orchard sprays which, in turn, have been reflected in a general improvement in fruit storage potential in this country.

Calcium concentrations are also supplemented by drenching the fruit immediately after picking with solutions of calcium chloride.

Other nutrients, including N, P and K also affect storage quality in one way or another and both leaf and fruit analysis are used to assist advisers and packhouse managers to identify the causes of persistent storage problems and also to categorize the crops from different orchards so that only fruit of good storage potential is allocated for long-term storage (Sharples, 1980). As storage periods were extended, rotting (e.g. by *Gloeosporium* spp. on Cox) became a major problem but most rots can now be effectively controlled by treatment with benzimidazole or related fungicides. Many apple cultivars, notably Bramley's Seedling, are susceptible to a physiological disorder called "superficial scald"; this is controlled by post-harvest applications of anti-oxidants. Chemical regulators of fruit ripening are now available; these are used at present to influence time of harvest (Sharples, 1973) but the future holds the prospect of chemical regulation of the rate of senescence after harvest.

EATING QUALITY OF STORED APPLES

The development of storage technology has so far been weighted towards the prevention of the wider aspects of senescence and wastage and the extension of the longevity of fruit has not yet reached its biological limits. However, there are other limits imposed by the growing sophistication and awareness of people as consumers of fruits. Hence there is now a need to investigate the way in which the eating qualities of fruit are affected by storage practices and, in this, we are immediately confronted by the inherent subjectivity of quality assessment. This can be offset by a dual approach in which, (i) subjective descriptions are gathered systematically and analysed statistically and, (ii) objective features of the fruit can be measured and correlated with quality (see also Williams, pp. 17-33, this Symposium). There is also a requirement for metabolic studies which reveal how post-harvest treatments affect processes related to quality changes and whether there is further scope for control or improvement.

THE DETERMINANTS OF QUALITY

FRUIT RIPENING

Ripening can be seen as a modification of the basic pattern of senescence in plant tissues. It is also apparent that the physiological changes which occur during ripening are associated with changes in quality as well as senescence leading to death through "physiological disorders". Ethylene is usually considered to be the immediate trigger for ripening but, because fruit is stored in sealed chambers, ethylene accumulates quickly and can reach levels in excess of 1000 $\mu l\ l^{-1}$. However, under CA conditions, ripening is not triggered and this must be because high CO_2 and low O_2 inhibit the processes, consequent upon ethylene action, leading to softening, colour change, aroma synthesis, etc. (Knee, 1976).

Recent research on the role of oxygen in senescence of plants and animals suggests a possible explanation of the mechanism of low O_2 storage. In senescence the organism succumbs to the "toxic" effects of O_2, which arise from its tendency to give rise to free radicals such as superoxide in biological systems (Fridovich, 1977). Preliminary results indicate that benzoate, a hydroxyl radical scavenger has similar effects to low O_2 treatment on chlorophyll loss, ester synthesis and softening of apples.

The most obvious role for free radicals in ripening is in the reactions leading to chlorophyll loss and aroma synthesis. The plastid membrane is rich in unsaturated fatty acids which are particularly labile to oxidative degradation involving free radical intermediates; this exposes chlorophyll to attack and provides precursors for aroma synthesis (see below). Lipid degradation can be accelerated by light treatment both of isolated chloroplasts (Heath and Packer, 1965) and intact leaves (Van Hasselt, 1974). Light accelerates chlorophyll destruction and ester synthesis in apples during storage (Table I). The way in which O_2 could accelerate the reactions involved in cell wall breakdown is less obvious; cell wall polysaccharides are labile to superoxide (Kon, 1978) but the role of O_2 could be more indirect; for example, it might influence the release of hydrolytic enzymes from a lysosomal compartment.

TABLE I

Chlorophyll and hexyl acetate content of Golden Delicious apples stored in 2·5% O_2 + 97·5% N_2 in the light (8500 lux) and dark for 128 days

Treatment	Chlorophyll ($\mu g\ cm^{-2}$ peel)	Hexyl acetate ($\mu g\ g^{-1}$)
Dark	3·70	2·05
Light	1·77	5·45

Data taken from Knee *et al.* (1980).

FRUIT COLOUR

Although intensity of the red colour of many red and bicoloured dessert apples is probably the most important aspect of colour quality, anthocyanin production in apple peel has an absolute requirement for light and is thus influenced by pre-harvest factors rather than storage.

Wholesalers equate longer shelf-life with a green rather than yellow ground colour and this tends to be reflected in grading standards and marketing requirements. However, there is little evidence for a clear preference among consumers of green or yellow-skinned fruits although imported green Golden Delicious apples command a higher price in some countries than more yellow-skinned home-grown fruits (Gormley and Egan, 1977).

During ripening, yellow pigments are synthesized but the dominant change determining colour is the loss of chlorophyll. This is inhibited by reducing storage temperatures, increasing CO_2 and reducing O_2 in the atmosphere of the store, and since, in long-term storage, apples are held in these inhibitory conditions fruit ground colour is largely determined before harvest, when rapid chlorophyll loss occurs on the tree (Fig. 1).

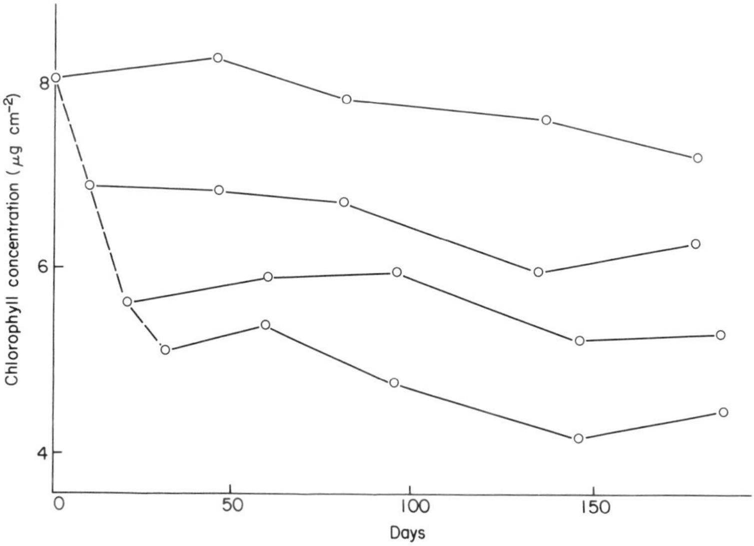

Fig. 1. Chlorophyll concentration in the peel of Cox's Orange Pippin apples harvested at 10 day intervals from 8 September 1973 (Day 1) and stored in 2% O_2 + 0·5% CO_2 for up to 178 days. The discontinuous line shows the decline in concentration on the tree, the continuous lines show the change during storage for successive harvests (redrawn from Knee, 1976).

TEXTURE

We attempt to assess textural properties of fruit by measuring the force required for a plunger of given dimensions to penetrate cortex tissue. Although simple to operate, interpretation of the penetrometer test is complex in that it is influenced by resistance to shear and compression and may involve cell breakage or cell separation. The same factors will presumably be involved in the consumer's appreciation of texture but there is unlikely to be a direct relationship between taste panel assessment and instrumental measurement (Table II). Other investigators have pursued alternative approaches with improved "analogue" instruments (denture tenderometers) or more refined measurements of separate mechanical features, and occasionally have achieved better correlations with sensory assessments of texture (Brennan *et al.*, 1977).

TABLE II

Correlation of physical and chemical measurements with sensory data for stored Cox's Orange Pippin apples

Physical/chemical measurement	Sensory description	Correlation coefficient
Firmness with penetrometer	Crispness	0·35[a]
	Toughness	0·26 n.s.
	Flouriness	−0·37[a]
Juice pH	Acidity	−0·41[b]
Titratable acidity	Acidity	0·56[c]

Mean ratings (1-5) for sensory features by 12 people on a taste panel were compared with physical/chemical measurements for 43 samples of fruit produced in various storage trials. n.s. = not significant, [a], [b], [c] significant at $P = 0{\cdot}05$, 0·01 and 0·001 respectively.

From first principles we would expect the texture of a tissue like apple to be influenced by the structure of its cell walls and the water relations of the cells they enclose. We can speculate about the effects of store humidity and changing compartmentation of solutes on wall pressure and hydration of structural polysaccharides but we have no direct experimental evidence for effects on the physiology or mechanical and sensory properties of fruit. These areas demand future detailed investigation.

In our work to date, we have concentrated on cell wall composition and structure. The parenchymatous cells of the apple cortex are large, highly vacuolated and bounded by relatively thin primary walls. These are low in lignin, xylan, xyloglucan and hydroxyproline-rich protein. A highly-branched pectin seems to be present in the matrix of the wall, while a less-branched pectin is concentrated in the middle lamella where its free carboxyl groups, interacting with calcium, provide cell cohesion. Microscopic observation shows that, during ripening, cells become more easily separated; cell separation can also be achieved *in vitro* by chelating agents, by enzymes which selectively attack the linear pectin, or by diazomethylation of its carboxyl groups which selectively solubilizes the same fraction (Knee, 1978).

Loss of galactose and a small loss of uronic acid residues occur early in storage and are unaffected by the O_2 or CO_2 concentration in the store; a variable amount of softening seems to be associated with these changes. The concentrations of O_2 and CO_2 have more influence on the release and/or synthesis of soluble pectin, which is associated with further softening.

In the absence of more refined correlations of physical and sensory assessment, we assume that firmer fruit are of better quality and have therefore stored better. However, the extent to which softening can be controlled varies greatly between different batches of fruit stored under the same storage conditions. The delay in the increase of soluble polyuronide is reasonably independent of season (Fig. 2) and it seems that the rates of change of other components must vary to account for the more rapid initial softening in some years than in others.

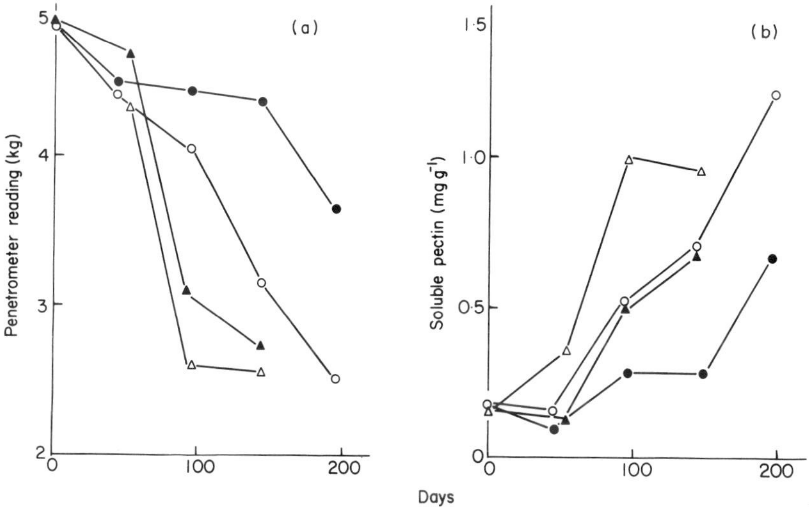

Fig. 2. (a) Firmness, and (b) soluble pectin concentration, of Cox's Orange Pippin apples stored in low oxygen concentrations in two different seasons.
○, 2% O_2, <0·5% CO_2 in 1974-75; ●, 1% O_2, <0·5% CO_2 in 1974-75;
△, 2% O_2, <0·5% CO_2 in 1976-77; ▲, 1·3% O_2, <0·5% CO_2 in 1976-77.

TASTE

Sweetness and acidity are the dominant taste attributes of dessert apples; the amounts of phenolic compounds are probably too low to contribute significant astringency or bitterness. The total sugar content of apples is essentially decided at harvest as fruit respiration during storage does not consume more than about 10% of the initially, high levels (Fidler and North, 1967). Sucrose is slowly hydrolysed during storage to glucose and fructose (Griffiths *et al.*, 1950) but this probably has little effect on sweetness. The major change is in the level of acid which may fall by as much as 50% during storage; this fall is inhibited by low O_2 conditions.

Comparison of sensory and analytical data relating to acidity reveals higher correlations than in other comparisons (Table II).

AROMA

Williams (1979) and his co-workers at Long Ashton are using both analytical and sensory assessments to identify the compounds which are important in apple aroma, particularly those responsible for the aroma of Cox's Orange Pippin. Research at East Malling is concentrated on the metabolism of aroma compounds, particularly esters, and the effects of storage treatments on their metabolism. The first three years of these complementary investigations have been summarized (Williams and Knee, 1977).

TABLE III

Acetylation of added hexanol by Cox's Orange Pippin apples in air and 2% O_2 + 98% N_2 at 3°C

Storage atmosphere	Hexyl acetate concn (μg g^{-1})	
	Control	Hexanol
Air	1·85	6·93
2% O_2 + 98% N_2	0	6·95

Whole apples were analysed after being supplied with hexanol (24 μg l^{-1}) as a vapour in a gas stream for 10 days at 3°C.

Storage in low O_2 atmospheres depresses production of esters, such as butyl and hexyl acetate, by apples (Patterson *et al.*, 1974). Since whole apples under these conditions are able to esterify added alcohols as rapidly as apples in air (Table III), we conclude that it is the supply of alcohols which is deficient in low O_2. It is likely that the alcohols are mainly products of fatty acid metabolism, an exception being 2-methyl butanol which is thought to be derived from leucine (Myers *et al.*, 1970). Oxidative degradation of linoleic and linolenic acids, initiated by lipoxygenase, can produce six-carbon aldehydes; subsequent saturation and reduction of these, yields hexanol (Eriksson, 1975). Oxygen is also required in the synthesis of these fatty acids at the desaturation steps (Harris and James, 1969) and thus it seems probable that storage in low O_2 atmospheres limits both the supply and further metabolism of the acids from which the alcohols are derived.

EFFECTS OF CONTROLLED ATMOSPHERES ON SENSORY PROPERTIES OF APPLES

We have seen how CA delays senescence changes and affects compositional changes related to aspects of fruit quality. Instrumental and chemical methods indicate that fruit texture (Fig. 2) and acidity (Table VI) are improved when O_2 concentrations are lowered but aroma volatiles (Table IV) are depressed. This is confirmed in general by sensory data.

TABLE IV

Concentrations of aroma compounds in Cox's Orange Pippin apples stored in 2% O_2+98% N_2 and 1% O_2+99% N_2 at 3°C for 80 days followed by air at 15°C for 14 days

Compound	Concn ($\mu g\ g^{-1}$) in:	
	1% O_2	2% O_2
Butan-1-ol	3·61	10·0
Hexan-1-ol	0·57	1·14
Butyl acetate	3·09	7·01
Hexyl acetate	1·98	4·32
2-Methyl butyl acetate	0·14	0·18

(Data derived from Knee and Hatfield, 1976)

Our approach to apple testing has been based on discrimination rather than detailed description; we have employed a modified "triangle test" in which the tasters are asked to state whether one of the samples differs from the other two presented and, if so, in what respects. Descriptions of differences are based on an agreed vocabulary (Holland, 1977). Occasionally the taster may be given three samples from the same treatment and the statements that "all three samples are the same or all are different" are also accepted (see footnote to Table V).

Concern about the flavour of stored Cox gave rise to the suggestion that the change from storage in 5% CO_2 and 3% O_2 to 2% O_2 with minimal CO_2 had led to a deterioration in flavour. The taste panel could usually discriminate between fruit stored in 5% CO_2 and in less than 1% CO_2 with 3% O_2. Moreover, as the storage period was extended, so they became able to distinguish between apples stored in 2% O_2 rather than 3% O_2, but they were unable to discriminate consistently between fruit stored in 2% O_2 (<1% CO_2) and that from 5% CO_2 with 3% O_2 (Table V).

TABLE V

Discrimination by taste panel between Cox's Orange Pippin apples stored in low oxygen atmospheres with and without 5% CO_2 (4°C)

Comparisons	Frequency with which "correct" sample was identified in "triangle" test (%)			
	3 Jan.	14 Feb.	10 April	9 May
5% CO_2 + 3% O_2 with 3% O_2 (<1% CO_2)	29	42[b]	47[c]	37[b]
5% CO_2 + 3% O_2 with 2% O_2 (<1% CO_2)	33[a]	29	27	37[b]
3% O_2 (<1% CO_2) with 2% O_2 (<1% CO_2)	16	42[b]	17	37[b]

[a], [b], [c], discrimination significant at $P = 0{\cdot}05$, $0{\cdot}01$ and $0{\cdot}001$ respectively. On 40% of occasions these samples were said to be indistinguishable (significant at $P = 0{\cdot}05$). N.B. All CA-stored fruit distinguishable from apples stored in air at 4°C until 3 Jan. ($P < 0{\cdot}001$).

TABLE VI

Effects of oxygen concentration on sensory attributes of Cox's Orange Pippin apples stored at 4°C (CO_2 maintained below 0·7%) until 21 March 1977 and at ambient for a further 9 days

		Sensory ratings	
		1·25% O_2	2·0% O_2
Texture (1-5)	Crispness	3·16[b]	2·26
	Juiciness	3·16[a]	1·58
	Flouriness	0·53	1·63[c]
Taste (1-5)	Acidity	2·37	1·95
	Sweetness	2·84	2·95
Aroma (1-3)		1·26	1·58

[a], [b], [c] significantly greater at $P < 0{\cdot}05$, $0{\cdot}01$, $0{\cdot}001$ respectively.

In recent years further extension of the storage life of Cox has been obtained by reducing the O_2 levels to 1-1·5% while maintaining CO_2 at about 0·7%. Under these conditions most disorders are inhibited until May or June; the apples are firmer (Fig. 2) but with lower ester contents (Table IV) than fruit stored in 2% O_2. The taste panel discriminated between fruit stored at the different oxygen levels successfully ($P < 0{\cdot}05$ or better) in

43-50% of the comparisons made on eight different occasions from January to May; the fruit from the lower O_2 concentrations was described in a profile test as being crisper, juicier and less mealy but with a tendency to be less acid and lower in aroma (Table VI). These differences were also recorded in a consumer survey involving 200 housewives from different parts of the country. Apples from 2% O_2 were scored higher for aroma and appearance but the texture of the fruit from 1·25% O_2 was preferred. Fruit stored in 1·25% O_2 was slightly preferable for all age groups, more definitely so for respondents under 35, and those living in the South.

CONCLUSIONS

We have seen how CA-storage prevents deterioration due to rotting and physiological disorders, inhibits compositional changes which occur during normal ripening, and influences the eating quality of the fruit. The inhibition of chlorophyll loss and synthesis of aroma compounds by low O_2 conditions may be due to the prevention of O_2-free radical attack on membrane lipid. Whether there is a common basis for O_2 effects on softening and other ripening changes and the mode of action of CO_2 remain to be determined. The developing hypothesis suggests that CA-storage necessarily reduces aroma production as it retards softening and colour changes.

The results of several recent UK consumer surveys suggest that the overall acceptability of apples is principally related to their texture. However, a significant proportion of the population rate other attributes of equal importance, and Williams (1979) has proposed that certain minimum thresholds for acidity and aroma apply when texture requirements have been met. These thresholds will, of course, vary for different individuals and it is therefore probable that certain markets will prefer apples with more aroma and acidity than others. Moreover, most consumer surveys are limited since they involve the comparison of a new product with the standard and do not necessarily establish whether the product is such that even more people will buy it. Purchasing habits are the ultimate test of consumer acceptance and, on this basis, it is noteworthy that the total volume of apple sales has not increased with improving living standards. If CA-storage is used simply to extend storage life and standardize a product of only average quality, apple consumption is unlikely to increase but, by careful selection of cultivars and the adjustment of CA conditions, it should be possible to meet different marketing requirements. The work reported here does, however, indicate that where aroma is of prime consideration some reduction in the storage period will probably be necessary if texture is to remain acceptable.

ACKNOWLEDGEMENTS

We thank our colleagues, Mrs J.T. Cockburn, S.G.S. Hatfield, D.A. Holland and S.M. Smith for providing information and advice which contributed to this paper.

REFERENCES

Anon. (1973). Report on "Refrigerated Storage", pp. 21. The Apple and Pear Development Council, Tunbridge Wells, Kent.

Brennan, J.G., Jowitt, R. and Mohamed, A.M.A. (1977). *Ann. appl. Biol.* **87**, 121-127.

Eriksson, C. (1975). *J. Agric. Fd Chem.* **23**, 126-128.

Fidler, J.C. (1973). *In* "Biology of Apple and Pear Storage" (J.C. Fidler, B.G. Wilkinson, K.L. Edney and R.O. Sharples, eds), pp. 1-61. Commonwealth Agricultural Bureaux, Farnham Royal, Surrey.

Fidler, J.C. and North, C.J. (1967). *J. hort. Sci.* **42**, 207-221.

Fridovich, I. (1977). *In* "Biochemical and Biomedical Aspects of Active Oxygen" (O. Mayaishi and K. Asada, eds), pp. 171-181. University Park Press, Baltimore, London and Tokyo.

Gormley, T.R. and Egan, J.P. (1977). *Irish J. Fd Sci. Technol.* **1**, 23-32.

Griffiths, D.G., Potter, N.A. and Hulme, A.C. (1950). *J. hort. Sci.* **25**, 266-288.

Harris, P. and James, A.T. (1969). *Biochim. biophys. Acta* **187**, 13-18.

Heath, R.L. and Packer, L. (1965). *Biochem. Biophys. Res. Commun.* **19**, 716-720.

Holland, D.A. (1977). *J. Sci. Fd Agric.* **28**, 436-442.

Kidd, F. (1923). *Rep. Fd Invest. Bd for 1922*, 28-30.

Kidd, F. and West, C. (1936). *J. hort. Sci.* **14**, 276-294.

Knee, M. (1976). *J. Sci. Fd Agric.* **27**, 383-392.

Knee, M. (1978). *Phytochemistry* **17**, 1257-1260.

Knee, M. and Hatfield, S.G.S. (1976). *J. Fd Technol.* **11**, 485-493.

Knee, M., Hatfield, S.G.S. and Ratnayake, M. (1980). *J. exp. Bot.* **30**, 1013-1020.

Kon, S. (1978). *Fd Technol. Champaign* **32**, 85-94.

Myers, M.J., Issenberg, P. and Wick, E.L. (1970). *Phytochemistry* **9**, 1693-1700.

Patterson, B.D., Hatfield, S.G.S. and Knee, M. (1974). *J. Sci. Fd Agric.* **25**, 843-849.

Sharples, R.O. (1973). *In* "Biology of Apple and Pear Storage" (J.C. Fidler, B.G. Wilkinson, K.L. Edney and R.O. Sharples, eds), pp. 173-225. Commonwealth Agricultural Bureaux, Farnham Royal, Surrey.

Sharples, R.O. (1980). *In* "Mineral Nutrition of Fruit Trees" ISHS Conference, 1979 (D. Atkinson, J.E. Jackson, R.O. Sharples and W.M. Waller, eds), pp. 17-28. Butterworths, London.

Van Hasselt, P.R. (1974). *Acta bot. neerl.* **23**, 159-169.

Williams, A.A. (1979). *In* "Progress in Flavour Research" (D.G. Land and H.E. Nursten, eds), pp. 287-305. Applied Science, London.

Williams, A.A. and Knee, M. (1977). *Ann. appl. Biol.* **87**, 127-131.

Section V
Changes during Processing

V.1 Changes of Flavour Compounds during the Processing of Fruit Juices

P. SCHREIER

Institut für Lebensmitteltechnologie und Analytische Chemie der Technischen Universität München, D 8050 Freising - Weihenstephan, West Germany

INTRODUCTION

The aroma of foodstuffs is influenced by husbandry and region. It can be further influenced by subsequent processing, packaging and preservation. Today more and more foodstuffs are sold after processing as "convenience foods" (Robbins, 1976) and the proportion of total food sold in this form is likely to increase still further over the next decade. Sensory quality may be adversely affected for various reasons during processing, especially where mercantile qualities of uniformity, long-keeping quality and durability are sought. Knowledge of changes of food constituents, including the volatile compounds, is therefore of basic importance to (substrate related) food technology.

This paper will briefly discuss the analytical techniques used by flavour chemists, and then consider some examples of food aromas of plant origin and the changes they undergo during food processing, as a result of non-enzymic reactions, e.g. heat treatment, or of enzymic reactions, e.g. after cellular disruption.

METHODS USED IN FLAVOUR RESEARCH

The more important techniques used today for the analysis of the complex aroma mixture present in foodstuffs (Teranishi *et al.*, 1971) are shown in Fig. 1.

The procedures begin with the separation of the aroma components from the foodstuff. Depending upon the nature of the material, one possibility is homogenization followed by distillation, extraction or combined distillation-extraction to remove and concentrate the aroma. A further procedure

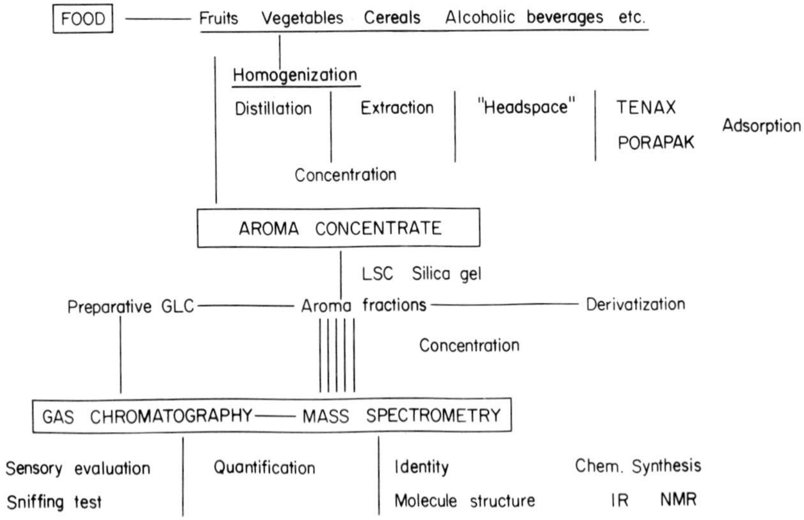

Fig. 1. Principle methods of flavour research.

consists of separation and enrichment of the aroma substances present in the air space above a substrate, either by direct "headspace" techniques, or after the enrichment of aroma components from the gas phase by adsorption on a polymer carrier like Tenax GC or Porapak Q. Concentration of these extracts allows a further enrichment of the separated aroma components. After the concentration step it may be appropriate to further separate the aromas, e.g. by adsorption chromatographic fractionation of the concentrate on silica gel; this procedure separates the aroma compounds into fractions of the same polarity. Instead of silica gel, column chromatography or, in combination with this method, preparative gas chromatography can also be used for the aroma pre-separation. Following such procedures, individual components are then usually finally separated by gas chromatographic methods. The derivatization of less volatile substances permits the use of gas chromatography for their separation.

Structural identification of the separated constituents is mostly achieved by means of mass spectrometry, in particular by coupled gas chromatography-mass spectrometry. Mass spectra give useful information about molecular structures of unknown substances. Where sufficient material is available, infrared or n.m.r. spectroscopy may also provide additional useful information. The chemical synthesis of the postulated compound and the comparison of its spectroscopic and chromatographic data with the unknown substance lead to a positive identification. Subsequently, the aroma constituents are determined quantitatively.

These physico-chemical approaches give data about the identity and the amount of a certain aroma constituent in a foodstuff, but they cannot give any indication of the sensory contribution of that component. To examine this aspect, a so-called "sniffing test" (Dravnieks, 1975) is carried out in which the eluate emerging from the gas chromatographic separation column is split into two parts and each analysed: one part with the GC detector (FID or selective detector) and the other by several people who judge the odour quality and intensity by sensory tests. By detecting components which the instrument cannot, this latter approach acknowledges the extraordinary sensitivity of the human nose in sensory work.

The qualitative identification of aroma substances in foods is difficult enough, but their quantitative determinations are even more so. This is evident from the number of papers published in the field of aroma research in recent years. Although 3000 different aroma substances have been detected in various foodstuffs (Van Straten *et al.*, 1977), only a small proportion of these publications deal with quantitative estimations. There are several reasons for this, one being that most food aroma components occur in traces, usually in the ppb range, and hence it requires sophisticated equipment for their separation and quantification.

In recent years, we have concentrated on the quantitative analysis of aroma compounds in fruits and processed fruit products (Schreier *et al.*, 1976, 1977a,b,c, 1978a,b). The following methods were developed specifically for this work (Schreier and Drawert, 1976).

After the addition of internal standards the material is homogenized with methanol. Using a press, a water-methanol filtrate is obtained, from which the aroma compounds are separated by liquid-liquid extraction using pentane/dichloromethane (2:1). Careful concentration is followed by separation of the aroma substances into classes by adsorption chromatography on silica gel using a pentane/diethyl ether gradient. According to the situation up to four or more aroma fractions may be eluted. These are concentrated and finally analysed using gas chromatography-mass spectrometry. For each silica gel fraction, one internal standard compound is added. After adding external standards, the individual aroma components are analysed by gas chromatography. The yields of the internal standards can be calculated from the external standards and these can then be used to calculate the amounts of substances in the original foodstuff. Calibration hence takes into account all the analytical parameters. Errors range from ± 8 to $\pm 15\%$, permitting the accurate determination of aroma substances isolated.

The use of these techniques to determine processing changes in oranges and tomatoes will now be discussed.

CHANGES BROUGHT ABOUT IN ORANGES BY PROCESSING

Table I lists the concentrations of some aroma constituents in oranges and orange juices, the latter being separated into fresh juice and pasteurized juice. Limonene is the main component in all the samples, although quantitatively only about 10% of the hydrocarbons found in the fruit are detectable in juices. Between 29 and 84% of the esters present in fruit were found in juices. Some components not present, or only present in trace concentrations in the genuine fruit aroma, increase markedly, especially during

TABLE I

Concentrations (μg l^{-1}) of some aroma compounds in oranges (I) and in fresh (II) and pasteurized (III) orange juices

Compound	Concentrations in: I	II	III
α-Pinene	178	22	20
Myrcene	690	68	88
Limonene	70000	4100	4500
β-Caryophyllene	64	10	5
Valencene	5000	200	120
Hexanal	20	260	230
Decanal	77	14	15
Citral	58	45	37
Ethyl butanoate	800	290	300
Ethyl hexanoate	52	15	15
Citronellyl acetate	12	10	10
Neryl acetate	20	9	12
Ethyl 3-hydroxybutanoate	650	110	93
Methyl 3-hydroxyhexanoate	100	34	30
Ethyl 3-hydroxyhexanoate	700	460	380
Linalool	215	150	130
4-Terpinenol	**45**	**42**	**120**
α-Terpineol	**55**	**93**	**270**
Carveol	—	**10**	**170**
Geraniol	11	9	10
Citronellol	20	15	12
Nerol	20	9	7
trans-p-Menthen-9-ol	22	15	12
***trans*-2,8-Menthadien-1-ol**	—	**44**	**166**
cis-Linalooloxide	10	+	+
trans-Linalooloxide	50	15	14
3-Hydroxy-2-butanone	—	—	**59**
3-Methyl-2-buten-1-ol	—	—	**455**
1-Hexanol	30	26	30
cis-3-Hexen-1-ol	54	50	45
1-Octanol	17	10	12

— = not detectable; + = <5 μg l^{-1}.

pasteurization, to give "secondary" aroma substances. Some of these, e.g. carveol, *trans*-2,8-menthadien-1-ol, 4-terpinenol and α-terpineol, are all known to be oxidation products of limonene under the acidic conditions prevailing in the orange juice (see compounds set in bold in Table I). The newly formed acetoin is a known product of the Maillard reaction, whereas methylbutenol probably originates from terpenoid compounds. These investigations revealed no increase in carvone, a substance which has been found in high amounts in some commercial orange juices (Koch, 1973).

A very different analytical picture was obtained following thermal concentration of orange juice by a thin film evaporator ("Centritherm" type Alpha-Laval Ltd.) (Heindze, 1978). This system involves evaporation from a disc-like centrifuge, the disc possessing a number of cavities to assist in evaporation. Figure 2 summarizes the percentage increase of some selected aroma constituents in the concentrate (Fig. 2a) and distillate (Fig. 2b) after

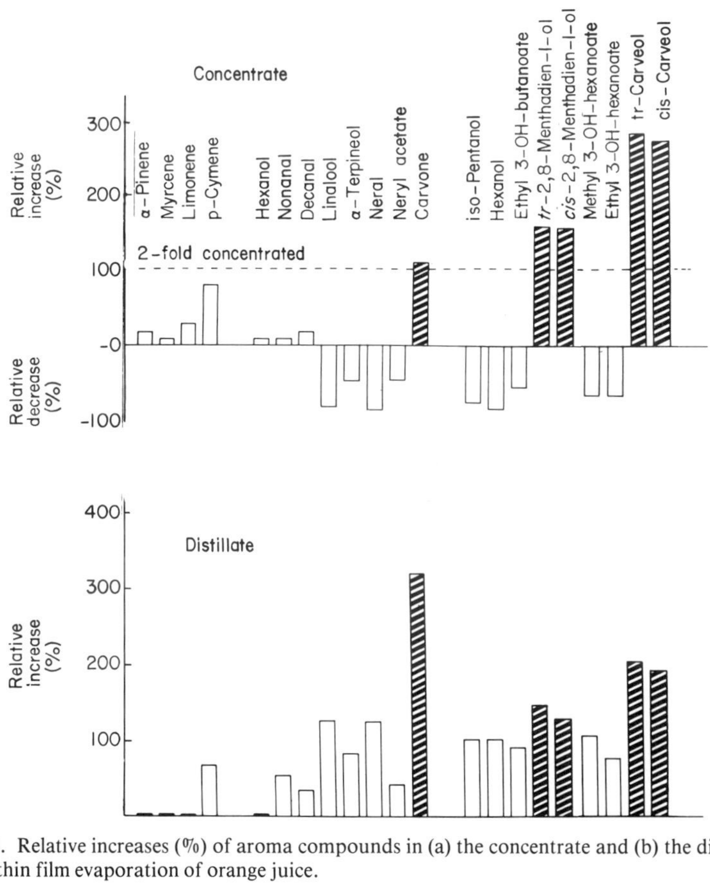

Fig. 2. Relative increases (%) of aroma compounds in (a) the concentrate and (b) the distillate after thin film evaporation of orange juice.

two-fold concentration. The values are expressed as a percentage of the aroma constituent in the original juice. Note that different zero points are chosen for the concentration of compounds in the concentrate and distillate. For aroma substances in the concentrate, concentrations are expressed as % increase (or decrease) with respect to their concentration in the original juice; in the case of the distillate, as a % of their amount in the juice. If the same amount of aroma substance is found in the distillate and juice, this corresponds to 100%. This diagram shows how the genuine aroma substances are distributed between concentrate and distillate. In general, an enrichment of aroma components was noted in the distillate but only in a few cases were the aroma substances of orange juice enriched in the concentrate. These latter components include hydrocarbons and some hydrophilic compounds, such as the aldehydes hexanal, nonanal and decanal. Carvone, *cis*- and *trans*-2,8-methadien-1-ol, and, in particular, *cis*- and *trans*-carveol are of special interest as their concentrations increase significantly in both the concentrate and distillate. All these are produced during thermal concentration. In contrast to the comparatively large amounts of terpineol formed during juice production, only relatively small quantities of this substance are formed during thermal concentration.

Because of the marked change in volatiles during heat treatment of orange juices, the gentler technique of freeze concentration (Thijssen, 1974) was investigated to determine whether fewer changes were induced and the genuine aroma substances better preserved. The plant built for this purpose is illustrated in Fig. 3 (Heindze, 1978). It consisted of an inclined crystal-

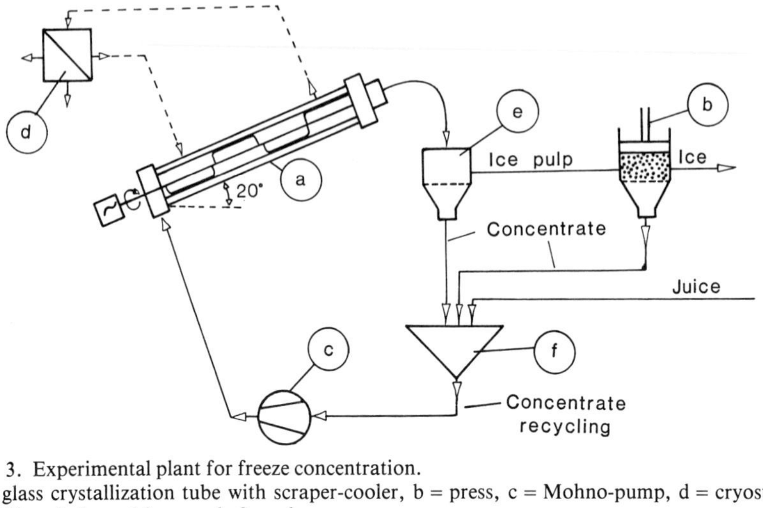

Fig. 3. Experimental plant for freeze concentration.
a = glass crystallization tube with scraper-cooler, b = press, c = Mohno-pump, d = cryostat, e = filter cloth, and f = supply funnel.

lization tube containing a scraper-cooler (a) and surrounded on the outside by a cooling jacket. The equipment also contained a press (b) for the separation of ice, and a pump (c) to convey the liquid phase through the crystallization tube hence enabling continuous operation. Methanol-water mixture supplied from a cryostat (d) allows indirect cooling. The ice pulp is collected in a filter cloth (e) prior to squeezing and thus a part of the concentrate can be drained into the supply funnel (f) from which a pump feeds the sample back to the crystallization tube. In this experiment the ice was not washed. Samples for aroma analysis were continuously removed from both the

TABLE II

Relative increases (%) in the amounts of aroma compounds during freeze concentration of orange juice (C = concentrate, I = ice)

Aroma compound	Extract enrichment 2-fold		Extract enrichment 3-fold	
	C	I	C	I
α-Pinene	74	4	142	5
Myrcene	42	2	88	2
Limonene	58	3	118	6
γ-Terpinene	38	2	79	5
***p*-Cymene**	**170**	**50**	**335**	**52**
Valencene	54	3	105	5
Hexanal	69	58	139	61
Octanal	39	4	105	4
Nonanal	99	11	196	13
Decanal	98	5	193	7
Linalool	84	14	167	20
4-Terpinenol	91	15	170	22
Neral	25	7	55	10
Geranial	42	9	83	11
Carvone	**393**	**60**	**770**	**100**
Perillaldehyde	62	11	126	15
Nerolidol	**127**	**37**	**247**	**39**
Isopentanol	98	23	188	27
Acetoin	54	13	108	22
Hexanol	85	20	170	29
Ethyl 3-hydroxybutyrate	96	7	191	15
Octanol	105	11	207	18
***trans*-2,8-Menthadien-1-ol**	**156**	**16**	**215**	**26**
Methyl 3-hydroxyhexanoate	115	8	227	18
Ethyl 3-hydroxyhexanoate	114	13	224	18
α-Terpineol	**297**	**22**	**610**	**41**
Citronellol	117	12	221	18
Nerol	101	12	208	20
***trans*-Carveol**	**180**	**18**	**300**	**28**
Isopiperitenone	112	12	207	13

concentrate and the separated ice, the temperature simultaneously being measured using thermocouples.

The enrichment of aroma compounds during a 2 - 3-fold freeze concentration was examined by gas chromatography and coupled gas chromatography-mass spectrometry. The volatile substances in the separated ice were also analysed to assess the losses of aroma components in the ice.

Table II shows the relative increases (%) of a selection of aroma compounds. The amounts of these substances in the concentrate and ice are related to the quantities occurring in the corresponding orange juice. In this way, the distribution of the aroma compounds between the concentrate and separated ice can be determined.

In spite of the low temperatures involved, terpene hydrocarbons were degraded during freeze concentration (Table II). This is probably due to chemical oxidation caused by the frequent intermixing of air with the acidic orange juice medium.

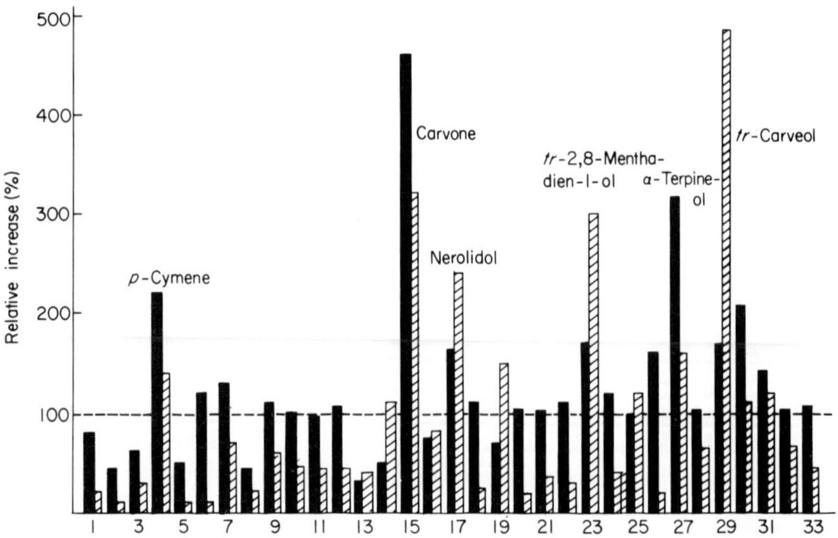

Fig. 4. Comparison of thermal and freeze concentration of orange juice: relative increases (%) of the amounts of aroma compounds in course of freeze concentration (concentrate + ice) (black) and thermal concentration (concentrate + distillate) (hatched) at 2-fold extract enrichment.

1 = α-pinene, 2 = myrcene, 3 = limonene, 4 = *p*-cymene, 5 = valencene, 6 = hexanal, 7 = heptanal, 8 = octanal, 9 = nonanal, 10 = decanal, 11 = linalool, 12 = 4-terpinenol, 13 = neral, 14 = geranial, 15 = carvone, 16 = perillaldehyde, 17 = nerolidol, 18 = isopentanol, 19 = acetoin, 20 = hexanol, 21 = ethyl 3-hydroxybutanoate, 22 = octanol, 23 = *trans*-2,8-menthadien-1-ol, 24 = methyl 3-hydroxyhexanoate, 25 = nonanol, 26 = ethyl 3-hydroxyhexanoate, 27 = α-terpineol, 28 = citronellol, 29 = *trans*-carveol, 30 = geraniol, 31 = isopiperitenone, 32 = 1, 8-*p*-menthadien-9-ol, 33 = nootkatone.

Among the other components in Table II, some show increases proportional to the degree of concentration. However, *p*-cymene, nerolidol, *trans*-2,8-menthadien-1-ol, α-terpineol, *trans*-carveol, and especially carvone experience higher enrichments.

A comparison of the results obtained using thin film evaporation with those using freeze concentration is given in Fig. 4. This shows the relative % increases of selected aroma compounds. For this comparison total concentrations were considered, i.e. the amounts of aroma substances in concentrate plus ice (freeze concentration) plus distillate (thermal process). It is again obvious that during thermal concentration carvone, carveol and 2,8-menthadien-1-ol are formed whereas only a small amount of α-terpineol was produced. In general, "secondary" aroma substances were formed more slowly during freeze concentration. However, relatively high concentrations of α-terpineol and carvone were formed, and *p*-cymene was

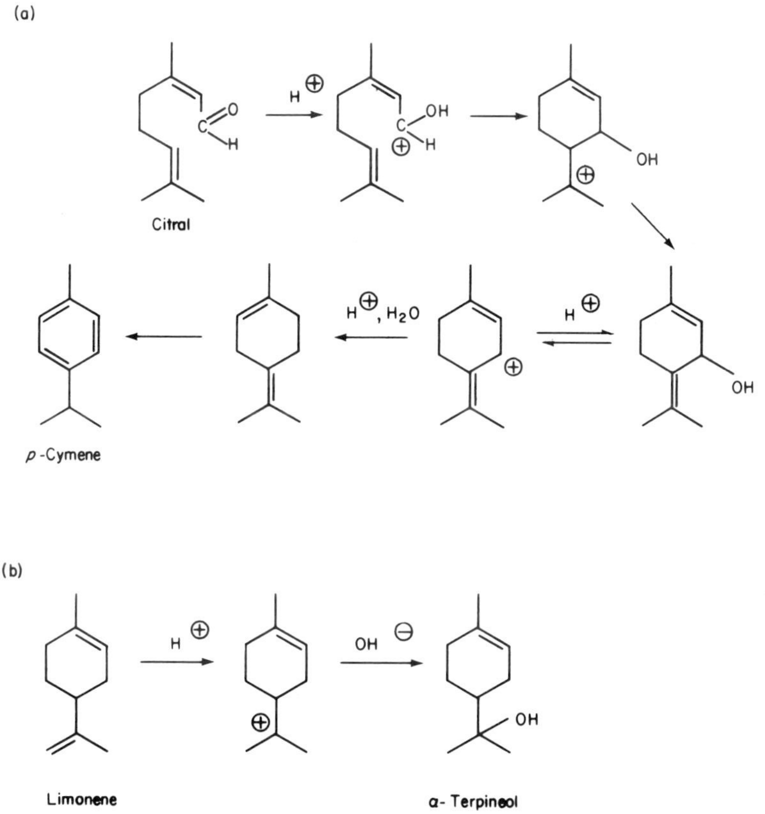

Fig. 5. Acid-catalysed formation mechanism of (a) *p*-cymene from citral (after Loori and Cover, 1964) and (b) α-terpineol from limonene (after Simonsen and Owen, 1947).

produced in comparable amounts to those obtained during thermal concentration.

The formation of different "secondary" aroma compounds results from distinct chemical pathways (Fig. 5). Both *p*-cymene and α-terpineol are formed from an "acid catalysed hydration dehydration (ACDH)" reaction (Simonsen and Owen, 1947). *p*-Cymene could be formed from citral (Loori and Cover, 1964) as shown in Fig. 5, or from γ-terpinene (Ikeda *et al.*, 1961). Considerable losses in the amounts of citral and of γ-terpinene were found during freeze concentration (Table II).

The chemical reactions outlined in Fig. 5 are independent of the oxygen content; however, the formation of the remaining "secondary" aroma substances such as carveol, carvone and 2,8-menthadien-1-ol, are oxidative processes (Fig. 6).

These compounds originate from D-limonene and involve radical structures as possible intermediates. The α-methylene groups behave as active

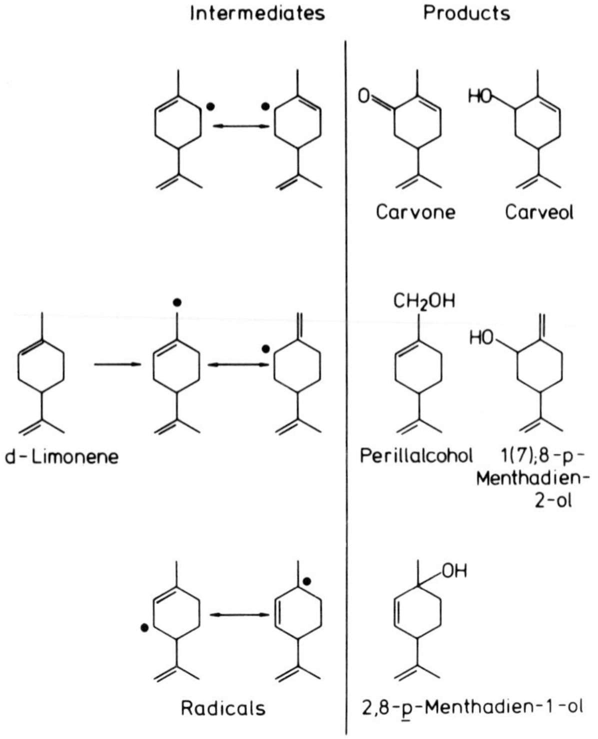

Fig. 6. Limonene oxidation products involving a radical mechanism (after Verghese, 1968).

centres which can easily react without the necessity for enzymic catalysis (Verghese, 1968).

It may be concluded that while the freeze concentration retains considerably more genuine aroma substances than the heat treatment method, it does cause changes and losses of aroma substances, in particular when considering the chemically unstable terpene hydrocarbons.

CHANGES BROUGHT ABOUT IN TOMATOES BY PROCESSING

The above-mentioned changes of aroma components during fruit juice processing can all be related to non-enzymic reactions. Some examples of changes in aroma constituents that are of enzymic origin will now be discussed. In this connection, attention is drawn to the method of fruit homogenization, in particular the use of methanol. Adding methanol serves as an inhibitor of the enzymes present in fruits. It is known that upon the destruction of cells, e.g. during mashing and juice production, very fast enzymic processes may lead to the composition of the aroma components changing. One such pathway is degradation of unsaturated compounds (Drawert *et al.*, 1973) to give rise to substances exhibiting strong "green-grassy" flavour characteristics. The formation of C_6-aroma constituents during tomato juice production will serve to illustrate this process (Fig. 7).

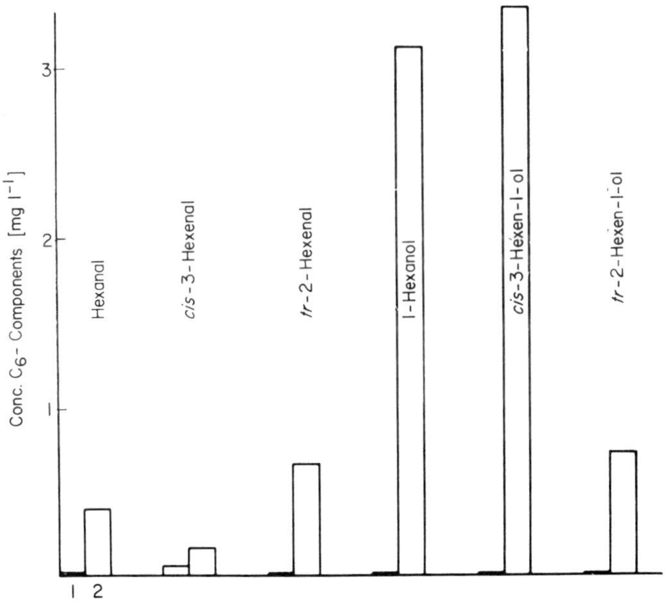

Fig. 7. C_6-aroma compounds in tomatoes (columns 1) and tomato juice (columns 2).

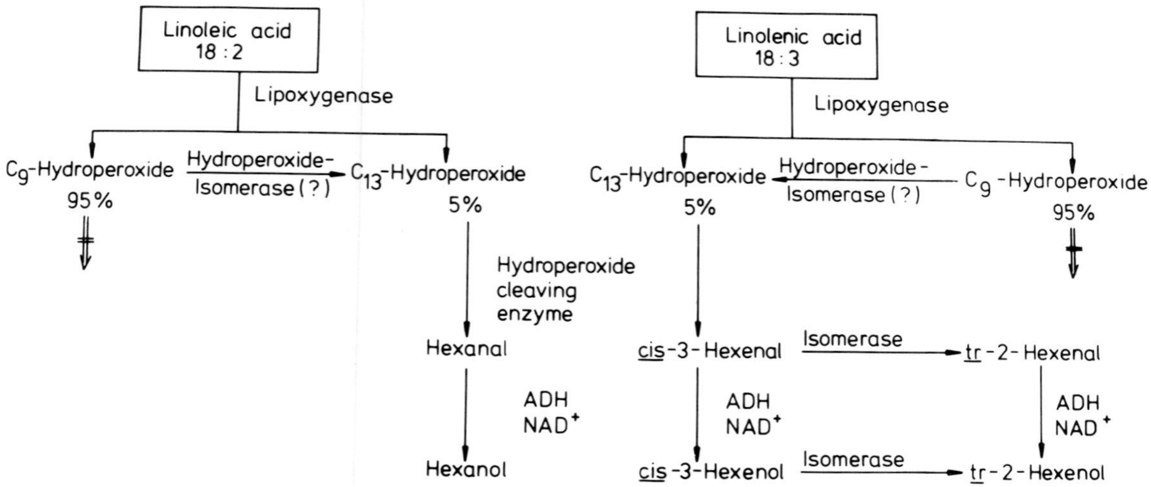

Fig. 8. Hypothetical scheme for the formation of C_6-aroma compounds in tomato homogenates by a lipoxygenase/hydroperoxide cleaving enzyme-system (Galliard and Matthew, 1977; Bhiwapurkar, 1978).

Intact tomatoes contain no C_6-components except *cis*-3-hexenal, whereas in juices there are several, notably *cis*-3-hexen-1-ol and 1-hexanol (Fig. 7). A mechanism for the enzymic formation of C_6-aldehydes in tomato fruits upon the destruction of cell structure involving the enzyme lipoxygenase and a hydroperoxide cleaving enzyme system (Fig. 8), has been recently postulated (Galliard and Matthew, 1977; Bhiwapurkar, 1978).

Free fatty acids possessing a *cis-cis*-1,4-pentadien system like linoleic or linolenic acid are transformed to hydroperoxides by the action of lipoxygenase, the tomato lipoxygenase favouring the formation of C_9-hydroperoxides. The ratio of C_{13}:C_9-hydroperoxides is about 5:95 (Galliard and Matthew, 1977; Bhiwapurkar, 1978). The C_9-hydroperoxides cannot be further degraded to volatile aldehydes, but the C_{13}-hydroperoxides are specifically cleaved to C_6-aldehydes by the action of another enzyme system active in tomatoes. By this mechanism, hexanal originates from linoleic acid and *cis*-3-hexenal from linolenic acid. An isomerase can now transform *cis*-3-hexenal to *trans*-2-hexenal. The reduction of the aldehydes to the corresponding alcohols shown in Fig. 8 is catalysed by an alcohol dehydrogenase with NAD^+ as cofactor, a reaction which has been investigated in detail by several research teams (Stone *et al.*, 1975; Hatanaka *et al.*, 1976).

The markedly poor formation of C_{13}-hydroperoxides in model experiments (Galliard and Matthew, 1977; Bhiwapurkar, 1978) makes it difficult to explain the production of rather large amounts of C_6-compounds during the juice production (Fig. 7). We therefore assume that an enzymic isomerase re-arrangement of C_9-hydroperoxides to C_{13}-hydroperoxides should take place, although no experimental evidence exists for this postulation.

The development of C_6-components as a result of destruction of cells can also be shown in several other fruits, the results of our recent investigations being summarized in Fig. 9. This gives average data for a series of fruits. The extent of formation of C_6-components depends upon the species. Thus, in comparison with homogenized redcurrants and oranges, apples, grapes and tomatoes form relatively high amounts of C_6-compounds. Concentrations of C_6-components are further influenced by cultivar and fruit ripeness (Drawert *et al.*, 1973).

The development of other "secondary" products in tomatoes during juice production is shown in Fig. 10. Components, such as the carotenoid degradation products methylheptenone, geranyl acetone, farnesyl acetone and β-ionone, are absent or occur only in traces in intact tomatoes (columns M = methanol homogenate), but are greater in the juices (J, HJ, PJ). The juice J was produced using pectolytic enzymes (2 h, 50°C) to avoid pulp separation (MacMillan and Sheiman, 1974); HJ indicates an HTST treatment (1 min, 100°C) of the juice, which is known to drastically reduce the

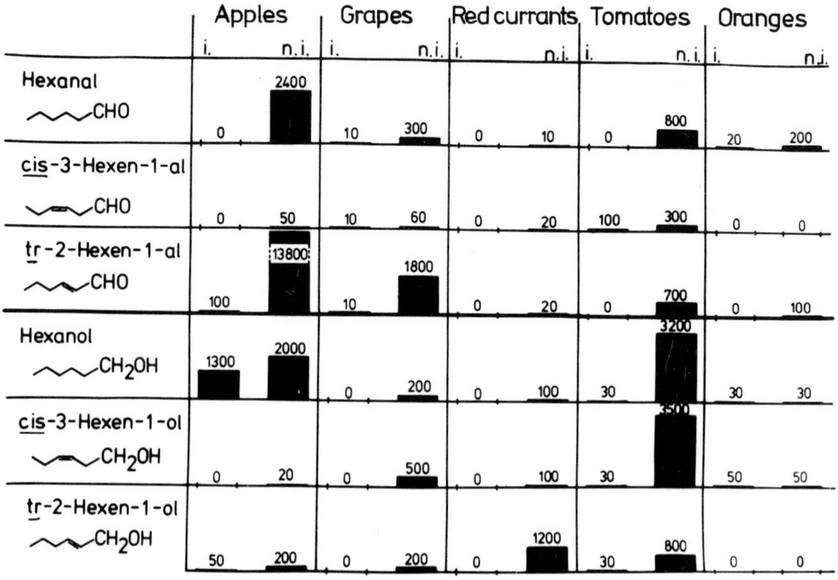

Fig. 9. Amounts (μg l^{-1}) of C_6-aroma components in different fruits after breakdown of cell structure.
i. = Concentrations in enzyme-inhibited experiment, n.i. = Concentrations in non-inhibited experiment.

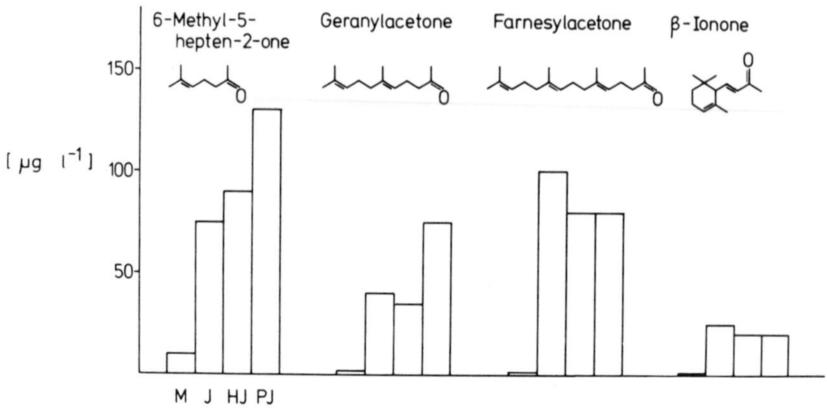

Fig. 10. Formation of carotenoid degradation products during tomato juice production.
M = Methanol homogenate (1 + 1) of tomatoes, J = Juice, HJ = HTST treated juice, PJ = Pasteurized juice.

formation of C_6-aroma substances in fruit pulps (Drawert *et al.*, 1976); PJ is a pasteurized juice.

In this case, HTST treatment does not reduce the amounts of these "secondary" aroma compounds; on the contrary, in the case of methylheptenone small increases of concentrations were detectable after HTST treatment.

These components may again arise as products of an enzymic pathway involving lipoxygenase. It is known that lipoxygenases from different plants show so-called co-oxidation activity. For example, β-carotene is enzymatically degraded in the linoleic acid/lipoxygenase system (Ikediobi and Snyder, 1977; Fischer and Grosch, 1977). However, using enzyme extracts from tomatoes and corresponding heat-treated material, it was found that the formation of the substances shown in Fig. 10 arise by a non-enzymic reaction. This result, in agreement with Grosch (1976), suggests that tomato lipoxygenase shows no co-oxidation activity. Figure 11 shows how the volatile components outlined in Fig. 10 can be derived from the carotenoids lycopene (the quantitatively dominating carotenoid in tomatoes), phytoene and β-carotene.

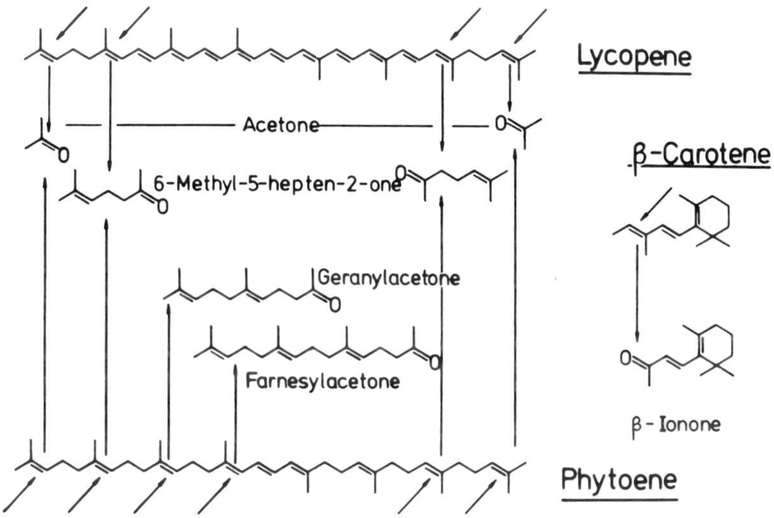

Fig. 11. Carotenoids as precursors for some thermally produced aroma components in tomato juice.

SUMMARY

Examples of two products, oranges and tomatoes, are used to illustrate technologically related changes of aroma substances during the processing of plant materials. All the results quoted are based on physico-chemical techniques which enable the aroma components in foodstuffs to be identified and their concentrations to be measured. No indications are given of the sensory importance of significant compounds. Like other researchers, however, we are currently engrossed in investigating the relationships between quantitative analytical data and the sensory role of the "secondary" flavour compounds formed during fruit juice production.

ACKNOWLEDGEMENTS

I would like to thank the co-workers involved in the reported investigations, Dr I. Heindze, Dr S. Bhiwapurkar and Mr A. Junker.

REFERENCES

Bhiwapurkar, S. (1978). Untersuchungen über primäre und sekundäre pflanzliche Aromastoffe und deren Genese. Ph.D. Thesis, TU München.

Dravnieks, A. (1975). *In* "Methods in Olfactory Research" (D.G. Moulton, A. Turk and J.W. Johnston Jr., eds), pp. 1-61. Academic Press, London, New York, San Francisco.

Drawert, F., Tressl, R., Heimann, W., Emberger, R. and Speck, M. (1973). *Chem. Mikrobiol. Technol. Lebensm.* **2**, 10-22.

Drawert, F., Kuchenbauer, F., Brückner, H. and Schreier, P. (1976). *Chem. Mikrobiol. Technol. Lebensm.* **5**, 27-32.

Fischer, K.H. and Grosch, W. (1977). *Z. Lebensm. Unters.-Forsch.* **165**, 137-139.

Galliard, T. and Matthew, J.A. (1977). Phytochemistry **16**, 339-343.

Grosch, W. (1976). *Mitt. Lebensm. u. gerichtl. Chemie* **30**, 1-5.

Hatanaka, A., Kajiwara, T. and Sekiya, J. (1976). Phytochemistry **15**, 1125-1126.

Heindze, I. (1978). Untersuchungen über das Verhalten der Aromastoffe bei Gefrierkonzentrierung sowie thermischer Behandlung von Fruchtsäften und Gewürzauszügen. Ph. D. Thesis, TU München.

Ikeda, R.M., Stanley, W.L., Vannier, S.H. and Rolle, L.A. (1961). *Fd Technol.* **15**, 379-380.

Ikediobi, Ch.O. and Snyder, H.E. (1977). *J. Agric. Fd Chem.* **25**, 124-127.

Koch, J. (1973). *Flüssiges Obst.* **40**, 42-48.

Loori, J.J. and Cover, A.R. (1964). *J. Fd Sci.* **29**, 576-582.

MacMillan, J.D. and Sheiman, M.I. (1974). "Food Related Enzymes" (J.R. Whitaker, ed.) pp. 101-130. Adv. Chem. Ser. No. 136, Am. Chem. Soc.

Robbins, P.M. (1976). "Convenience Foods. Recent Technology". Noyes Data Corp., Park Ridge, New Jersey.

Schreier, P. and Drawert, F. (1976). *Z. Analyt. Chemie* **279**, 141-142.

Schreier, P., Drawert, F. and Junker, A. (1976). *J. Agric. Fd Chem.* **24**, 331-336.

Schreier, P., Drawert, F. and Junker, A. (1977a). *Lebensm. Wiss. u. Technol.* **10**, 337-340.

Schreier, P., Drawert, F. and Junker, A. (1977b). *Z. Lebensm. Unters.-Forsch.* **165**, 23-27.

Schreier, P., Drawert, F., Junker, A. and Mick, W. (1977c). *Z. Lebensm. Unters.-Forsch.* **164**, 188-193.

Schreier, P., Drawert, F. and Mick, W. (1978a). *Lebensm. Wiss. u. Technol.* **11**, 116-121.

Schreier, P., Drawert, F., Steiger, G. and Mick, W. (1978b). *J. Fd Sci.* **43**, 1797-1800.

Simonsen, J.L. and Owen, L.N. (1947). "The Terpenes", Vol. 1 (2nd edn). Cambridge University Press, Cambridge.

Stone, E.J., Hall, R.M. and Kazeniac, S.J. (1975). *J. Fd Sci.* **40**, 1138-1141.

Teranishi, R., Issenberg, P., Hornstein, I. and Wick, E.L. (1971). "Flavor Research". Dekker, New York.

Thijssen, H.A.C. (1974). *In* "Advances on Preconcentration and Dehydration of Foods" (A. Spicer, ed.) pp. 115-149. Applied Sci. Publ., London.

Van Straten, S., de Vrijer, F. and de Beauveser, J.C. (1977). "Volatile Compounds in Food" (4th edn). CIVO, TNC, Zeist.

Verghese, J. (1968). *Perfum. essent. Oil Rec.* **59**, 876-886.

V.2 Changes during Processing of Frozen Vegetables and Fruit

M. KALBASSI

Technical Department, Findus Ltd., Grimsby, South Humberside, DN32 8HP, England

INTRODUCTION

Frozen vegetables and fruits are now making a substantial contribution to the daily diet of most Western and many other nations throughout the World. It is no longer necessary to wait 12 months for the "pea" season or to consume the same fresh vegetable every day to comply with the forces of nature. Thanks to the advancements in frozen food production, we are able to preserve fruit and vegetables to a very high degree of consumer acceptability and without a significant change to their overall nutritional value.

As with all industrial practices, the final quality always depends on the successful handling and the technological capabilities of the processor. There are inevitably variations that can be discovered on analytical or sensory examination of commercial samples. This is nowhere better illustrated than in the scientific and technical publications available to the Food Technologist. In this paper I shall combine my theoretical as well as practical observations as a Vegetable Technologist in order to re-emphasize the need for caution when making statements about the final quality of frozen foods. Not all the changes in quality or character occur as a direct result of processing; nor, to preserve the accepted indicators of quality such as appearance and taste, is it necessary to sacrifice certain other characteristics either intentionally or by chance.

To discuss such possible alterations to the natural product, I shall refer to the main features of processing techniques, then discuss the expected changes in quality during various stages of production, and finally look briefly at selected common product groups.

PROCESSING STEPS

The stages of process which influence final quality for most products include:

Raw material, transport, storage, handling.

↓

Initial cleaning using air, water,
chlorinated water or manual.

↓

Trimming and size adjustment either mechanical
or manual and in certain cases, peeling and colour sorting.

↓

In most cases, blanching and use of additives.

↓

In some cases, frying followed by cooling (15° to 3°C).

↓

Final freezing to −18°C followed by packing
and cold storage at −25°C.

Not all stages in the process influence final quality nor have all stages been exhaustively studied except in isolated cases. It is well known that with perishable goods, large companies achieve very short post-harvest delays and that generally, the process itself takes less than half an hour.

GENERAL CHANGES AT EACH STAGE OF THE PROCESS

RAW MATERIAL HANDLING

Changes after harvest are numerous and dependant on the type of raw material, temperature of storage and the amount of damage inflicted. Certain vegetables can withstand long delays prior to freezing. Fruits such as strawberries and vegetables such as peas and sweetcorn require the shortest possible delay and immediate cooling to prevent bacterial growth or detrimental activation of enzymes. Loss of sweetness, wilting, discolouration and off-flavours can also occur in the absence of optimal post-harvest handling.

FACTORY CLEANING

As the first stage in the processing of fruits and vegetables, cleaning in fresh/chlorinated water, air or detergents serves to remove unwanted materials, reduce bacterial load and highlight defects. It also cools the product and starts the commercial preservation of quality. Few data are available on nutrient losses or effect of chlorine on such products although our experience has shown this to be insignificant and the quality benefits of factory cleaning far outweigh any losses.

TRIMMING AND SIZE ADJUSTMENT, PEELING AND SORTING

Changes during these stages are directly related to the tastes and traditions of the market for which product is prepared. Whole beans as opposed to cut beans, pepper strips instead of diced peppers, and chopped spinach in place of leaf spinach, are all examples of where quality is not always altered. However, there is evidence that peel removal from roots and fruits results in a loss of skin nutrients (Henshall, 1973) although it does increase the consumer appeal of all such products. During peeling a reduction in surface flora as well as blemishes or discolouration occurs. Steam-peeling requires less subsequent washing and may benefit from lower losses as opposed to lye-peeling.

Size adjustment is achieved by direct grading or by segmentation (potato chips, cauliflower florets and peach halves). Losses may follow if cells are damaged. Also, as more internal surfaces are exposed to air, the need for immediate blanching or freezing becomes paramount and storage under water or ascorbic acid and other treatments, may be necessary. Therefore, depending on the processor's possibilities, the final quality can be vastly different as a result of this treatment which is often an expensive part of a process.

Sorting, whether manual or electronic, removes unsightly pieces as a direct service to the consumer. In many Third World markets, and indeed in the fresh market, far less effort is placed on provision of defect-free items. However, frozen products are bought for their convenience and as such, producers use various methods to remove unwanted pieces and this they normally do prior to blanching to save on energy costs involved.

The degree of peeling, trimming, size adjustment and defect sorting depends entirely on the quality aspirations of the producers and their customers. Reject material is of commercial importance and in the case of blemishes, a significant quality improvement is achieved when up to 95% of initial blemishes are inspected out by most quality producers.

BLANCHING, USE OF ADDITIVES, AND FRYING

Prior to blanching, fruit and vegetables can be considered comparable to fresh and in many instances, better than the fresh produce sold in shops. For prolonged shelf-life and all year availability, it is often necessary to blanch these products before freezing. Aims of blanching include:

(i) Inactivation of enzymes.

(ii) Texture stabilization or adjustment.
(iii) Colour stabilization or adjustment.
(iv) Reduction of bacterial load.
(v) Removal of raw flavours/colours.
(vi) Removal of intercellular gases.
(vii) Addition of desirable additives.

The extent of heat treatment varies from product to product and the amount of blanching necessary for each vegetable is dependant on many variables. Known enzymes such as peroxidases, catalase, polyphenoloxidase, ascorbic acid oxidase, lipase together with many unknown agents are thought to be responsible for changes during storage of fruits and vegetables. Off-flavours, rancidity, discolouration such as browning of cut apples, internal browning and a general loss of nutrients can all result from activity of such enzymes during cold storage and their inactivation is necessary for prolonged shelf-life.

Experience shows that heat destruction of peroxidase (measured by routine tests) is sufficient to indicate a successful inhibition of other enzymes, and in most cases, heat treatment is beneficial in preserving most quality values. Many changes then occur, such as:

(i) Inactivation of enzymes resulting in shelf-life of up to five years.
(ii) Adjustment of texture—softening as in peas or firming as in chips or beans.
(iii) Reduction in bacterial number.
(iv) Reduction in nutrients (heat-labile or water-soluble).
(v) Loss of raw colours to bright processed colours.
(vi) Loss of some soluble solids and a general reduction in flavour.
(vii) Pick up or loss of water.
(viii) Loss of undesirable flavours.

The extent of these changes depends on the product and the blanching techniques used and research into commercial methods is currently underway in many countries. Main blanching systems in use include:

(i) Water blanchers. Cheapest method and easy to control, versatile for use of additives but can result in osmotic movement and loss of solids and nutrients; obviously more damaging to diced root vegetables or low solids fruits.

(ii) Steam blanchers. Expensive and reputedly slower to complete enzyme destruction. In the absence of water, the loss of solids may be less, though longer heat treatment can be detrimental to vitamins and colour.

Also used are individual quick blanchers, microwave systems, combined water/steam blanchers and hot air systems, all of which attempt to reduce solids losses, but often result in poorer colour, flavour and nutrients. In the

UK, most vegetables are water-blanched and the changes in quality depend on domestic temperatures and methods.

Little information is available on individual practices, although a short high-temperature blanch is noted to give the best colour and flavour for most products and a low-temperature long blanch results in firmer texture of chips, beans etc.

The effect of blanching on the texture of fruit and vegetables is related to their cell structure (Dewdney, 1978). Heat treatment destroys cell walls, allowing air and cell sap to escape and this normally results in softening. With potatoes and beans, low-temperature heat-treatment increases firmness, possibly due to a slower rate of increase in insoluble pectic acid content. Calcium ions in the blanching water also affect firmness of fruits and vegetables, especially at lower blanching temperatures.

Degree of inactivation, quality of water, use of additives (ascorbic acid, $CaCl_2$ etc.), type of cooling (whether in water, cold air or both), delays and handling before and after blanching, blanching method and control, pH of steam or water, level of solids in blancher or transportation water, and the state and condition of the product, all have significant effects on the changes that occur during processing of fruit and vegetables. Generally speaking, as in cooking of fresh material in the home, changes and losses do occur during heat treatment and some of the changes are in fact desirable.

The degree of change depends on the vegetable and our technology may be adequate for a particular product (as in peas), but may require advancement to bring final quality closer to the consumer's expectations (as in sprouts and fruit).

Following blanching, only a limited number of vegetables are parfried at high temperature. The aim here is to reduce moisture content, parcook the product and by hardening the case to produce a protective barrier after freezing. Potato products receive various amounts of frying and it has been reported that frying can adversely affect the vitamin C content although frozen chips require less "home" frying which compensates for early losses. There is normally a degree of surface browning due to frying and colour variation which can be attributed to surface sugar content of raw potatoes stored at freezing temperatures.

FREEZING AND COLD STORAGE

Freezing and cold storage at temperatures of $-18°$ to $-30°C$, combine to significantly retard chemical, biochemical and microbial development without a proven detriment to quality (Brown, 1977), except as a result of bad practice. Certain products suffer little textural damage during freezing whereas low solid fruit invariably lose their texture as a result of irreversible cell damage. General changes during freezing/cold storage include:

(i) Reduction in bacterial count.
(ii) Reduction of available water, down to 10% at −20°C.
(iii) Cell damage, higher due to slow freezing.
(iv) Possible change in colour, conversion of chlorophyll to pheophytin during cold storage.
(v) Surface dehydration either mild or severe (freezer burn) if not protected.
(vi) Possible cracking of surface due to fast freezing especially in fruit.
(vii) Long-term storability at temperatures below −20°C and preservation of nutrients.
(viii) Welding and subsequent damage during post-storage handling.
(ix) Loss of fluid during defrosting of fruits such as strawberries (Dewdney, 1978).

In general, provided fast freezing, whether in flo freezers, cryogenic or plate freezers is employed, changes during freezing do not detract from the quality of most fruit and vegetables and the consumer now fully appreciates the convenience of most products on the market. Along with all the physical and chemical changes summarized above, the price per unit also increases and this must influence consumer loyalty. Different fruits and vegetables exhibit their own particular characteristics and cell structure and moisture content play an important part in the freeze stability of such products.

Cultivar, maturity and rate of freezing all influence the final texture. Slow freezing results in the formation of large ice crystals and irreversible damage to the cell structure and texture. Also, punctured cell walls in frozen fruit can lead to destruction of pectic substances and toughening of the product. Freezing damage, however, is not so serious to products containing gelatinized starch, though here again quick freezing is more suitable.

CHANGES DURING PRODUCTION OF SELECTED FROZEN FRUIT AND VEGETABLES

PEAS

Representing the most accepted frozen vegetable, peas belong to a highly perishable and seasonal fruit and vegetable family and require intensive and highly efficient post-harvest treatment. Some producers seem to have achieved the seemingly impossible task of harvesting peas at their prime, ice cooling and subsequently blanching and freezing them in 2-3 hours. As such, frozen peas from such sources are always better than any sold fresh in the summer.

Peas, and to the same extent sweetcorn, broad beans and some fruit require the full blanch/freeze combination which aid their preservation but result in irreversible changes that include:

(i) The usual removal of unwanted materials, reduction in microflora, inactivation of known enzymes and loss of sugars and soluble solids.
(ii) Frozen peas retain some 80% of their vitamin C and 90% of their original thiamine, this being better than most fresh produce (Morrison, 1974).
(iii) Reduction in sweetness due to delays or unsuitable blanching, especially in water.
(iv) Brighter colour, though colour variation has been noted possibly due to localized blancher pH or temperature fluctuations.
(v) Softening of texture and picking up of "blancher" flavours.
(vi) General quality reduction if under high temperatures in cold store or poor handling by retailer or customer.

POTATOES AND ROOT VEGETABLES

Unlike the leguminous vegetables just mentioned, roots require extra treatments such as peeling and cutting as well as frying and blanching. Potatoes are often parfried because despite low temperature blanching and use of calcium chloride few non-fried frozen potato products are edible. Carrots and other roots are not usually parfried and they too suffer extensively during blanching and freezing. Fried potato products are popular on their own merit and not because of their closeness to the fresh chip. Other roots appear popular only when mixed with several vegetables, possibly because of their poorer taste. Observations of changes during processing of such vegetables indicate that:

(i) Peeling loses about 20% yield and generally improves their surface quality. Peeling, followed by trimming, blanching and frying, reduce the preparation time for chips from over 30 to 2-6 minutes.
(ii) The whole process loses up to 40% in yield when steam/lye peeling (more if abrasive peeled) and up to 60% when including frying.
(iii) Blanching softens the texture of non-starchy roots, but can improve that of potatoes (Brown and Morales, 1970); low temperature blanching is particularly helpful.
(iv) Loss of surface starch, sugars, dry matter content during peeling, blanching and cooling reduces the flavour and texture of most root vegetables except potatoes; results in a more uniform surface sugar content leading to improved fried appearance; and changes the vitamin and nutrient content especially on slow blanching of potatoes.
(v) A brighter colour in most root vegetables and a cooked appearance in potatoes.

(vi) Case hardening, reduction of moisture content by some 20% as a result of frying; and prevention of cold storage darkening (Voirol, 1972).

(vii) Development of colour variation in potato products of non-uniform raw quality. Retention of some 5% fat.

(viii) Freezing does appear to soften the texture of most root vegetables. Carrots lose their crispness and slow freezing can cause major damage.

Many reports (e.g. Brown and Morales, 1970) have been published on the merits of different blanching conditions and additives. Water blanching to adjust colour and texture of chips appears the best available method, although steam may be more suitable where sugars need to be retained. There is little risk of vitamin A loss except during handling and exposure to light.

Individually Quick Frozen chips normally have a higher dry matter content than fresh and as such, they do not exhibit the same succulence as the latter.

Although therefore, the overall nutritional quality of root vegetables is good, the eating quality of carrots, swedes and some potato products is inferior to peas and other legumes. Also, little knowledge is available on the seasonal and annual changes in such vegetables.

LEAFY VEGETABLES (BRASSICAS)

Sprouts, cabbage, spinach, cauliflower and leeks exhibit different processing characteristics to all the above mentioned groups. Normally, a high degree of trimming is followed by blanching and freezing. Delays are not so serious as with peas, except for spinach and generally, their raw quality is good.

Sprouts

These suffer yield losses during processing. Blanching in water or steam damages larger sprouts and to date full enzyme inactivation is recommended, this causing softened outer leaves. Softened leaves often result in high freezer damage. Sprouts are rich in vitamin C and minerals, and the long blanching times may reduce their nutritional quality. Water blanching produces the brightest green colour. Surface dehydration during cold storage can be serious in frozen sprouts.

Spinach

Though not highly popular in the UK, spinach is sold extensively as chopped, pellofrozen or leaf in many countries. Its large leaf surface makes it susceptible to oxidation and loss of vitamins. Blanching greatly reduces its volume and English spinach often breaks up during the process. Despite this, it is highly nutritious, retaining most of its nutrients and minerals to a higher degree than fresh due to quick commercial handling (Table I). As a result of good freezing techniques, frozen spinach is an excellent source of nutrients.

TABLE I

Percentage of US recommended daily allowance for an adult male in a 3 oz serving of cooked frozen and fresh spinach

	Frozen	Fresh
Protein	4·28	2·58
Calcium	9·02	6·83
Iron	9·39	16·23
Vitamin A	160·32	99·60
Vitamin C	11·77	2·03
Vitamin B_1 (thiamine)	2·65	·80
Vitamin B_2 (riboflavin)[a]	2·26	4·35
Niacin	1·49	0·08
Calories (number)	23·80	15·23
Fat (%)	0·114	0·204

Data from Martin (1977). [a]Exposure to light may explain the fall in riboflavin content.

Cauliflower/Cabbage/Leek

Rich in minerals and vitamin C, their final nutritional quality is unlikely to be much lower than the normal fresh product cooked in the home. Cabbage requires little blanching, and in the shredded form, it resembles the cooked quality of freshly prepared product. Water or steam blanched and often plate frozen, it is highly underrated.

Cauliflower is cored and trimmed during which it loses some 15% in yield. Blanching water of about pH 4·5-5·0 gives the best final colour. Cauliflower must be fully blanched and loses some flavour in the process. Flo freezing damages its florets and smaller florets produce the best eating quality.

Leeks require much cleaning and trimming, but keep for a long time with little blanching.

FRUITS

Frozen fruits are produced extensively in the USA and some European countries. Not all fruits are suitable for direct freezing, some are peeled, some blanched and some receive ascorbic acid or syrup treatment, prior to quick freezing. Low solid fruits such as strawberries, can suffer extensive cell damage during freezing and special cultivars need to be selected for this process. Liquid nitrogen or CO_2 freezing usually produces a better final texture.

Post-harvest changes in some fruits, coupled with bruising and damage can adversely alter colour and general quality during processing and when thawing. Any delay after peeling can cause discolouration in apples and cause dehydration and cell fluid losses in tomatoes and similar products. Where surfaces have to be cut as in diced apples and quartered tomatoes, the exposed surfaces can dehydrate and discolour in the presence of air and such remedies as ascorbic acid treatment or vacuum-packing are often employed.

Becuase of their fragile cell structure, most fruit are damaged by blanching and freezing. Also, as with vegetables, fluctuations in storage temperatures cause permanent deterioration in their quality.

CONCLUSIONS

The physical and chemical changes that occur during production of frozen fruit and vegetables depend on the species, and the conditions and methods employed by the processors. Avoidance of delay, adequate control of peeling, blanching, frying and freezing together with suitable protection and constantly low temperature storage, all assist in preservation of quality. Some products comply with the image the consumer has of the final quality, and peas in particular must be regarded as the superior frozen vegetable. Nutritionally, the information published varies but generally, the frozen substitute is often as good as and more convenient than the freshly picked cultivar. Where the product quality does not match the real thing, either the consumer has to be induced to accept the offered quality or alternative processes should be sought by the industry. The latter would only be possible by research into first principles and improved knowledge of cell structure, varietal handling characteristics and special process techniques.

REFERENCES

Brown, M.S. (1977). *J. Texture Stud.* **8**, 391-404.
Brown, M.S. and Morales, J.W. (1970). *Am. Potato J.* **47**, 321-325.
Dewdney, P. (1978). *Scient. tech. Surv. Br. Fd Mfg Ind. Res Ass.* No. 107, pp. 36.
Henshall, J.D. (1973). *Proc. Nutr. Soc.* **32**, 17-22.
Martin, S. (1977). *Quick froz. Fds int.* (Nov), 34-53.
Morrison, M.H. (1974). *J. Fd Technol.* **9**, 491-500.
Voirol, F. (1972). *Food Process. Ind.* 27-33.

V.3 Changes during Processing—Preserves

D. E. CROSS

James Robertson & Sons Ltd., Bristol BS4 5AP, England

INTRODUCTION

Preserves, as understood in the UK, consist of jams, citrus marmalades, mincemeat and fruit curds. The last two items are little known outside the UK and this paper will consider only jams and marmalades.

The origin of preserve-making goes back many centuries but the concept is that of taking fruit in season and preserving it so that it can be used over a longer period of the year. The basic requirements in a preserve are good flavour, fruit texture, colour and set.

MANUFACTURE

PROVISIONAL PRESERVATION

In large-scale manufacture, the practical and financial aspects of converting fruit during the harvesting season has led to the general adoption of semi-processing (provisional preservation) at harvest time and conversion later. There are three basic systems of provisional preservation—sulphiting, freezing and canning.

Sulphiting

This is the most economical method and consists of adding sulphur dioxide (SO_2), gas or liquor at a concentration of up to 2000 mg/kg of fruit. With citrus peels there is often an addition of salt. Storage is generally in drums or bulk tanks, the wooden keg having now disappeared from commerce. Whilst the use of SO_2 is criticized in some quarters, it is still the most widely used method for preserving fruit products, particularly juices. One obvious problem in sulphited fruit is the bleaching of anthocyanins, although most of the colour returns when the SO_2 is later removed.

Freezing

This is used to produce free flow, block or plate frozen fruit from whole or puréed material. Whilst most fruits survive the freezing and thawing regime fairly well, strawberry is a well known exception, as the rupturing of cells results in poor texture when the fruit is thawed. Quality changes can, of course, occur if correct storage conditions are not maintained. Whilst freezing may appear to have advantages as a process, there are also disadvantages. It is expensive as a process, and also in storage and transport costs. An additional point is that the restricted harvesting periods put heavy load factors on the freezing plants for short periods.

Canning

This is generally confined to imported tropical fruits, e.g. apricots, peaches and pineapple. Another method allied to canning that is becoming more popular, is the use of aseptically filled drums—the main problem, as in canning, is one of blown or damaged containers. Problems can also occur in emptying the container through the small bunghole if the product is viscous.

BOILING

In simple terms, preserve manufacture consists of a boil of fruit and sugars with or without the addition of pectin and fruit acid. The particular characteristics of the fruit and the manner in which it has been provisionally preserved can affect the nature of the final product. There are, however, a number of common factors.

Flavour

Changes in flavour have several causes. The first change that occurs is a considerable increase in the sugar level, well above the natural levels present in the fruit and varying with the type of fruit. Taking pineapple as an example and looking at the raw fruit, fruit in light syrup and jam, three quite different flavour profiles are obtained. Most fruit flavours are related to the raw state and when cooked, the flavour profile is altered. In the boiling of preserves we are cooking an acidified sugar solution and developing a caramelized background.

There are two basic approaches to jam boiling in current practice—the traditional open pan method and the vacuum boiling system. In the former method, the boiling temperature rises as water is removed to a maximum at approximately 105°C. In the vacuum system, the boiling temperature is held at approximately 60°C under reduced pressure until the desired level of soluble solids is reached. The pressure is then brought back to atmospheric and the temperature allowed to rise to 80-90°C in order to obtain sterilization and a suitable temperature for filling. The differing process temperatures obviously produce different flavour profiles. In both cases, steam distillation of flavours also takes place. It is feasible, of course, to collect these using flavour recovery systems, but it is questionable whether the cost of this equipment is justified in relation to the market value of the product.

The last basic change to flavour is that due to the control of acidity. Jams and marmalades are adjusted to about pH 3·0, the optimum for setting of pectin, but different from the natural pH level of the original fruit.

Colour

Changes may occur during storage of the fruit, but there is a significant change during heat processing due to degradation of red colours to brown. This is usually countered in manufacturing practice by the addition of artificial colouring materials.

Texture

Changes are influenced by the original nature of fruit, the method of preparation, the degree of mechanical handling and the effect of cooking. In the cooking of fruits, one of the changes which takes place is a depolymerization of the intercellular pectic substances (see Hall, M.A. p. 53 this Symposium). This has a double significance in jam manufacture because this effect produces soluble pectins to provide the setting agent for the finished product and, at the same time, this weakening of the intercellular structure also causes some disintegration of the fruit. Also, the physical action of boiling is extremely violent.

Stresses are introduced because one is raising the dissolved solids level from less than 10% to almost 70% in a matter of minutes. In one direction we are trying to diffuse sugar syrup into the fruit, whilst osmotic effects are tending to remove water and cause collapse. In addition the surrounding matrix is being raised to the boiling point of the syrup at which temperature the cell contents are being super-heated, resulting in elevated internal pressure. Intercellular air is also being replaced by syrup. With such violent changes taking place we are perhaps fortunate in retaining at least some recognizable structure.

STRAWBERRIES

With strawberries our basic requirement is to retain whole or recognizable pieces of fruit in the finished product. If the fruit has been frozen beforehand, there will be rupturing of mainly parenchyma cells with resultant loss of structure. Some cultivars suffer more than others. There may also be some enzymic breakdown of pectic substances before freezing is actually achieved.

If the strawberries were sulphited we may again have enzymic breakdown even after sulphiting (see Dennis, C. and Harris, J.E., p. 275, this Symposium). Some hydrolysis of pectic substances may occur in fruit stored in sulphite liquor due to the acid conditions. This hydrolysis can be used to advantage by the addition of calcium ions which will then react with the resulting soluble pectins to form insoluble calcium pectates, and thus provide firmness in the berries. By varying the quantity of calcium added, one can obtain some control over the texture of the sulphited berries. To a lesser extent calcium salts can also be added to fruit which is to be frozen.

One other factor is that in the boiled product the remaining structure is related to the vascular bundles which interconnect the achenes to the pith of the strawberry. Since the quantity is directly related to the number of achenes, then there is a relationship between structural integrity and number of achenes per unit area of the surface (Jewell *et al.*, 1973).

The emphasis in commercial practice is on retaining the integrity of the fruit, especially Cambridge Favourite, the most widely used cultivar in the UK. There are cultivars, however, which produce remarkably tough fruit and if these are used the end product can be composed of firm entire berries suspended in almost clear syrup. This shows if one uses quaternary berries from Cambridge Vigour which have been stored in calcium bisulphite solution.

RASPBERRIES

With raspberries, the situation is totally different. In this case the fruit structure will simply not survive the rigours of jam-making. The main concern with raspberry is the visual appearance and number of seeds. The endocarp of the raspberry seed normally provides a secure case and in raspberry jam the seeds show up clearly against the background matrix. If the endocarp is damaged, this allows the interstitial air to be replaced by syrup. When this occurs the differential visual effect is lost and when the seeds are no longer readily discernible, they are described as ''blind''. There are a number of factors involved, not least being varietal differences. There is a greater tendency to blindness in sulphited fruit than in frozen fruit, but one of the main factors affecting blindness in commerce is that of mechanical damage. It is normal practice to pump and stir raspberry pulp and yet this can have a direct effect on the degree of blindness.

OTHER FRUITS

With other fruits such as blackcurrants and citrus fruit, the pectic substances in the skins cause the products to be unduly tough unless preparatory heat treatment is carried out. The degree of pre-treatment necessary has to be judged in relation to the prevailing situation, and whilst there are a number of methods available for measurement of toughness we find that this is still an area which owes more to art than to science.

In the case of citrus peels, ginger and pineapple, the fibrous structure is such that sugar syrup is taken in without undue breakdown and hence these fruits stand up well to the boiling process. These are, of course, the items which are often prepared as candied fruits.

SUMMARY

The quality of preserves is related not simply to the state of the original fruit, but can be greatly affected by how the fruit is processed at both the preparation and final product stages. New cultivars and new methods mean that the procedures are under constant review, but despite recent advances, preserve manufacture is still a very inexact science.

REFERENCE

Jewell, G.G., Rantsios, A.R. and Scholey, J. (1973). *J. Texture Stud.* **4**, 363-370.

V.4 Apple Pies

M. G. DARGUE

British Bakeries Ltd.,
Westpoint P.O.Box 152, The Grove, Slough, Berks SL1 1QF, England

INTRODUCTION

Fruit pies have long been a popular item in the baker's shop. The many different types of fruit which are available during the year enable the small baker to offer his customers variety. He can also take full advantage of market price and offer real value for money. However, one constraint is that his products are short shelf-life items, just as with home-made products.

Fruit pies sold in larger outlets are, however, a different matter requiring, for example, a degree of shelf-life, thereby allowing time for national distribution. To be economically viable, products sold through such outlets usually have a shelf-life in excess of 14 days. Manufacturers of these products therefore have to carefully design both pastry and fillings to produce quality items having the full designated shelf-life.

THE APPLE PIE

Of fruit pies sold in this way, the apple pie is by far the biggest seller. I want therefore to consider what manufacturers look for in the prime raw material, mainly apple, to achieve a consistent product throughout the year. I shall outline the manufacture of a small 2 oz product, consisting of short pastry base and lid with apple filling contained in a foil patty pan of about 2½ ″ diameter.

FILLING CONSISTENCY

Because the product made in the home or by a small baker has short shelf-life, usually only one or two days, the minimum amount of sugar is added to the apple filling, usually just enough to counteract any excessive sharpness of flavour associated with the apple. If such a filling were used in a pie designed for at least 14 days shelf-life several problems would ensue.

Firstly, even if the apple were diced into cubes of 10 mm size it would be impossible to deposit into the pastry cups at the rate encountered on modern high-speed, multi-deposit machines. For instance a 15-nozzle deposit machine runs at 35-40 strokes a minute or 500 pies a minute giving an hourly output of 30-35 000 pies! At these speeds a degree of fluidity is needed and hence fillings must consist of part pieces, part liquid.

Secondly, a pie which consisted solely of apple pieces would not meet the desired shelf-life parameters. Moisture would soon migrate from the apple to the pastry resulting in a dried-up filling of leathery texture.

The pastry would turn soft and become a potential mould hazard, particularly during warmer weather. To counteract this tendency, a certain soluble solids content is essential and this is usually achieved by adding sugar in one form or another, that is, a quantity of sucrose plus other non-sweet sugars, for example, wheat sugar or low dextrose equivalent glucose products. This ensures the necessary soluble solids content without being excessively sweet tasting. Apple pulp is also used as this imparts a certain flavour and this also helps in stabilizing the filling.

APPLE-TYPE AND AVAILABILITY

Having decided upon the consistency of the filling the question of apple type is important. With a branded apple pie for sale nationally, a year-round supply is desirable so availability is important. Although certain dried apple products are on the market, usually of Italian origin, they are generally of poor flavour, and when reconstituted and used in a filling of high soluble solids content the apple pieces tend to become tough and rubbery during shelf-life.

Bramley's Seedling apple is usually the primary source for apple pies of this type, the reasons being its flavour, texture and availability.

The flavour of the apple used in a pie with high soluble solids content is important. Bramley apples provide, it seems, the best possible apple flavour. Because of their fairly vigorous mixing during preparation of the fillings the apple must also be of sufficiently firm texture to withstand possible damage and remain as discrete apple pieces rather than disintegrate into pulp. With a pie plant mixing at high speeds, the method of filling preparation must also be quick. On such a mixer, the blades revolve at several thousand r.p.m. and mixing of a 500 lb batch of filling can be accomplished in less than half a minute. A pie plant, running at 40 000 pies an hour uses in excess of one ton of filling per hour.

Bramley apples are available out of the growing season from gas stores and also in the canned form. Sealed in air-tight storage chambers at controlled temperatures, the Bramley produces CO_2 which once at the desired level is controlled by venting. Apples thus stored can then be peeled, cored, diced and treated with either calcium or sodium metabisulphite solution at any time. The product can then be packed in polythene bags and delivered in plastic crates to factories. In this way, apple of one or two days age from peeling is available and this has certain flavour and textural advantages over canned apple.

Bramleys are generally gas-stored only up to the middle or end of June when they tend to deteriorate. To cover the gaps before the new season's crop, in about September, some apples are removed from gas store in June and bulk processed. This is then stored at 35°-38°F (1·7-3·3°C) until required and referred to as "laid down" apple.

Canned Bramley apple is used as an alternative to the "fresh" product at certain times. The cans contain apple pieces (3-5 cm^3). These must be further diced at the factory to the desired cube size (10 mm^3). Pre-diced canned apple has given problems in the past with blown cans, because the cubes tend to pack down in the cans and heat treatment is less effective than with larger pieces.

Canning involves a certain heat treatment which softens the apple so that when received at the factory, further heat treatment prior to baking is unnecessary. With the "fresh" apple, however, a heat treatment prior to baking is necessary otherwise the apple cubes become tough during shelf-life. The best treatment involves blanching as this causes less damage to the produce than other bulk heating methods.

QUALITY CHARACTERISTICS

The final baked product must contain discrete pieces of apple, and to achieve this, fairly rigorous specifications regarding cube size and quantity are laid down. A representative 5 kg sample is taken from which four 100 g samples are extracted from four separate portions, mixed together and then two 100 g sub-samples taken for sorting. Product dimensions are divided into four categories:

(i) Cubes. Not less than 60% by weight of sample to consist of 10-8 mm^3 (3/8-5/16″) cubes.

(ii) Part cubes. Not less than 85% weight of sample to consist of 10-8 mm^3 cubes *plus* part-size cubes (i.e. at least two dimensions shall be not less than 8 mm).

(iii) Chippings. Pieces of apple less than 8 mm in two dimensions to be not greater than 15% by weight.

(iv) Oversize cubes. Pieces exceeding 12 mm (½″) in length to be absent.

Similarly, product defects are classified in terms of the following maxima: pips (6); core pieces, including septum (+24); peel, bruised or discoloured pieces and other minor blemishes, for normal diced apple (+70), and for "rough trimmed" diced apple (+300); and stalks (nil). The number of "foreign" objects which could cause possible consumer complaints is thereby minimized.

In addition to ensuring that the filling is "fluid" in nature, it is also necessary to control the amount of liquid emanating from the apple otherwise the balance of the filling and subsequent shelf-life could be altered. The specification for apple therefore contains reference to "free liquid" such that:

(i) wash liquor and bisulphite solution used in the process are eliminated as far as practicable by thorough draining before packing for transport; and

(ii) the soluble solids content of the liquor obtained by drainage from the packed product at any stage after preparation, as measured by refractometer, shall not be less than 8%.

BAKED PRODUCT QUALITY

During the entire production sequence, further routine quality checks are done. For instance, after the apple filling has been mixed a portion is blended in a liquidizer and soluble solids content determined by pocket refractometer; after baking and cooling the filling from six pies is also tested. This ensures that the correct amount of the various sugars have been added and that the pies will last for the full shelf-life.

Further samples of filling are taken from the baked product, washed and the pieces of apple counted. This ensures that the desired amount of discrete pieces of apple are present and also checks that the filling operation has been successful and that undue separation of the filling in the depositor hoppers is avoided.

OTHER FRUIT PIE PRODUCTS

Other fruit pie products are made on the same scale, for example, apple and blackcurrant, and raspberry and redcurrant pies. These raw materials are bought in the frozen state. Similar raw material specifications exist as for apple. With a frozen raw material, however, the filling has to be pre-heated in bulk mixers to ensure adequate softening and production of uniform colour.

I hope this account illustrates the care which is taken with a simple raw material, such as apple, to produce a consistent end product on a mass scale.

Subject Index